HANDBOOK OF STRATA-BOUND AND STRATIFORM ORE DEPOSITS

Volume 1
CLASSIFICATIONS AND HISTORICAL STUDIES

HANDBOOK OF STRATA-BOUND AND STRATIFORM ORE DEPOSITS

Edited by
K.H. WOLF

I
PRINCIPLES AND GENERAL STUDIES

1. Classifications and Historical Studies

2. Geochemical Studies

3. Supergene and Surficial Ore Deposits ; Textures and Fabrics

4. Tectonics and Metamorphism

II
REGIONAL STUDIES AND SPECIFIC DEPOSITS

5. Regional Studies

6. Cu, Zn, Pb, and Ag Deposits

7. Au, U, Fe, Mn, Hg, Sb, W, and P Deposits

ELSEVIER SCIENTIFIC PUBLISHING COMPANY
Amsterdam — Oxford — New York   1976

HANDBOOK OF STRATA-BOUND AND STRATIFORM ORE DEPOSITS

I. PRINCIPLES AND GENERAL STUDIES

Edited by
K.H. WOLF

Volume 1
CLASSIFICATIONS AND HISTORICAL STUDIES

ELSEVIER SCIENTIFIC PUBLISHING COMPANY
Amsterdam — Oxford — New York   1976

ELSEVIER SCIENTIFIC PUBLISHING COMPANY
335 Jan van Galenstraat
P.O. Box 211, Amsterdam, The Netherlands

AMERICAN ELSEVIER PUBLISHING COMPANY, INC.
52 Vanderbilt Avenue
New York, New York 10017

ISBN: 0-444-41401-0

Printed in The Netherlands

LIST OF CONTRIBUTORS TO THIS VOLUME

A.J. BERNARD
Laboratoire de Géologie Appliquée de l'Ecole Nationale Supérieure de Géologie, Nancy,
France

J.W. GABELMAN
U.S. Energy Research and Development Administration, Rockville, Md., U.S.A.

P. GILMOUR
Consulting Mining Geologist, Tucson, Ariz., U.S.A.

H.F. KING
Department of Earth Sciences, University of Waterloo, Waterloo, Ont., Canada

J.D. RIDGE
The Pennsylvania State University, University Park, Pa., U.S.A.

J.C. SAMAMA
Institut National Polytechnique de Nancy, Ecole Nationale Supérieure de la Géologie Appliquée et de Prospection Minière, Nancy, France

K.H. WOLF *
Watts, Griffis and McQuat Ltd., and Directorate General of Mineral Resources, Jeddah,
Saudi Arabia

---

* Formerly: Laurentian University, Sudbury, Ont., Canada.

CONTENTS

Chapter 5. DEVELOPMENT OF SYNGENETIC IDEAS IN AUSTRALIA
by H.F. King

Chapter 6. ORIGIN, DEVELOPMENT, AND CHANGES IN CONCEPTS OF SYNGENETIC
            ORE DEPOSITS AS SEEN BY NORTH AMERICAN GEOLOGISTS
by J.D. Ridge

Chapter 7. SUMMARY OF THE FRENCH SCHOOL OF STUDIES OF ORES IN SEDIMEN-
TARY AND ASSOCIATED VOLCANIC ROCKS–EPIGENESIS VERSUS SYN-
GENESIS
by A.J. Bernard and J.C. Samama

Chapter 1

INTRODUCTION

KARL H. WOLF

The investigation of stratiform/stratabound ore deposits associated with sediments and volcanics and ranging in age from Precambrian to Recent, has become a specialized branch within the wider scope of ore petrology. It is, by all means, not an entirely new field of endeavour and it has always had a place in economic geology, but an accelerated shift in both individualized and collective efforts in the study of the stratiform mineral accumulations has resulted in a large amount of dispersed data. Although during the past few years many publications were devoted to these ores, a more comprehensive summary of their specific aspects remained a challenge. Many phases of ore petrology are in a state of flux as new hypotheses and/or refinements of theories are to be expected, but the purely "building stages" or "accumulative/descriptive stages" have long been strongly supplemented by the at present so prominent "interpretive/genetic phase" in theoretical and applied geology. A synthesis stressing genetic concepts related to stratiform mineralization appeared to be an immediate necessity to enable an efficient dispersal of the presently held hypotheses and theories; in particular to establish if it were possible to combine singular geologic elements into one meaningful entity and reach useful conclusions. By so doing, it was hoped, not only would the summaries offer a collection of known and generally available information (otherwise widely distributed), but the constructive reviews would result in new concepts and new approaches through a critical examination of available data. This indeed, as it turned out, was the case, as is demonstrated by many of the chapters in the present multi-volume publication. The more than fifty contributions vary in both manner of approach and style of presentation, which is unavoidable even with the best of intentions on the part of the authors and the editor. Some chapters are based mainly on a summary and/or review of existing information; others offer some new ideas; whereas in some instances the syntheses of data have resulted in material that is predominantly based on discussions involving new — occasionally even unique — proposals.

Increasingly, there is a trend towards specialization which is reflected not only in the narrow concentration of research by individuals, but also in national and international congresses dealing exclusively with specific topics. The present multi-volume publication on stratiform ores is another inescapable consequence of such specialization. On the other hand, as is illustrated more fully in appropriate chapters, it has become obvious that the

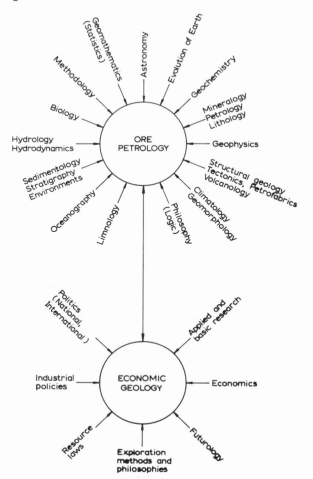

Fig. 1. Diagram depicting the sub-disciplines employed in ore petrology and economic geology.

boundaries between the numerous scientific sub-disciplines, which were artificial and
unfortunate at any time in the past in the earth sciences, have become increasingly
intolerant. Therefore, any trends in specialization are accompanied by an integration of
all relevant disciplines (Fig. 1), and this has to continue progressively in the future. A
conscientious effort (even if it takes some prodding) has to be made to eliminate any
illusive, illusionary and illogical barriers between geological fields — the boundaries are
not natural or factual, they only exist in our mind as a consequence of training and
attitude or possible reluctance to accommodate change. All this is self-evident to most of
us, as is demonstrated by new approaches being employed in solving old geological
problems. The combination and integration of sub-disciplines has usually resulted in a
"fresh look" and occasionally even a breakthrough has been an unexpected consequence,

although generally it has been found that developments of new theories progress slowly and are formulated step-by-step in an accumulative fashion, i.e., occurring not suddenly, but over a period of many years.

The increasing demand on the individual to work consciously in opposition to the ever-mounting departmentalization of knowledge, requires a certain amount of courage from each researcher in venturing outside his own field(s) of experience and interests. It is, no doubt, an emotional and "political" risk to leap over the walls into other people's territory and it may cause some "conflicts of interests",[1] but under no circumstances should one discourage attempts to draw information from sister disciplines to find the answer to scientific queries.

It does not take a large mental leap from the above to the sensitive situation, pointed out by Routhier (1968) so succinctly and in a friendly manner, namely that many of the English-speaking geological fraternity have ignored many pertinent foreign publications (with numerous exceptions). No doubt, this situation has been discussed verbally by the international community of earth scientists on innumerable — mostly informal — occasions. Routhier (p. 19) challenged all of us to have the courage to examine the ratio of English to non-English references in publications, realizing that we would find that many of the European papers (as well as others) have been neglected, thus increasingly arousing intellectual frustration. As Routhier has put it, "geology is a science only if it is world-wide" (p. 19) and a linguistic barrier is no excuse. Such neglect is, of course, contrary to our own philosophy, but it nevertheless has existed, and still does exist, with an intellectual isolation as the consequence. As the chapters on the evolution of geologic thought will demonstrate, such isolation has given rise to a neglect of genetic theories proposed many years earlier which were then "rediscovered" more recently. Under these circumstances, genuine acknowledgement to originators of specific concepts is made impossible. The present compilation of contributions by authors from fourteen countries is in accordance with the axiom that scientific thoughts are of an international character. Intellectual cross-fertilization should be made paramount.

All contributions to this publication had to be written in English, which placed an additional requirement on those authors whose mother tongue is other than English. Although all measured up remarkably well, a few of the manuscripts necessitated a good deal of rewriting to assure precise geologic meanings. A full and complete rephrasing was not possible, so that a few of the chapters are presented in a certain quaint, but not at all objectionable style — using the editorial position that the scientific information is paramount to perfect conventional grammar. The readers will also note that little attempt has been made to restrict the authors to one particular mode of English spelling, thus deleting the national grammatical straightjackets that separate, for example, the American from

---

[1] It seems to be only more recently that sociologists are seriously examining the rather complex human conflicts between researchers that have existed from the Middle Ages and earlier as a result of simultaneous discovery, for example.

the British way of writing. The same reasoning was applied to personal preferences in punctuation, for example.

Attempts were made during the original editorial planning (1971–72) of the present project to assure coverage of the most important aspects related to strata-bound ore genesis. Nevertheless, discrepancies between the readers' requirements and the authors' plus the editor's presuppositions may have introduced some disproportions. A treatise or handbook that deals with so large a subject as ores in sedimentary and volcanic rocks, invites some minor lopsidedness for which only the editor is responsible. It is hoped, however, that each chapter lists the most pertinent references, but inasmuch as here too judgement was involved, some omissions are unavoidable. For future revisions, suggestions are welcome.

The editorial responsibilities included prevention of extensive overlap or duplication as far as convenient from the readers' point of view. That is to say, several chapters may consider related subject matters, but the preliminary – rather extensive – friendly correspondence between authors and editor assured a compromise to prevent undue overlap. After arranging the more than fifty individual chapters in a logical sequence, the resulting collection as available now in the volumes is hopefully free from undesirable major discontinuities. A full chapter-by-chapter integration was, of course, not possible as each contribution was written independently, but the reader will encounter occasional directive statements within the text as well as in footnotes to consult specific pertinent contributions. However, a more efficient manner of obtaining the full benefit of all the details from the various volumes would result from a preliminary familiarization with the content of all chapters by first examining the above-average comprehensive contents, followed by a cursory page-by-page check of the individual chapters' material. Based on such a quick overview, the reader can pick out the most relevant chapters, or sections thereof, to satisfy his momentary personal and professional needs.

The volumes have been divided into Part I, composed of chapters of a more general nature, and Part II with contributions dealing with specific ore districts and/or ore types. In all instances, editorial attempts were made to present ideas useful to the international readers, i.e., information applicable in general in any part of the world wherever the geology appears similar. Therefore, even if a specific ore district is being discussed, the purely descriptive data of local interest were kept at a minimum and the genetic aspects were emphasized. The proportionality between descriptive and genetic information differs, of course, from one contribution to another, but editorial requests were made to let each chapter culminate in up-to-date genetic discussions – this in accordance with accepted standard procedures. Beyond this, particular requests were made to: (1) make attempts to formulate conceptual models, and (2) use a comparative approach by showing similarities and differences between two or more types of ore deposits. A large number of chapters, therefore, use these two approaches.

In particular, the first chapters in Part I are of a "philosophical" nature in spite of the possibility that some readers may consider them as unnecessary "sermons". This may be granted to be true for the experienced geologist, but one should recall that the writers are also addressing the younger generation of readers. Seasoned geologists may prefer "hard facts" often to the exclusion of "philosophy". This appears to be partly the result of the separation of science, art and philosophy since the Middle Ages and the Industrial Revolution, although science and certain fields of philosophy remain complementary to each other. On the contrary, as several authors have demonstrated in their reviews of genetic theories, for example, the historical changes in geologic thoughts should convince even the most reluctant reader that a periodic — if not continual — "dabbling" in philosophical aspects is not a dilatory activity. There is no escape from the fact that all scientific reasoning is based on some fundamental philosophical principles and these "truisms", "paradigms", or "axioms" are often not given their deserved deliberate attention — the reasons being numerous and varying from situation to situation and from one person to the next.

Without a chronology of the development of scientific thoughts, there can be no perspective and without perspective there can be no comprehension of the history of geology. Provision has been made, therefore, to give glimpses of the evolution of geological hypotheses as viewed by certain countries or individuals thereof (see in this Volume the chapters by Ridge, and King, for example, and in Volume 6 the chapter by Vokes).

Another brief comment on the topics chosen may be in order. Oceanology and limnology have become part of the repertoire of each ore petrologist who wishes to understand the origin of economic deposits in marine and lake sediments. It should come as no surprise, therefore, that the reader will find several chapters dealing with mineral concentrations in Recent sedimentary milieux.

One particular subject that has not been given full consideration is the economics of exploration[1], property evaluation, feasibility studies, and such, inasmuch as it has been decided to concentrate exclusively on ore petrology per se.

The books have been prepared with a whole spectrum of requirements in mind, but they address academic geologists and explorationists alike, including teachers, researchers and students. It is hoped that the volumes will serve as a reference and a handbook in both theoretical and applied ore petrology. The editor strongly believes that compilations such as this one will become increasingly necessary to accommodate the day-by-day demands of earth scientists, to give them the opportunity to keep up to date by merely having to refer to one set of books. Anyone who requires facts and information instead of surmises and opinions will agree that a well-organized library is the most efficient research tool offering the greatest economy in time and effort. As A.I.Levorsen put it in 1946

---

[1] As already pointed out in a recent publication (Wolf, 1976), the commonly used phrase "economic geology" for ore petrology is somewhat misleading, inasmuch as too frequently one does not refer specifically to the economic aspects of mineral resources. It would be an ill-sounding expression to speak of "the economics of Economic Geology"!

(then Dean of the School of Mineral Sciences, Stanford University, California — quoted by Gribi, 1973), "the first place to start in the hunt for oil, is the library". No doubt, this maxim applies also to mineral exploration.

As among the three basic purposes of a library or publications in a civilization, this set of books on strata-bound ores will, hopefully, assist in further discovery of knowledge as well as in the conservation and transmission of information. For example, all chapters follow the motto that "Half the Knowledge is the Knowledge where to find it" by serving as a starting point from which the reader can select certain pertinent references from which he can "spread out" to cover other fields of interest. The chapters thus fulfill three functions: they offer well-summarized and new information, they increase the speed of obtaining data, and they allow to widen one's horizon.

Woodward (1975) mentioned the impending crisis in libraries due to: (a) the "information explosion" which makes the buying, storing, retrieval and circulation of publications a formidable task, and (b) the financial crunch which has led already to the cancellation of hundreds or thousands of periodicals and made expansion impossible. Woodward also pointed out that this national crisis can be solved only by a "major revolution in the way bibliographical materials are organized, resources are shared and libraries are financed". It is the contention of the editor, however, that "old-fashioned" books will not be replaced by the specialized information-retrieval systems developed either recently or in the future. As Woodward said, the reading of microfiche cards cannot substitute totally for the pleasures derived from reading a book.

One often hears about the high rate of obsolescence in knowledge and that, for example, a publication is "out of date" the moment it appears in print. This is an exaggeration unworthy of further elaboration as it is self-evident that ideas continue to develop uninterruptedly, but this does not nullify the demands for books — on the contrary, periodic reviews and compilations remain important. Indeed, the editor believes that, although the rate at which the raw geological data is being made available may continue to increase, its "digestion" may *not* necessarily result in the development of new hypotheses or theories but fit into ones already developed. That is to say, more and more, it is predicted, we will narrow down the choices we have between several genetic interpretations until we have adopted *the most plausible* or *most factual* one. This indeed sounds like "intellectual arrogance", but all we have to remember is the development of the hydrothermal theory which has become refined to the extent where we now recognize several types of "hydrothermal" ore-forming solutions. For several of these hydrothermal sub-types we have located areas of present-day processes forming mineral concentrations. The raw data on these will pour out of many laboratories and will result in *refinements* of established theories, but on the whole the *basic* genetic concepts may remain. Thus, if we eventually can hope to agree on theories, debates among geoscientists will shift from the basics to the details. This in turn means that publications dealing with the basics will less rapidly and less frequently be out of date, which is in contrast to many books written in the past in which both the basics and the details were short-lived.

Synopses of particular topics, especially if relatively all-inclusive and based on constructive and evaluative critical syntheses, are recognized to have become increasingly necessary. Russell (1954/55, pp. 50–56) is one among numerous geologists who have correctly recognized that to do proper and useful geological syntheses requires particular attitudes and talents. Inasmuch as the synthetical field in geology has lagged far behind the analytical studies, despite being of tremendous importance, Russell goes as far as to propose that specially talented students should be educated for synthetic studies so that we may have a trained group for "fundamental" rather than routine thinking in geology. The present writer considered this problem when he first saw Russell's book in 1974 and found that synthetic research can be divided into at least three parts, each more complex and more useful than the preceding one: (1) collective synthesis → (2) comparative synthesis → (3) creative synthesis. Indeed, it seems that each synthetic study has to be based on this sequence to lead to any acceptable logical conclusion. Merely collecting data (which is often done in analytical work) is only the first step, and unless data from many different studies are compared (i.e. all differences and similarities established), the bank of pure numbers and observations are without meaning and remain irrelevant unless followed by step 2 in the above sequence. Many studies do not consider comparative work and, thus, fall short of utilizing a powerful mode of thinking.

From the above point of view, there seems to be little doubt that "synthetic" work constitutes research — it may even be called basic research if new theories are formulated. Synthetic and analytical studies are research of a particular kind, and one should not exclude one from the other; they must supplement each other. Although synthesis commences with the mere collecting of data, often supplied from published sources, the gathering of material may easily be misconstrued as "para-research" or "pseudo-research". This is definitely not the case, in particular if the data are properly employed in comparative and creative work. Trotter et al. (1973) have pointed out that it is increasingly the practice to subdivide the research function into "Research I", involving the discovery of new knowledge and new applications of knowledge, and "Research II", involving "keeping up in the discipline", assimilating new knowledge and relating it to existing knowledge. Taking the field of theoretical and applied geology as a whole, both kinds of research are necessary, although individual geologists may have a preference and abilities for one or the other. Nevertheless, differences of opinion as to the relative importance of "Research I" versus "Research II" and synthetic versus analytical research will remain. As Weinberg (1967) has stated: "As I see it, at least part of the conflict amounts to a philosophic judgement whether Science is the search for new knowledge or the organizer of existing knowledge." Each one of us engaged in research and exploration is using science as a method of approach to solve problems and not as a final body of hard-and-fast doctrine. "Each individual can only hope by his efforts to come a little nearer than his predecessors to a full comprehension of the processes he is studying" (Waddington, 1948, pp. 97–98). "In science men have learnt consciously to subordinate themselves to a common purpose without losing the individuality of their achievements. Each one knows that his work depends on that of his predecessors and colleagues, and that it can only reach its fruition through the work of his successors. In science men collaborate not because they are forced to by superior authority or because they blindly follow some chosen leader, but because they realise that only in this willing collaboration can each man find his goal" (Bernal in "The Social Function of Science").

Inasmuch as research in ore petrology is vital to the survival of our present industrial civilization, the collection, preservation, transfer, and refinement of geological knowledge is of direct practical importance. Rosenzweig (1971) has made a good case in convincing us that "geology constitutes one of the fundamental elements of industrial development in the past, present, and future of all nations ... the geologic substratum is the foundation on which the present world is built" (p. 2174) and that "geology is development" (p. 2176). Such a dependence on mineral wealth has been effective only since the Industrial Revolution — our ancestors of the not-too-remote past depended but little on minerals and ores which did not play any role in the world economy (Meyer and Strietelmeier, 1968). Even today, many countries remain unaffected by the Industrial Revolution; consequently, they do not consume large amounts of mineral resources, although some of them are great exporters. On the other hand, many of these countries are now entering the Industrial Revolution and their factory systems, together with the increasing demands for consumer goods, result in an unprecedented consumption of raw material. Nevertheless, as recently as 1971, the per capita resource demand and environmental impact of the western world was about 50 times that of an underdeveloped country (Prof. Paul Ehrlich, Stanford University; quoted by Pecora, 1971, p. 1716), but such disproportionalities will change rapidly in favour of the developing countries. The combined participation of the state and private, national and foreign spheres will be required to fulfill world demands for energy and minerals. The geologist and mining engineer will be the creators of wealth because through their hypotheses they will be instrumental in locating and developing the solid and liquid resources. The exploration for new deposits, however, will increasingly tax our ingenuity. The Canadian Geoscience Council has issued a report (cf. review in *The Northern Miner*, Jan. 16, 1975) which stresses that "probably no other group of disciplines will be so vitally important in the 25 years remaining in this century". Among several pertinent proposals made, it was urged that new conceptual models and syntheses should be provided to the exploration and research geologists.

The editor has been asked on occasions about the "sponsor" of the present project and it seems proper, therefore, to point out that no individual, group, society or institution has acted in this capacity. The project of summarizing pertinent ideas on the genesis of stratiform/strata-bound ores originated with the editor who had hoped to make it an inter-departmental or an inter-university effort together with authors from several continents. In the final analysis, however, the latter approach was used exclusively. As early as 1964, an urgent need was felt to write a comparative synopsis of the application of both "soft-rock" and "hard-rock" geology in the investigation of stratiform ore deposits, but it was not until 1971 that definite steps were taken to bring such a review project to fruition. After the Elsevier Publishing Company agreed to act as publisher, numerous prospective authors were invited to participate. At the time, the editor was Associate Professor of Geology at Laurentian University and held research grants awarded by the Geological Survey of Canada and the Canadian National Research Council. He gratefully acknowledges all assistance rendered to him that enabled him to maintain his professional interest and allowed the planning and execution of the present undertaking — not to forget the many researchers who have over the past 15 years responded to reprint requests. Particular thanks go to the editor's family who have given their generous encouragement in this and all other research endeavours; without their devotion and kind patience the work would have been impossible.

This Introduction would be incomplete without an expression of thanks to the contributors whose selfless cooperation made this collection of chapters possible — collectively, this series of books is *theirs*. As editor, I hope to have served them well and to their satisfaction. The authors' patience and understanding in the face of numerous editorial requests was indeed very much appreciated. If any shortcomings are present in the publication, the editor takes sole responsibility.

REFERENCES

Gribi, E.A., 1973. Geologic maxims. *Am. Assoc. Pet. Geol.*, 57: 215–216.
Meyer, A.H. and Strietelmeier, J.H., 1968. *Geography in World Society*. J.B. Lippincott, New York, N.Y.
Pecora, W.T., 1971. Uniqueness of man and his environment. *Am. Assoc. Pet. Geol.*, 55: 1715–1718.
Rosenzweig, A., 1971. Geology and industrial development. *Am. Assoc. Pet. Geol.*, 55: 2174–2176.
Routhier, P., 1968. America, Europe and geology. *Geotimes*, 1968: 19–20.
Russell, G.A., 1954–55. *Some Philosophical Aspects of Geology*. University of Manitoba, 171 pp.
Trotter, B., McQueen, D.L and Hansen, B.L., 1973. The ten o'clock scholar? What a professor does for his pay. *Bull. Can. Assoc. Univ. Teachers*, 21: 4–10.
Waddington, C.H., 1948. *The Scientific Attitude (Pelican, A84)*, Penguin Books, Middlesex, 175 pp.
Weinberg, A.M., 1967. *Reflections on Big Science*. Pergamon Press, London, 3 pp.
Woodward, K.L., 1975. On borrowed time. *Newsweek*, Feb. 24, p. 85.
Wolf, K.H., 1976. Influence of compaction on ore genesis. In: G.V. Chilingar and K.H. Wolf (Editors), *Compaction of Coarse-grained Sediments*, 2. Elsevier, Amsterdam, in press.

Chapter 2

CONCEPTUAL MODELS IN GEOLOGY [1]

KARL H. WOLF

Modelling of natural phenomena has become a widely accepted tool in the earth sciences and inasmuch as it has also proved to be a powerful procedure in economic geology (in ore petrology, writing reports, and outlining exploration programmes, for example), a few models from the published literature, as well as some prepared by the editor, will be made available below. The study of ore genesis is not strictly confined to ore-mineral precipitation but encompasses nearly all geological disciplines, including sedimentary petrology and geochemistry, stratigraphy, environmental reconstructions, volcanism, and tectonism, so that a number of diagrammatic models will be presented which go beyond the actual realms of ore petrology.

Conceptual models are nothing new in geology, except possibly the phrase itself which is rapidly becoming a "household" word among earth scientists. Bowen's theory and his familiar three-pronged diagram (with olivine and plagioclase at the top and K-feldspar and quartz near the bottom), is nothing but an idealized model established by both laboratory and field observations. What may be of more recent vintage, however, are the *types* of modelling performed as well as the *extent* to which conceptualization has been employed, in particular during the past 10–15 years. This trend will continue and practicing geologists have requested analogues or models that will assist them in comprehending the complexities of natural phenomena to be unravelled in their search for mineral resources. A report on the state of geosciences in Canada issued by the National Geoscience Council stated that several weaknesses exist in the earth sciences that require immediate attention – among these problems was the need for new conceptual models and syntheses (*Northern Miner*, Jan. 16, 1975).

Modelling is not confined to the geological disciplines, but has also been employed by the sister sciences, especially in biology, and it has also found many applications in

---

[1] The following authors, in their contributions, offer conceptual models: Dimroth on ore pretrology; Rackley on uranium mineralization in sandstones; Glasby and Read on Mn-precipitation; Roy on the genesis of ancient Mn-deposits; Pretorius on placer ore genesis; Ruitenberg on the genesis of an ophiolite; Ineson on the formation of Pb and Zn ores; Veizer on numerous aspects of ore genesis from the Precambrian to the Recent; Zuffardi on ore formation in karsts; Mookherjee on ore metamorphism; Gilmour on transitional types of ores; to mention only a few examples.

physics and chemistry. The construction of conceptual models, however, has also been adopted in Law. Lloyd (1972), for example, has a chapter entitled "Conceptual Thinking in Law", and it seems to the writer that in legal studies the "precedent cases" could be compared with the "analogues" or "models" used in geology. The warnings issued by Lloyd (pp. 284, 295) against a too-restricted use of concepts or models also applies to the natural sciences: ". . . concepts . . seem to exist as ideas in the human mind rather than as concrete entities" and "Concepts . . . are excellent servants, but not always good masters." Nevertheless, if employed with reason and good judgement, models serve as theoretical analogues. As Amstutz (1971, p. 251) put it: "The search for congruency is the integrative fundamental 'sine-qua-non method' of all science. We do not identify anything except by analogy."

As will become obvious by examining critically some of the models given below, diagrammatic models can depict the entire natural system, can determine how the individual parts are related to each other and can allow us to find an explanation of how some parts of the whole operate in nature. The fabric of such flow-chart types of model is composed of tightly interwoven threads connecting parameters, factors and processes. Certain processes in nature are shown to be more complex than usually depicted (e.g., see models on soil genesis) as can be visualized by altering a number of the variables, so that causes become consequences and causes again. In describing such complex systems, no particular description is uniquely correct; much depends on where one breaks into the matrix of the model and on which parameters are varied and to what degree.

Before proceeding, let us consider what "models" are by quoting some recent publications (cf. also Wolf, 1973a, b):

(1) Krumbein and Graybill (1965, p. 13) pointed out that ". . . a model provides a framework for organizing or 'structuring' a geologic study . . . the geologist studies multivariate problems by selecting and integrating from innumerable details those elements that appear to control a given geological phenomenon. The internal consistency of the data and the comparison of several lines of evidence simultaneously commonly provide a basis for selecting some single set of conditions that most satisfactorily accounts for the phenomenon."

(2) Potter and Pettijohn (1963) referred to sedimentary models, and some of these can be applied in the study of ores within sedimentary-volcanic piles, as "intellectual construct . . . based on a prototype." The model concept relates all "geologic elements to one another to provide an integrated description of a recurring pattern of sedimentation."

(3) Chorley (1964) mentioned that the need for models, as well as for classification schemes in general, comes from the "high degree of ambiguity" presented by the various subject matters of the different scientific disciplines, and "the attendantly large 'elbow room' which the researcher has for the manner in which this material may be organized and interpreted. This characteristic is a necessary result of the relatively small amount of available information which has been extracted in a very partial manner from a large and multivariate reality and leads not only to radically conflicting 'explanation' . . . but to

differing opinions regarding the significant aspects of . . . reality." "Where such ambiguity exists, scholars commonly handle the associated information either by means of classifications or models." "Model building, which sometimes may even precede the collection of a great deal of data, involves the association of supposedly significant aspects of reality into a system which seems to possess special properties of intellectual stimulation."

(4) As to the types of models, Kendall and Buckland (1960) stated that a model is ". . . a formalized expression of a theory or the causal situation which is regarded as having generated observed data. In statistical analysis, the model is generally expressed . . . in a mathematical form, but diagrammatic models are also found." The "initial model is commonly simplified to retain the essential features of the phenomenon, with extraneous details omitted" (Chorley, 1964), and the models are based on observations, expressing some segment of the real world in idealized form. These initial models are often qualitative, but quantitative mathematical models are being increasingly developed in the various geological fields. Nevertheless, qualitative models will remain useful and will never be replaced entirely by quantitative ones, because the latter are often developed from the qualitative types. "Conceptual models in diagrammatic form are largely qualitative, but they point the way toward a choice of observations and measurements that can be used to implement the model, either to test its validity in terms of real-world phenomena or to use the model as a predicting device" (Krumbein and Graybill, 1965).

Two parameters enter earth-science models that deserve special consideration, namely, the "time" and "space" variables. The time factor (over millions of years in geology) may result in a sequence of events, and as a consequence it may lead to a succession of models. The same arguments apply to space. Indeed, time and space may be interrelated as in the development of a geosynclinal cycle. A particular model, applied to a specific location during the early geosynclinal stages, may prevail for a certain length of time. With a change of rate and style of tectonism, new patterns of sedimentation, volcanism and possibly ore-forming conditions, are established which are represented by a new conceptual model. This development continues to the termination of the tectonic cycle. Some models may succeed in an orderly predictable fashion, as, for example, the molasse-type follows the flysch-type of sedimentation. On the other hand, several models may coexist, e.g., fluvial, deltaic and off-shore sedimentological models depicting the distribution of placer-ore accumulations may be present in a particular basin more or less pcne-contemporaneously.

Among the numerous types of models only three varieties will be considered here, namely the diagrammatic/pictorial, the flow-chart and the tabular types of conceptualizations. They are particularly useful inasmuch as they can be utilized at any time and at any stage of geological investigations. All that is required is a pencil and a notebook for the more simple models, although the more complicated ones require some efforts. Once prepared, the models can be employed for a number of situations by slightly modifying them. The efforts of many researchers will eventually have covered all demands to the benefit of everyone. A shift from the purely qualitative to the more quantitative models should then be increasingly possible. The three models just mentioned can be profitably used during the planning stage of an investigation, as well as for descriptive and genetic work on any scale from the macro- and mega-scale during field work to the microscale in the laboratory (cf. models below).

Some of the advantages of employing models during the various stages of the scientific method (Wolf, 1973a, pp. 153—159; 1976) are:

(1) To organize and structure data[1] and to demonstrate interrelationships between two or more factors or parameters, processes, and situations, and thereby compel the investigator to consider all variables and all possible interpretations.

(2) To discover new problems by formulating correct, meaningful questions because "meaningless questions beget meaningless answers" (Griffiths, 1967).

(3) To improve the critical appraisal and testing of ideas in research and exploration work, to improve problem-solving procedures, to find some workable solutions, and to form prognostications (Griffiths, 1967).

(4) As a "mental crutch" assisting in learning complex phenomena, and to improve the "visualization" of phenomena during teaching; the conceptual models give a bird's eye view and at the same time summarize a complex subject mater. (Through personal communications with individuals from industry, both mining and oil, and from universities and research institutions, support has been given to the above-mentioned advantages of using models.)

(5) Conceptualization is the first phase in the preparation of flow diagrams in computor geology and in the formulization of mathematical models.

(6) Potter and Pettijohn (1963, p. 226) go so far as saying that with a proper choice of models one should be able to make more successful predictions about those parts within a sedimentary-volcanic pile, for example, which are concealed and unexplored during the search for petroleum and ore deposits. Many examples might be cited from the oil exploration industry where stratigraphic-environmental analyses based on model concepts have resulted in new discoveries. Indeed, this is the main function of applied sedimentology (with its numerous sub-disciplines). In metal exploration, the best example that comes to mind immediately is the work done on the Witwatersrand gold—uranium fields (see, in Vol. 7, the chapters by Pretorius, and details below).

A sequence of conceptual models or analogues will be made available below, all of which should be of interest to ore petrologists and economic geologists, inasmuch as both the ore per se as well as the host rocks have to be critically examined to comprehend fully the genesis of ore deposits. In the flowchart-like type of models, all the complexly

---

[1] One of the earliest attempts with which the writer is familiar is that by Koch (1949) who used the Tetraktys or four-fold spatial arrangement in organizing geological data into a natural unit and illustrated various interconnections as determined through any complete investigation; namely, as a result of Observation (through sense-perception), Determination (through discrimination), Classification (through grouping), and Description (after Naming).

interrelated parameters are connected by arrows. Insofar as in most instances no quantitative data is available as yet, as pointed out earlier, the diagrammatic models are purely qualitative — nevertheless, they fulfill an obvious function. Eventually, by extending our field and laboratory observations and by developing methods for extrapolation in geologic time and space, we may accumulate sufficient numerical information to, first, make certain parts of a particular model quantitative, and then to expand the numerical treatment step-by-step until the whole model has been covered.

Literally volumes could be written on the genetic relationships between variables listed in the models; this is, however, not possible because of lack of space. But every geologist with a sufficient background can "visualize" for himself what the genetic controls are that have been illustrated by the arrows.

Fig. 1 is a model relating all important variables controlling the origin of soils and is, therefore, applicable also in the investigation of laterites and bauxites as well as supergene

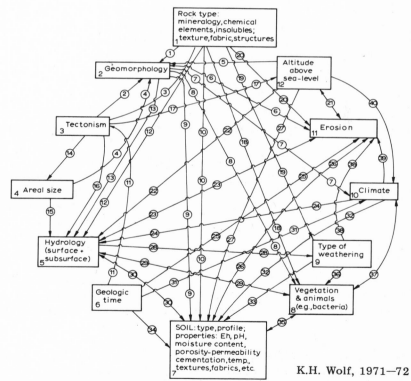

Fig. 1. Model relating all parameters important in the genesis of soils, and supergene ore concentrations such as bauxite and laterite; also useful in provenance reconstructions. (After Wolf, 1973a, fig. 9.)

ores. In sedimentological studies it is useful for provenance reconstructions. Let us consider some of the genetic connections.

(1) Using "rock type" as a starting point or main center of concern, the following relationships exist (see respective arrows): the rock type and its properties, such as mineralogy, texture, structure and strength, will determine the type of soil formed; the rate of degradation or erosion; the type and amount of vegetation and organisms; the mode of water run-off or hydrology in general; and as a result of controlling the mode of erosion, the rock type will determine the general topography.

(2) Many of the above-mentioned variables, of course, are also controlled by climate; the type of weathering, i.e., chemical versus mechanical; tectonism; geologic time; as well as others. All these influence all others in complex ways depicted by the arrows. If progress is to be made by detailed geochemical, geological and biological observations in the study of supergene ores and bauxite, for example, these interrelated complex connections have to be considered; in particular in quantitative investigations.

In petrographic and petrologic studies of volcanic lithic sandstones, tuffaceous sand-

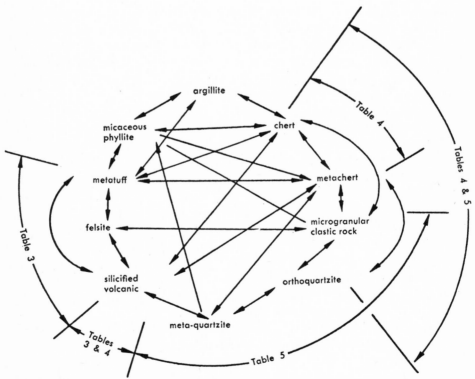

Fig. 2. Circular diagram to show ten lithologies between which textural and compositional transitions are possible, that make identification and discrimination in sand grains and pebbles often difficult or impossible. (After Wolf, 1971, fig. 1; for tables 3 to 5, see original publication.)

stones, and pyroclastics, ranging in age from Tertiary to Precambrian, it has been found that difficulties are encountered in discriminating the various types of lithic sand and pebble grains (e.g., Wolf, 1971). In order to facilitate the interpretations, a circular diagram (Fig. 2) has been prepared showing the textural and compositional transitions between the numerous petrographic types that are liable to be confused with each other. The transitions to be expected in microscopic work are indicated by the arrows. One may, on first consideration, wonder about the reasons for including such a model here in a book on ore petrology. By way of justification, the best example that comes to mind is the controversy around the genesis of the Mt. Isa (Australia) ores where it was only recently that detailed work has revealed some evidence of volcanicity through the identification of tuffaceous sediments. Wherever microscopic work is being applied in ore genesis, all techniques available to the petrologist that enable him to recognize and differentiate individual grains, should be utilized.

In the search for ore bodies, exploration work has been increasingly concentrated on sedimentary—volcanic piles as one major host-rock environment. Anyone who has done field work in such a setting can vouch for the intricate stratigraphic complexities that can be encountered. The consequent associated difficulties in interpreting the petrography and petrology of the rocks are especially acute in metamorphosed Precambrian belts where the criteria usually handled to distinguish between volcanic and sedimentary rocks have been obliterated, leaving only a few criteria or none at all. In younger rock sequences, e.g., of the Kuroko- and Cyprus-types of unmetamorphosed host rock, the petrography may be much easier to work out. On the other hand, independent of the degree of preservation of the primary details of the sedimentary and volcanic deposits, whenever genetic interpretations are attempted in reconstructing the larger-scale processes and depositional environments, visualization is enhanced by either flowchart-like sequential process-response models or pictorial analogues. The model presented here (Fig. 3) is from Wolf (1971, adapted from Whitten, 1964) and applies to the accumulation of both epiclastic and pyroclastic material in a large Tertiary lake in which the deposits were secondarily reworked prior to incorporation in the stratigraphic column. Table I and Fig. 4 present information on the derivation of the numerous types of material, in support of the conceptualized processes in Fig. 3. There is little doubt that similar stratigraphic work on the associated volcanic and sedimentary accumulations will eventually be performed on the piles containing ore bodies in order better to comprehend the environmental setting that gave rise to the stratabound ores.

Another two general conceptual models might be mentioned before presenting those that are more directly related to ore genesis, namely a model of sedimentary environments and one on stratigraphic sequences. In regional geologic analyses forming the basis in the search for natural resources, the utilization of models such as those of Figs. 5 and 6 (pp. 21—24) will indeed enhance one's interpretations (aside from being a proper tool in presenting a convincing, imaginative and comprehensive report to management, for example, as has been related to the writer by individuals working in the mining and petroleum

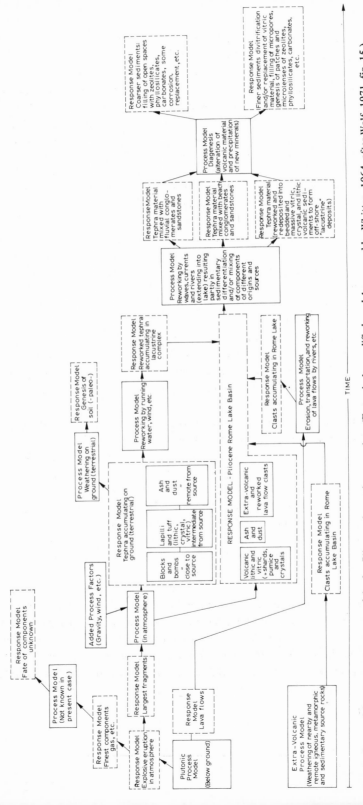

Fig. 3. Sequential process-response model of a Pliocene lacustrine complex. (Extensively modified model proposed by Whitten, 1964, after Wolf, 1971, fig. 15.)

TABLE I

Source rock and mode of origin of various types of epiclastic, volcanic and pyroclastic components in a Pliocene complex

Lithology of source	Age of source	Location of source	Mode of origin of grains
(1) Sedimentary	old	extrabasinal	epiclastic
(2) Metamorphic	old	extrabasinal	epiclastic
(3) Plutonic	old	extrabasinal	epiclastic
(4) Volcanic	(a) old	extrabasinal	epiclastic
	(b) penecontemporaneous	extrabasinal	epiclastic
	(c) contemporaneous	extrabasinal	pyroclastic: extensive re-working and transportation into Rome basin
	(d) contemporaneous	intrabasinal	pyroclastic: (i) no reworking (ii) slight reworking (iii) extensive reworking

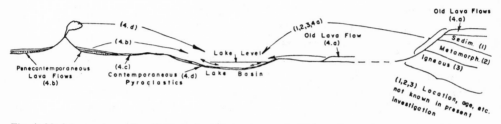

Fig. 4. Model, supplementing Table I, depicting sources of a Pliocene lacustrine complex. (After Wolf, 1971, fig. 16.)

industry). The two figures are selfexplanatory, but the reader should be warned that these two models are *idealized* and do not represent natural occurrences, as demonstrated by the numerous "contradictions" a critical examination would reveal. These are unavoidable in such models that are all-inclusive, but can and will be eliminated as soon as the models are adapted by individuals to their specific natural setting.

In most of the chapters in the present multi-volume publication, reference is made to carbonate and clastic sedimentary rocks (or their metamorphosed equivalents) which constitute two major groups of host rocks for stratiform/strata-bound ores. The environmental reconstructions in detailed investigations necessitate an intimate familiarization with the origin of sedimentary deposits of all types, and the numerous variables in the natural system which have been modellized by a number of researchers and explora-

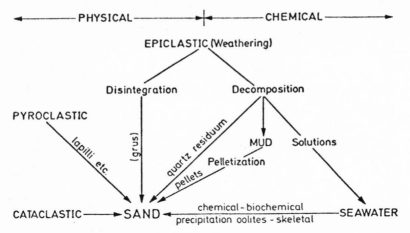

Fig. 7. Processes of sand formation. (After Pettijohn et al., 1972, fig. 8-1, with permission of Springer-Verlag.)

tionists. The parameters in the land (= continental = terrestrial) source area, from which sand, silt and clay particles are derived and transported to a spectrum of possible sedimentary depositional milieux, are depicted in Fig. 6. Pettijohn et al.[1] (1972, fig. 6-1) illustrated diagrammatically the provenance and evolution of the non-carbonate components of sandstones and conglomerates as shown in Fig. 8, whereas the same authors have broadened their model in Fig. 7 by including purely chemically precipitated particles. With a progressive increase in the latter, clastic sediments grade into limestones — which are characteristically the host rocks of the Mississippi Valley-type ores. This is sufficient reason to include here conceptual models of the origin of limestones of which Fig. 9 (Wolf, 1973c, fig. 3) shows the numerous carbonate particles and their possible derivation. Figure 10 (Wolf, 1973a, fig. 7) summarizes the limestone-forming environments in a very generalized and simplified fashion, and Fig. 11 (Wolf, 1973a, fig. 8) is a synopsis of the parameters controlling the origin of limestone reefs.

King (1973, p. 1373) stated that "it is time we outgrew the tacit concept that in the context of ore, geological events happened one at a time, e.g., sedimentation, and/or extrusion, hardening, folding, granite intrusion, etc. I venture to visualize that there is considerable overlapping of these, even that the formation of granite may provide the ingredients for a contemporaneous stratiform deposit at a higher level." King's above proposal has been followed in many of the chapters and a number of diagrammatic conceptual models have been offered by the respective authors in support of their ideas

---

[1] Note that the term "greywacke" is used by sedimentary petrologists in different ways, so that whenever this term is employed, a definition plus the author of the classification scheme used must be provided to the readers.

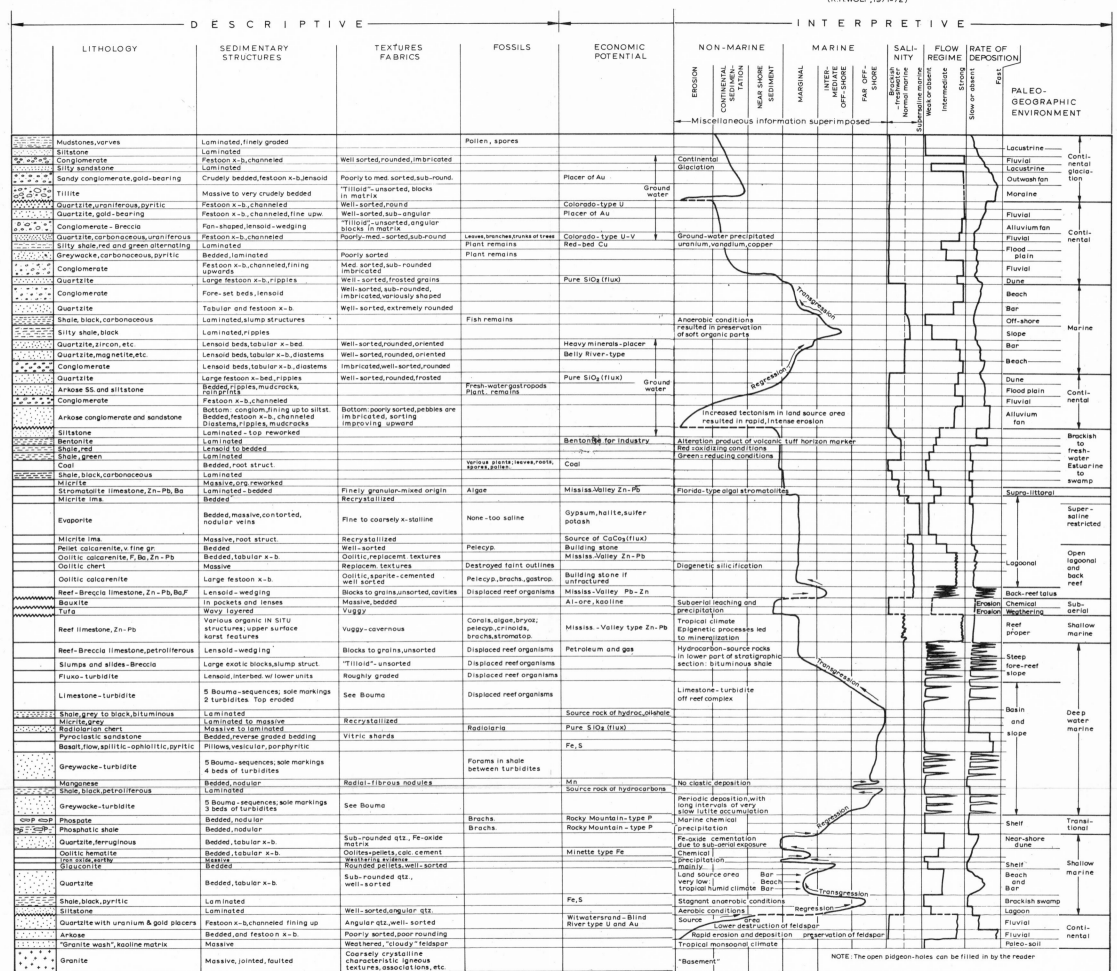

Fig. 6. Idealized model of a sedimentary–stratigraphic–environmental sequence. (After Wolf, 1973b, fig. 14.)

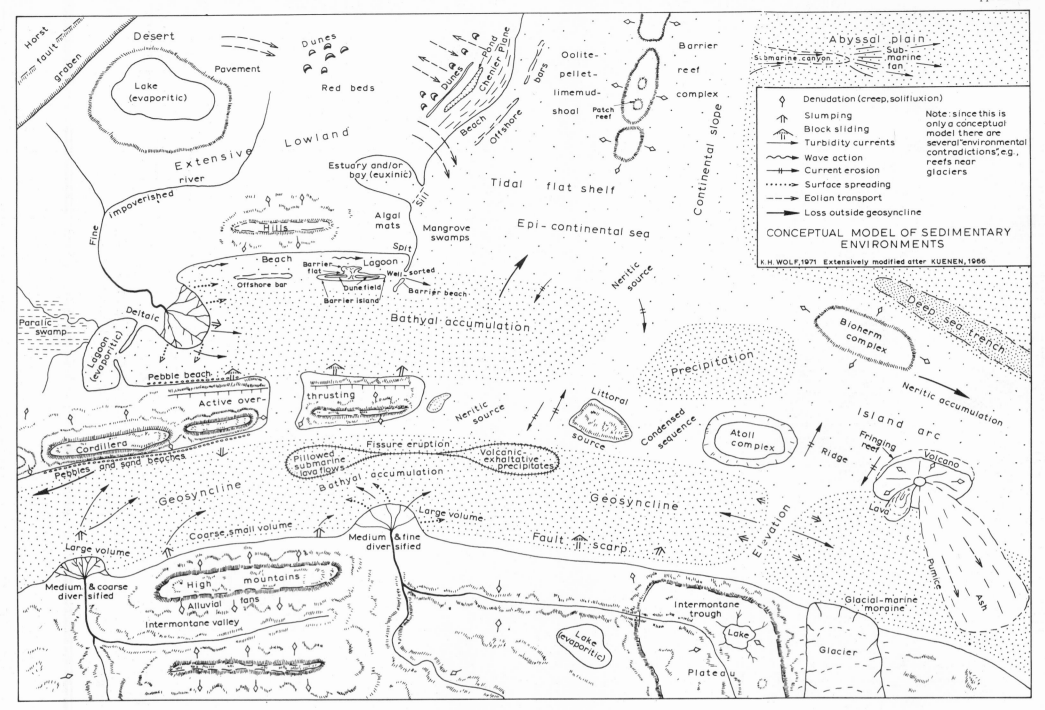

Fig. 5. Conceptual model of sedimentary depositional environments. (Modified after Kuenen, 1966, with permission; from Wolf, 1973a, fig. 6.)

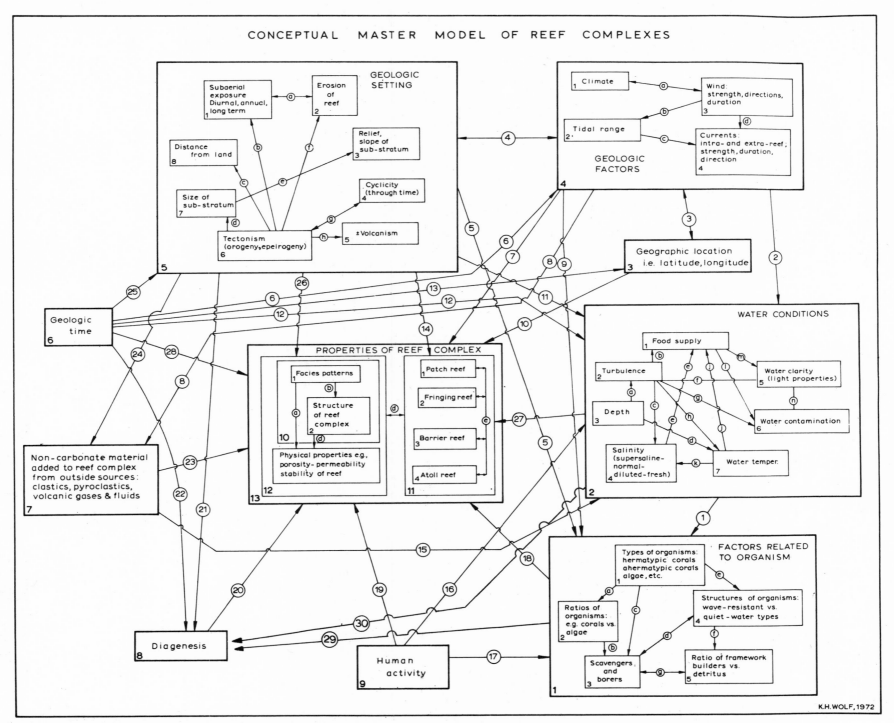

Fig. 10. Conceptual model presenting the factors and processes controlling the origin of limestone reefs. (After Wolf, 1973a, fig. 8.)

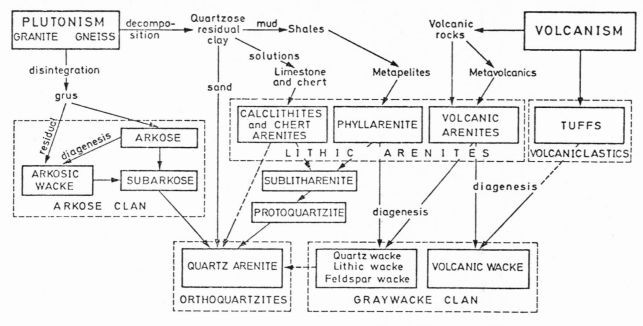

Fig. 8. Provenance and evolution of the noncarbonate sands. (After Pettijohn et al., 1972, fig. 6-1; with permission of Springer-Verlag.)

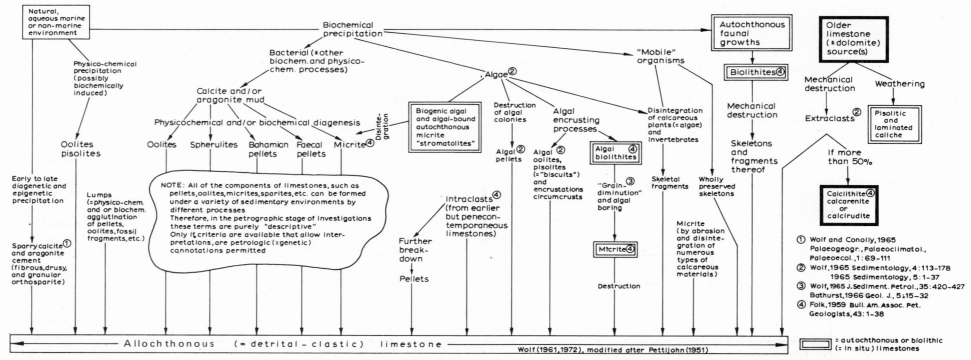

Fig. 9. Model depicting the genesis of carbonate particles, matrix material and cement of limestones. (After Wolf, 1973c, fig. 3.)

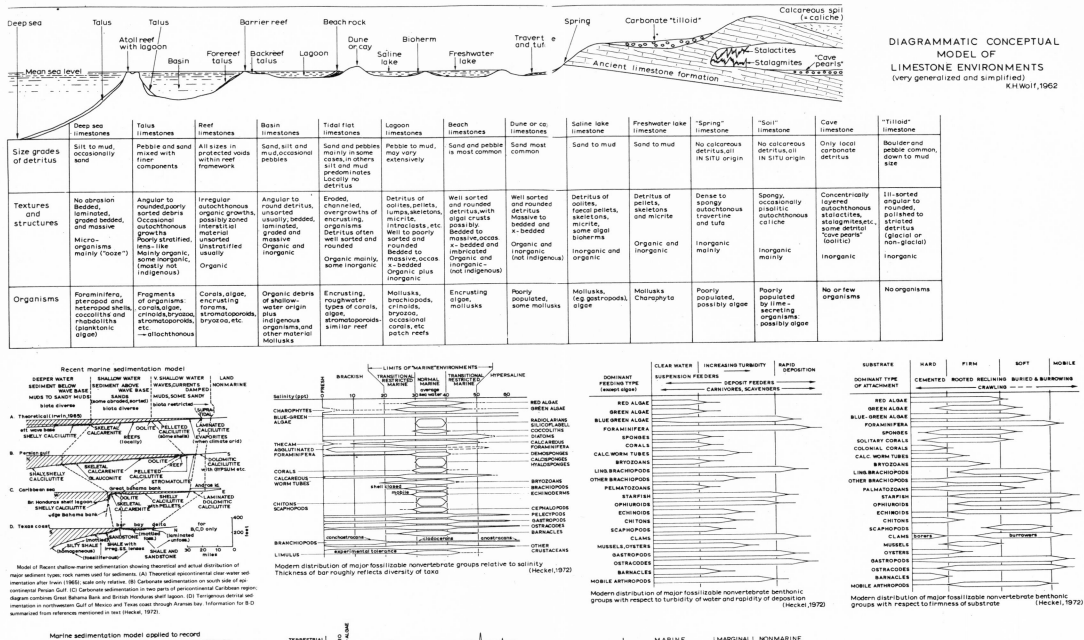

DIAGRAMMATIC CONCEPTUAL
MODEL OF
LIMESTONE ENVIRONMENTS
(very generalized and simplified)
K.H.Wolf, 1962

	Deep sea limestones	Talus limestones	Reef limestones	Basin limestones	Tidal flat limestones	Lagoon limestones	Beach limestones	Dune or cay limestones	Saline lake limestone	Freshwater lake limestone	"Spring" limestone	"Soil" limestone	Cave limestone	"Tilloid" limestone
Size grades of detritus	Silt to mud, occasionally sand	Pebble and sand mixed with finer components	All sizes in protected voids within reef framework	Sand, silt and mud, occasional pebbles	Sand and pebbles mainly in some cases, in others silt and mud predominates Locally no detritus	Pebble to mud, may vary extensively	Sand and pebble is most common	Sand most common	Sand to mud	Sand to mud	No calcareous detritus, all IN SITU origin	No calcareous detritus, all IN SITU origin	Only local carbonate detritus	Boulder and pebble common, down to mud size
Textures and structures	No abrasion Bedded, laminated, graded bedded, and massive; Micro-organisms mainly ("ooze")	Angular to rounded, poorly sorted debris Occasional autochthonous growths Poorly stratified, lens-like Mainly organic, some inorganic, (mostly not indigenous)	Irregular autochthonous organic growths, possibly zoned Interstitial material unsorted Unstratified usually Organic	Angular to round detritus, unsorted usually; bedded, laminated, graded and massive Organic and inorganic	Eroded, channeled, overgrowths of encrusting organisms Detritus often well sorted and rounded Organic mainly, some inorganic	Detritus of oolites, pellets, lumps, skeletons, micrite, intraclasts, etc. Well to poorly sorted and rounded Bedded to massive, occas. x-bedded Organic plus inorganic	Well sorted and rounded detritus, with algal crusts possibly. Bedded to massive, occas. x-bedded and imbricated Organic and inorganic- (not indigenous)	Well sorted and rounded Massive to bedded and x-bedded Organic and inorganic (not indigenous)	Detritus of oolites, foecal pellets, skeletons, micrite, some algal bioherms Inorganic and organic	Detritus of pellets, skeletons and micrite Inorganic mainly	Dense to spongy autochthonous travertine and tufa Organic and inorganic	Spongy, occasionally pisolitic autochthonous caliche Inorganic mainly	Concentrically layered autochthonous stalactites, stalagmites, etc., some detrital "cave pearls" (oolitic) Inorganic	Ill-sorted angular to rounded, polished to striated detritus (glacial or non-glacial) Inorganic
Organisms	Foraminifera, pteropod and heteropod shells, coccoliths and rhabdoliths (planktonic algae)	Fragments of organisms: corals, algae, crinoids, bryozoa, stromatoporoids, etc. → allochthonous	Corals, algae, encrusting forams, stromatoporoids, bryozoa, etc.	Organic debris of shallow-water origin plus indigenous organisms, and other material Mollusks	Encrusting, roughwater types of corals, algae, stromatoporoids- similar reef	Mollusks, brachiopods, crinoids, bryozoa, occasional corals, etc patch reefs	Encrusting algae, mollusks	Poorly populated, some mollusks	Mollusks, (e.g. gastropods), algae	Mollusks Charophyta	Poorly populated, possibly algae	Poorly populated by lime-secreting organisms: possibly algae	No or few organisms	No organisms

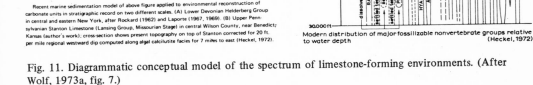

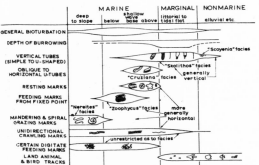

Fig. 11. Diagrammatic conceptual model of the spectrum of limestone-forming environments. (After Wolf, 1973a, fig. 7.)

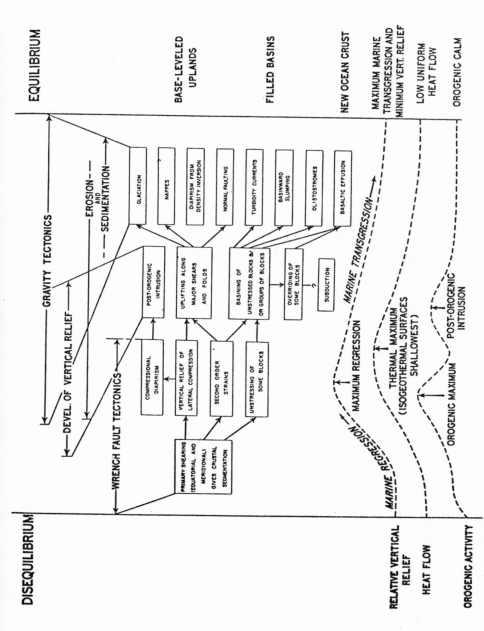

Fig. 12. Schematic model relating tectonics, faulting, sedimentation, and other variables in space and time. (After Moody, 1973; with permission of Am. Assoc. Petrol. Geol.)

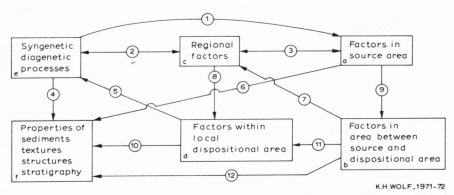

K.H.WOLF,1971-72

Fig. 13. General model of factors that determine properties of sediments. More complicated all-inclusive models were developed from this one, e.g., Fig. 15 and 16. (After Wolf, 1973a, fig. 4.)

on ore genesis. One further, more general example is made available in Fig. 12, which combines orogenesis, tectonics, magmatism and volcanism, formation of basins and sedimentation. Similar conceptualization should prove valuable in the investigation of the distribution of ore-forming episodes in the context of regional geologic developments.

A model that includes all natural variables to be considered in the study of clastic sediments has been summarized in Figs. 13 and 15; the former giving all mega-factors from which the more detailed model was prepared in Fig. 15. It should be noted that from these diagrams the models on placer and chemical ores were prepared - see sections below. It may be wise to use Fig. 15 to demonstrate "how to read" such flowchart-like models which use the "nesting" technique, for once understood all others can be easily deciphered.

The "starting point" of this conceptual diagram (Fig. 15) lies near the lower left-hand corner, i.e., "box" No. 1: *Particle properties*. These properties, such as composition, colour, roundness, etc., determine the textural characteristics and these in turn determine the fabrics. Both the textures and fabrics are also closely interrelated and are, therefore, combined to form box No. 2. Inasmuch as the study of particle properties, textures and fabrics comes under the heading of petrography and lithology, they are grouped together to form box No. 3. It is clear that the properties of the variables in box 3 determine those of No. 4: *Sedimentary structures* and these then control No. 5: *Stratigraphic properties*. Together with all syngenetic-diagenetic factors and processes, boxes Nos. 1–6 constitute petrography and petrology (box No. 7), which are directly controlled by all the factors in the local sedimentary environment. The latter are complexly interrelated among themselves, as shown by the arrows connecting the twelve variables. There are many complex relationships between the factors in boxes No. 7 and 8, but any additional arrows would obliterate the diagram. For this reason, just one arrow (i.e., No. VI) has been used between the two boxes. It is up to the reader to insert additional arrows to adapt such models to his own investigative problems. By expanding the models step-by-step — which, incidentally, is progressive from a local 4th-order scale through a 2nd-order to a large regional 1st-order scale — one can include *all* natural variables of direct and indirect influence in producing a sedimentary unit.

The following models are directly concerned with the petrology of ores and the economics of resources:

(1) a master diagram from which many of the flow-chart models have been derived by inserting all available details (e.g., Figs. 13 and 14);

(2) a model of chemical sedimentary ores (Fig. 16) (in support of many chapters in this multi-volume publication);

(3) a mastermodel of sedimentary placer ores (Fig. 17) (supplemental to Hail's Chapter 5 in Vol. 3);

(4) a model of the Witwatersrand-type uranium ores as an example of continental-terrestrial placer ores (Fig. 18) (partly in support of Pretorius' two chapters in Vol. 7);

(5) a sequence of tabular conceptual models on the Witwatersrand ore deposits (by Pretorius, 1966, reproduced with permission; Figs. 19–26);

(6) several diagrams conceptualizing the geochemical cycles of a number of chemical elements (Figs. 27–34);

(7) a generalized schematic model of ore-genesis theories according to the conventional and the new patterns of thought, the latter of which are well represented in the present volumes on stratiform/strata-bound ores (after Amstutz, 1964, with permission);

(8) a master model (Wolf, 1976, fig. 79) depicting metal sources and origin of uranium deposits in clastic sandstones of the Western States-type (i.e., Colorado Plateau and Wyoming types of ores; cf. Rackley's chapter 3 in Vol. 7) and Texas-type of uranium (i.e., near-shore sandstone milieux) (Fig. 36);

(9) a master model showing multiple metal sources, mode of transportation and origins of ores within four major host-rock types (Fig. 37; Wolf, 1976, fig. 92) (in support of several chapters);

(10) diagrammatic and tabular models (Wolf, 1976, fig. 52 and table VIII) illustrating possible metal sources, mode of transportation, and origin of some specific Cu and Pb–Zn deposits with their subdivision according to host-rock types (Fig. 38 and Table II) (supplemental to the chapters on the Mississipi Valley-type of ores in Vol. 6);

(11) a model of ore depostion in the geotectonic cycle (Laznicka, 1973, fig. 2; Figs. 39 and 40)) (in support of several chapters dealing with the relationships between tectonism and metallogeny); and

(12) a model (Fig. 41) summarizing the parameters or variables of the policy environment of the mineral systems as applied, for example, to Canada (Austin, 1974, fig. 12).

Inasmuch as the above-mentioned models are addressed to geologists with sufficient experience to comprehend the underlying principles, the models will be accompanied only by some superficial comments as a brief guide.

Figure 14 merely outlines the seven major groups of geologic settings of ore deposits: two subsurface hydrothermal milieux, one associated with the oceanic crust and the other

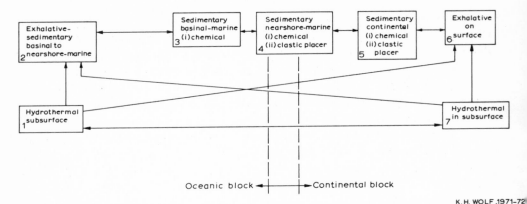

Fig. 14. Master diagram relating the various types of ores in sedimentary and sedimentary-volcanic piles, demonstrating their transitional nature. (After Wolf, 1973a, fig. 10.)

with the continental crust; two surface and sub-aqueous environments where volcanic—exhalative processes can occur; the same environments may offer conditions for purely physicochemical and biochemical ore precipitation; and the transition zone between marine and continental environments (supergene deposits have not been included here). Placer accumulations are usually confined to the nearshore and continental localities and a separate model has been drafted for them (cf. Fig. 17). For the chemical ores of various types, refer to the model in Fig. 16. The arrows joining the seven "boxes" in Fig. 14 merely indicate that a number of transitional types of ores have been recognized as being based on source, precipitation mechanisms, location of origin, and tectonic setting (for a discussion on gradational ore types, see in this Vol., Chapter 4 by Gilmour).

Fig. 16 modellizes the genesis of several types of physicochemical and biochemical ores in an all-inclusive fashion, i.e., it is a generalized model applicable to ores with components derived from a landmass (boxes 11 and 12), from volcanic-exhalative sources (in box 9) as well as from the sea water itself (boxes 7, 9 and 10). Secondary diagenetic and metamorphic modifications have been considered also in boxes 7 and 13. For any particular geologic setting under investigation, the researcher merely has to isolate those boxes pertinent to his problem, i.e., delete those not applicable. The exploration geologist, for example, can then use his model to phrase meaningful questions about an area, to prepare a research programme, and write a research/exploration report that has to be delivered in written and/or verbal form. By employing such models in both the planning stages at the commencement and during the interpretive phases at the end of a geological/geochemical/geophysical programme, no possibility and no alternative need be inadvertently overlooked. The same arguments, of course, apply to all other models offered below. Hallberg's (1972) "energy circuit system approach" applied to sedimentary sulfide-mineral formation should prove to be valuable, possibly in conjuction with the model given here.

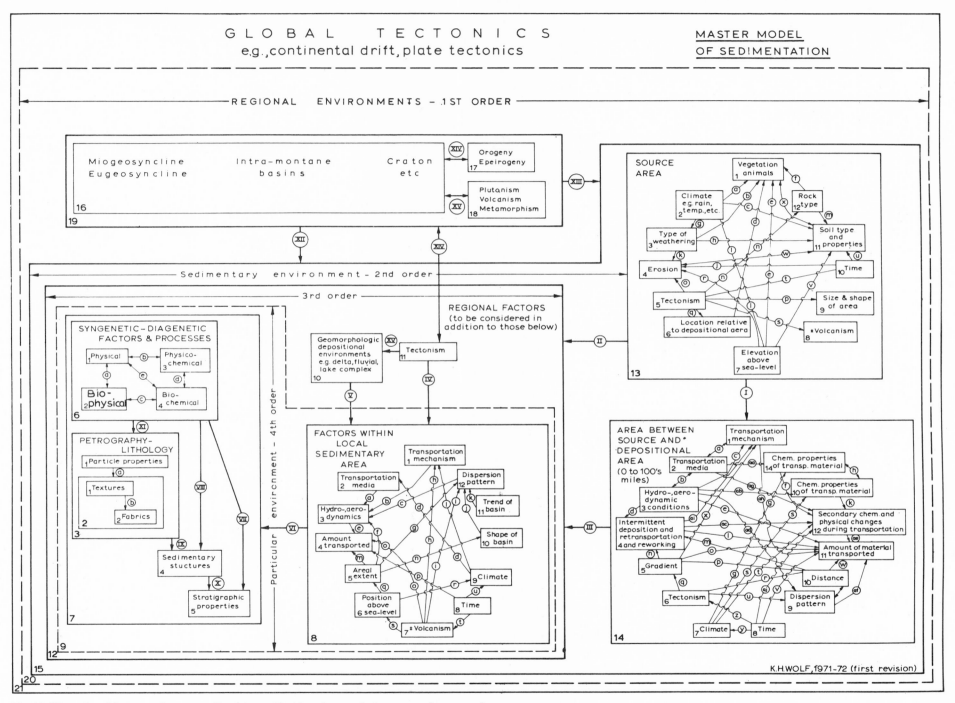

Fig. 15. Hierarchy of factors and processes (local to world-wide; microscopic to meso- and megascopic; simple to complex) that singly or in combination determine the properties of sediments (clastics, pyroclastics and sedimentary ores). (After Wolf, 1973a, fig. 5.)

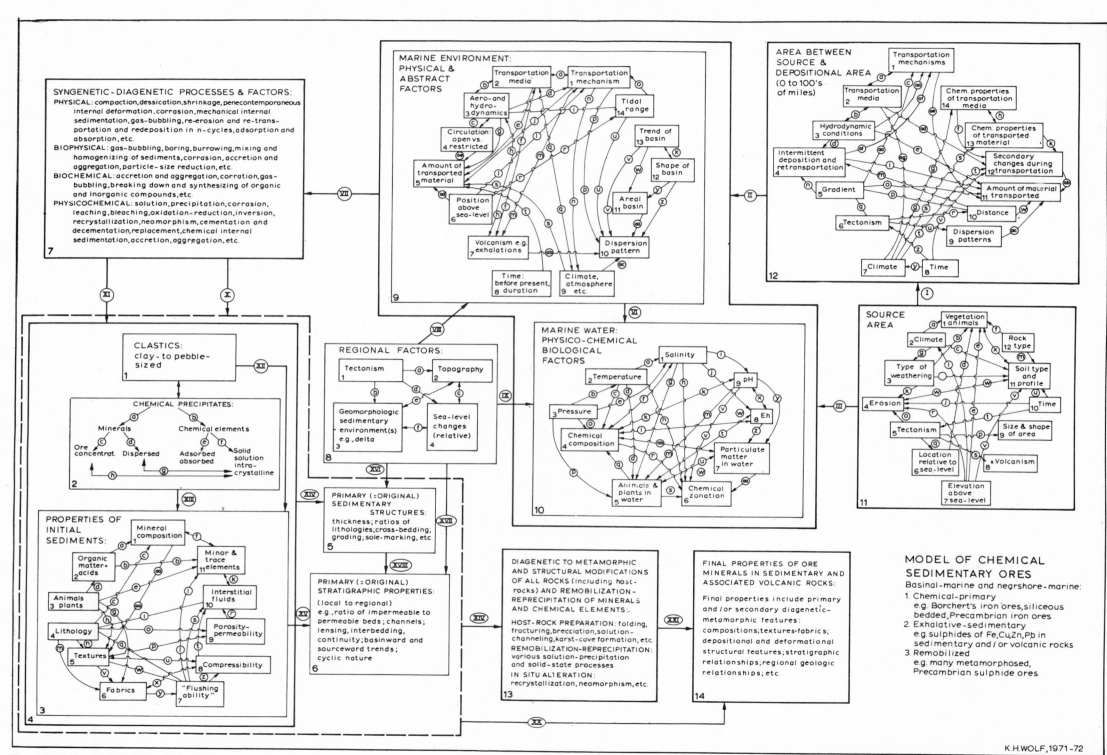

Fig. 16. Master model of chemical sedimentary ores (i.e., primary chemical, biochemical, volcanic exhalative-sedimentary, and remobilized). (After Wolf, 1973a, fig. 11.)

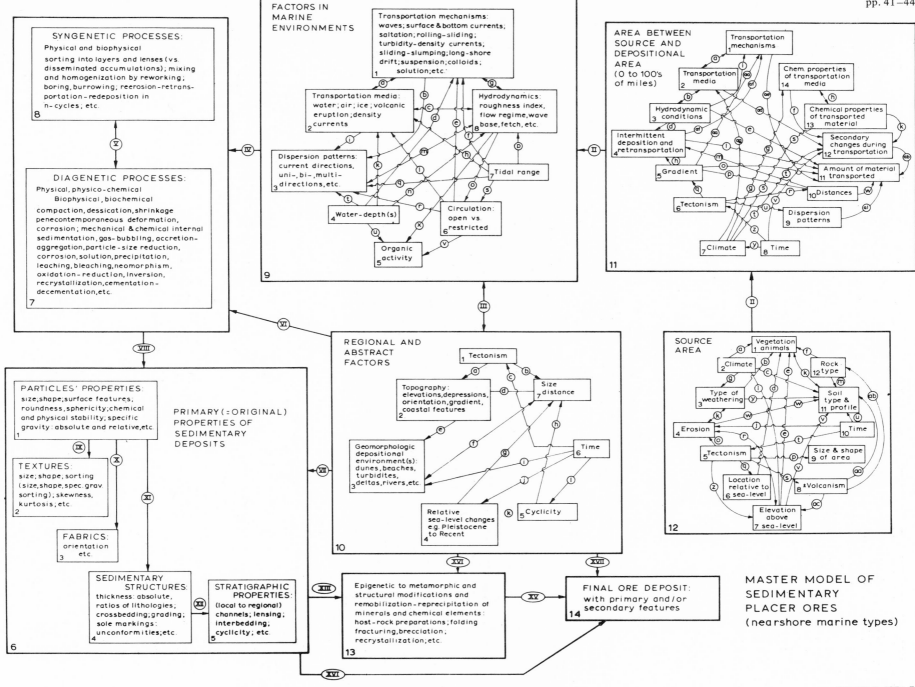

Fig. 17. Model of placer-type of sedimentary ores. (After Wolf, 1973a, fig. 12.)

MASTER MODEL OF
SEDIMENTARY
PLACER ORES
(nearshore marine types)

K.H.WOLF,1971-72

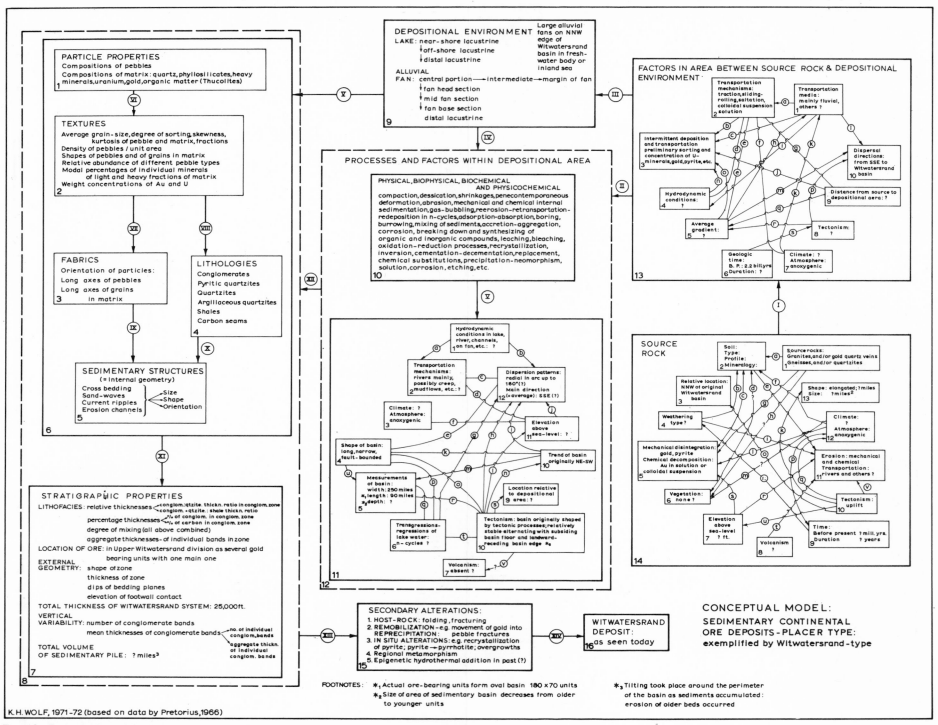

Fig. 18. Conceptual model of the placer-type Witwatersrand uranium-gold deposits. (After Wolf, 1973a, fig. 13; based on data by Pretorius, 1966.)

The two conceptual diagrams in Figs. 17 and 18 are established for ores formed by mechanical erosion-transportation-accumulation processes, i.e., the so-called placer accumulations described by Hails (Chapter 5, Vol. 3) and in two chapters on the Witwatersrand by Pretorius (Vol. 7). Several alternative source areas included: (1) a source area on land wich may be hundreds of miles remote from the final depositional milieux; (2) a land situated close to the accumulative environment; and (3) a source within the marine environment itself which furnished the placer minerals by long-shore drift or from farther offshore by shoreward currents and waves. In the latter case, one merely deletes boxes 11 and 12. Inasmuch as Fig. 17 is a generalized diagram, another one was prepared for the Witwatersrand-type of uranium—gold ores. Similar models should prove to be valuable for genetically related Precambrian U-containing conglomerates, such as those of the Blind River—Elliot Lake type and Jacobina type in Canada and Brazil, respectively.

With some pleasure, the editor presents the following eight figures by Pretorius (1966)[1], in particular because these were employed successfully in the first phases of computerization of the exploration programmes that resulted in extending known ore deposits and in the localization of new ones (Figs. 19—26). Each of the subscribes of the figures provides enough information to allow the reader to recognize the purpose of the individual models. Instead of arrows, merely lines were used to connect the boxes that are related to each other. The reader is well advised to refer to these models either while or after reading the chapters by Pretorius (Vol. 7, Chapter 1 and 2).

Two aspects are of particular interest to the editor, namely (1) the very detailed sedimentary petrology undertaken (together with stratigraphic, structural and environmental reconstructions) (see Fig. 25), and (2) the list of statistical methods employed, as given in the right-hand column in Fig. 26.

The *geochemical cycles* of all the major chemical elements and isotopes have been modelled in different ways by numerous researchers, whereas more quantitative data has to be made available before the intricacies of all the parts of the cycle of the rarer elements can be fully apprehended. In some instances, the geochemical and rock cycles are synonymous, whereas in other cases the former is more inclusive, insofar as the geochemical cycle incorporates the hydrosphere and atmosphere in addition to the lithosphere. To illustrate one mode of the conceptualization of chemical elements, eight diagrams are reproduced here (with the kind permission of the authors) from a more extensive collection prepared by Borchert and his associates (Figs. 27—34 depicting the geochemical distribution of In, Cd, Ga, Te, Se, Tl, Ge, and P.)

Amstutz (1964) compared in an elegant way the older conventional theories of ore

[1] The editor is grateful to Dr. J.E. Gill of McGill University, Montreal, Canada, who in 1968 drew his attention to the publications by Pretorius. The latter, upon being contacted, supplied a number of publications from the University of Witwatersrand. Together with earlier reading, Pretorius' models formed the "seed" that gave rise to the editor's own efforts in modelling diagrammatically geological phenomena related to ore genesis, including that on the Witwatersrand (Wolf, 1973a and b).

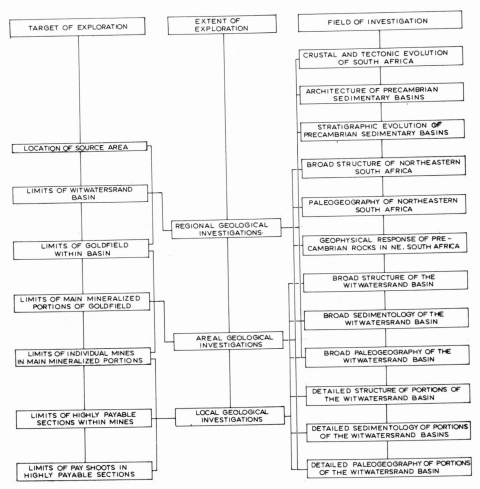

Fig. 19. Conceptual model of scales, targets, and fields of geological exploration for goldfields, mines, and pay shoots in the Witwatersrand basin. (After Pretorius, 1966; reproduced with permission.)

genesis, on the one hand, with the more modern concepts, on the other. These latter, "new patterns of thought" (Fig. 35), as he called them, are now well entrenched in the field of ore petrology, as can be deduced from the chapters in this multi-volume synopsis. With only minor exceptions, the ore-forming mechanisms accepted today agree with the genetic hypotheses offered by Amstutz in Fig. 35, and this classificatory scheme is, therefore, provided in support of the other schemes in the chapter on terminology and classifications. Amstutz's circular diagram is superior to the more common pigeon-hole classifications as no artificial boundaries are set that may convey the false impression that ore types are distinct varieties. The circular approach is more realistic inasmuch as any

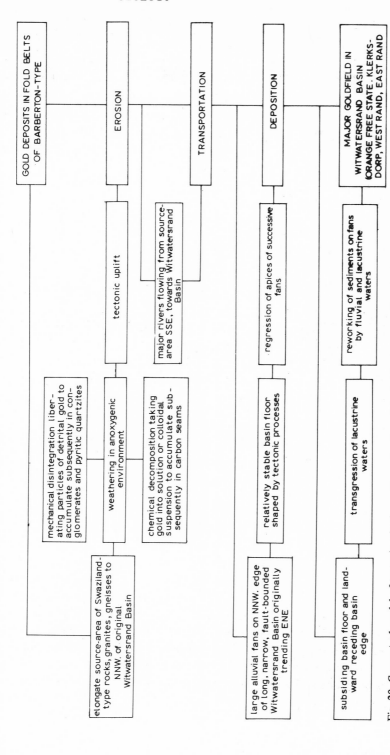

Fig. 20. Conceptual model of tectonic, physiographic, and sedimentological factors controlling the development of a major goldfield in the Witwatersrand basin. (After Pretorius, 1966; reproduced with permission.)

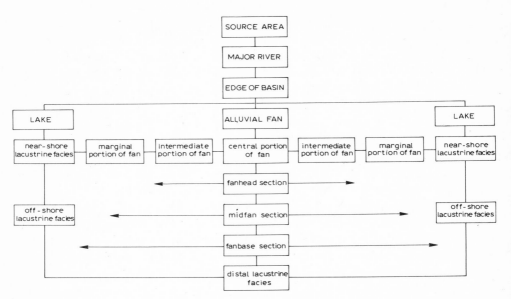

Fig. 21. Conceptual model of the radial and lateral geometry of an alluvial fan located on the edge of a lake to form a major goldfield in the Witwatersrand basin. (After Pretorius, 1966; reproduced with permission.)

number of transitional genetic ores can be visualized between main end-member types, as discussed at lenght by Gilmour in Chapter 4, this Vol. (For details on the four possible combinations of syngenetic, epigenetic, endogenous, and exogenous ore-forming processes, see also Fig. 35.)

In a brief review of the origin of uranium in sandstones, as part of a synopsis of the influence of compaction on the origin of ores in sediments and volcanics, Wolf (1976, fig. 79) conceptualized the numerous genetic variables, as demonstrated here in Fig. 36. This model has the following features:

(1) The possibility of at least two uranium-concentration episodes has been shown, i.e., one occurring early during the filling of the sedimentary basin and prior to tectonism (i.e., intra-basinal stage) and the second one occurring after the deformation of the sedimentary and volcanic sequences (i.e., post-deformational stage).

(2) A sedimentary depocenter, receiving detritus, has an extra-basinal (= outside) terrestrial source (box I) that upon weathering, supplies lithic grains and pebbles, various sand and silt grains (boxes III and IV) and, depending on the degree of chemical weathering, various clay minerals, to the basin of accumulation (boxes VI and VII). Volcanism may take place on land (box II) and pyroclastic debris would then be moved directly through passing via the air, or indirectly by rivers, to the sedimentary basin (box V). All the types of detritus described above may contain a certain amount of elements, such as uranium and vanadium, for example.

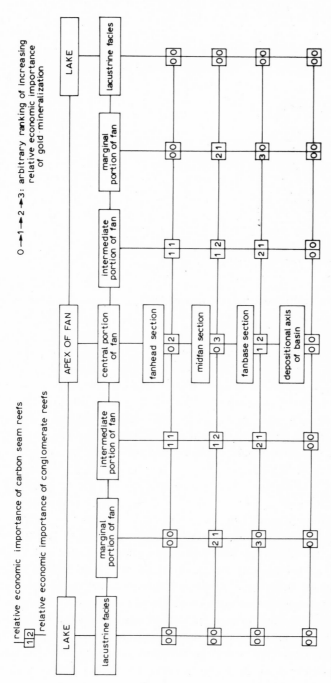

Fig. 22. Conceptual model of the geometry of optimum conditions for gold mineralization in conglomerate and carbon seam reefs in a major goldfield in the Witwatersrand basin. (After Pretorius, 1966; reproduced with permission.)

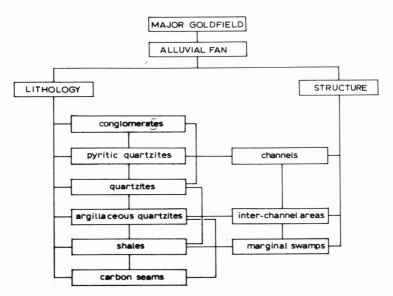

Fig. 23. Conceptual model of the lithology and structure of an alluvial fan constituting a major goldfield in the Witwatersrand basin. (After Pretorius, 1966; reproduced with permission.)

(3) In addition to the non-volcanic (= epiclastic) and volcanic (= pyroclastic) debris delivered from the land, the same two sources may provide chemical elements to surface and subsurface fluids (line 1A) whenever deep weathering and leaching takes place. These elements may or may not be precipitated upon reaching the sedimentary milieu.

(4) The deposits, which accumulated in the depocenter (i.e., boxes III to VII), could have been composed of grains that contain traces of uranium in arkoses, lithic sediments, and pyroclastics and/or clay minerals in muds and claystones with adsorbed and absorbed uranium.

(5) Compaction waters (plus some meteoric fluids and during late diagenesis also water of crystallization released as a result of clay-mineral transformations) moving from the fine-grained muds interbedded with the sandstones and conglomerates in a fluvial sedimentary complex, as well as from deeper basinal sediments, may carry uranium from volcanic ash and clay minerals. Additionally, compaction fluids moving through permeable sandstones, conglomerates and coarse pyroclastics may leach out uranium. Within the coarser-grained aquifers, these compaction solutions may precipitate uranium wherever organic matter is present and provides a local reducing environment. The uranium mineralization is either in the form of fine disseminations or as ore-grade concentrations (box VIII). The process may have been cyclic in nature, namely, dissolution—transportation--reprecipitation, and could have occurred many times before the flow of subsurface water was terminated for a number of possible geologic reasons. Theoretically, this would

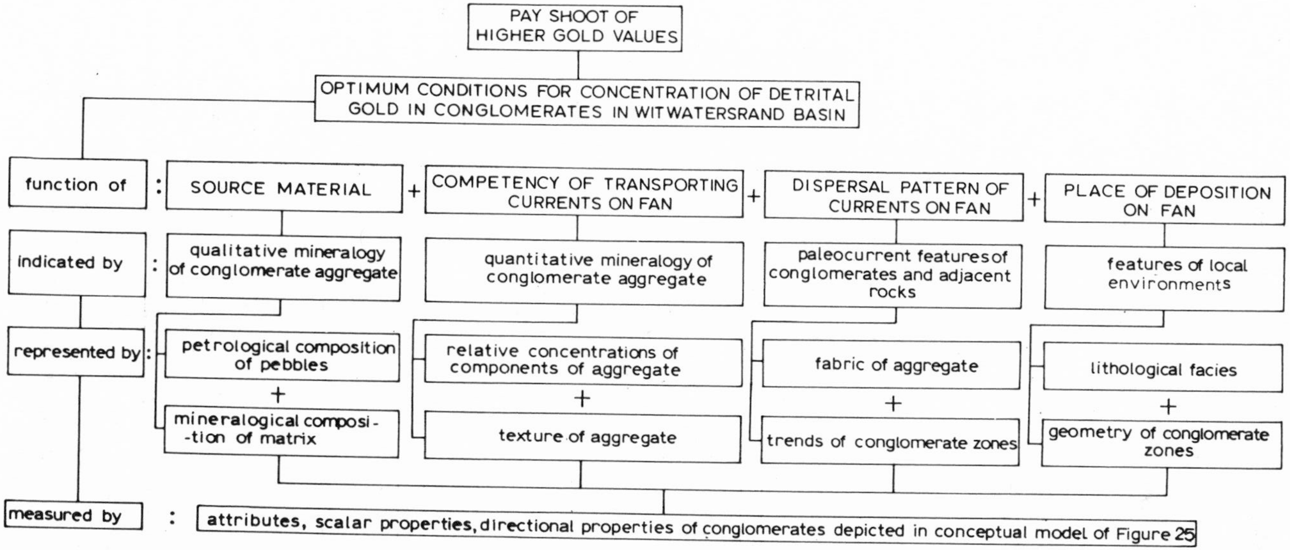

Fig. 24. Conceptual model of geological factors controlling the development of pay shoots in conglomerate reefs of the Witwatersrand basin. (After Pretorius, 1966; reproduced with permission.)

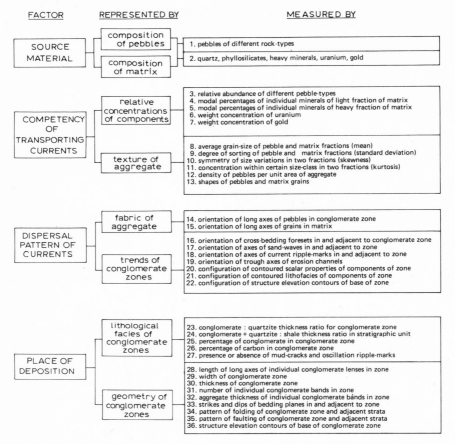

FACTOR   REPRESENTED BY   MEASURED BY

SOURCE MATERIAL
- composition of pebbles — 1. pebbles of different rock-types
- composition of matrix — 2. quartz, phyllosilicates, heavy minerals, uranium, gold

COMPETENCY OF TRANSPORTING CURRENTS
- relative concentrations of components
  - 3. relative abundance of different pebble-types
  - 4. modal percentages of individual minerals of light fraction of matrix
  - 5. modal percentages of individual minerals of heavy fraction of matrix
  - 6. weight concentration of uranium
  - 7. weight concentration of gold
- texture of aggregate
  - 8. average grain-size of pebble and matrix fractions (mean)
  - 9. degree of sorting of pebble and matrix fractions (standard deviation)
  - 10. symmetry of size variations in two fractions (skewness)
  - 11. concentration within certain size-class in two fractions (kurtosis)
  - 12. density of pebbles per unit area of aggregate
  - 13. shapes of pebbles and matrix grains

DISPERSAL PATTERN OF CURRENTS
- fabric of aggregate
  - 14. orientation of long axes of pebbles in conglomerate zone
  - 15. orientation of long axes of grains in matrix
- trends of conglomerate zones
  - 16. orientation of cross-bedding foresets in and adjacent to conglomerate zone
  - 17. orientation of axes of sand-waves in and adjacent to zone
  - 18. orientation of axes of current ripple-marks in and adjacent to zone
  - 19. orientation of trough axes of erosion channels
  - 20. configuration of contoured scalar properties of components of zone
  - 21. configuration of contoured lithofacies of components of zone
  - 22. configuration of structure elevation contours of base of zone

PLACE OF DEPOSITION
- lithological facies of conglomerate zones
  - 23. conglomerate : quartzite thickness ratio for conglomerate zone
  - 24. conglomerate + quartzite : shale thickness ratio in stratigraphic unit
  - 25. percentage of conglomerate in conglomerate zone
  - 26. percentage of carbon in conglomerate zone
  - 27. presence or absence of mud-cracks and oscillation ripple-marks
- geometry of conglomerate zones
  - 28. length of long axes of individual conglomerate lenses in zone
  - 29. width of conglomerate zone
  - 30. thickness of conglomerate zone
  - 31. number of individual conglomerate bands in zone
  - 32. aggregate thickness of individual conglomerate bands in zone
  - 33. strikes and dips of bedding planes in and adjacent to zone
  - 34. pattern of folding of conglomerate zone and adjacent strata
  - 35. pattern of faulting of conglomerate zone and adjacent strata
  - 36. structure elevation contours of base of conglomerate zone

Fig. 25. Conceptual model of attributes, scalar properties, and directional properties as measures of geological factors controlling the development of pay shoots in Witwatersrand conglomerate reefs. (After Pretorius, 1966; reproduced with permission.)

complete the ore-concentration episode No. 1 in the idealized conceptual model presented here.

(6) The coarse-grained sediments that upon leaching can supply the uranium to the subsurface fluids, can also act penecontemporaneously as "reservoir" rocks for the ore mineralization. For this reason, a clear separation of "host rock" (boxes III and VIII, Fig. 36), on the one hand, and "source rock", such as arkoses (box IV), for example, on the other, is not possible. In other words, the separate boxes should not mislead the reader into assuming that the host rock of the uranium ore is different from the coarse-grained sediments listed in the model.

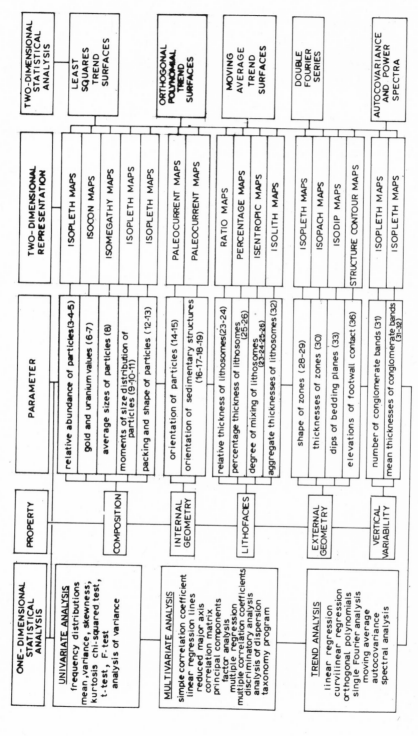

Fig. 26. Conceptual model of two-dimensional representation and tectonics of statistical analysis of geological parameters of conglomerate reefs of the Witwatersrand basin. (After Pretorius, 1966; reproduced with permission.)

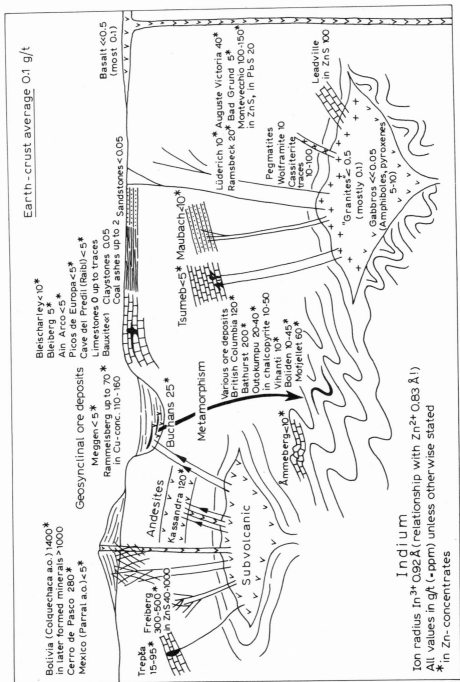

Fig. 27. Geochemical distribution of indium. (After J. Feiser, H. Borchert, and G. Anger, 1965, unpublished; reproduced with permission.)

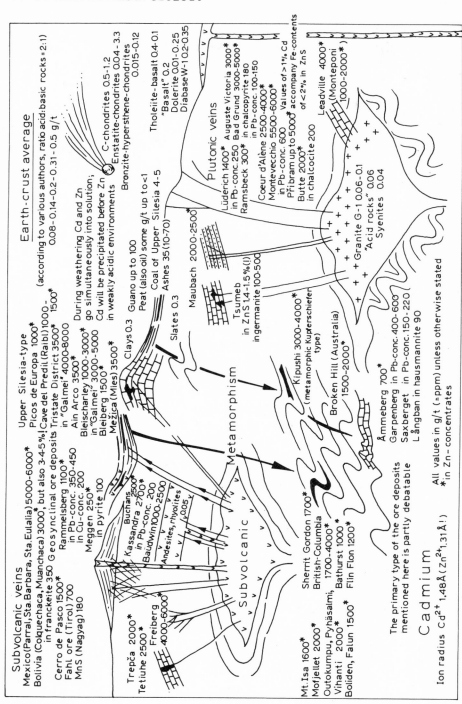

Fig. 28. Geochemical distribution of cadmium. (After J. Feiser, H. Borchert and G. Anger, 1965, unpublished; reproduced with permission.)

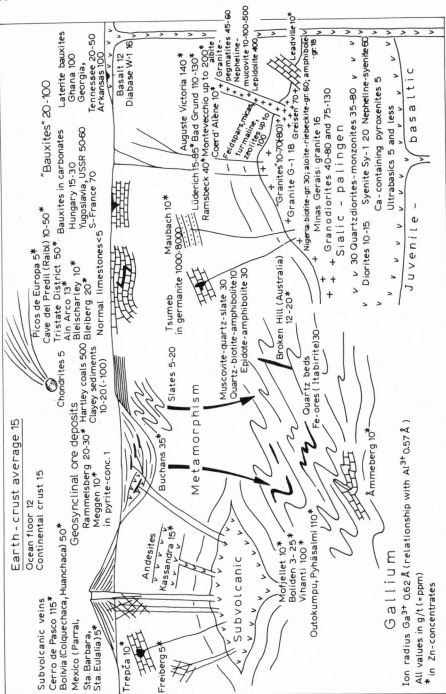

Fig. 29. Geochemical distribution of gallium. (After J. Feiser, H. Borchert and G. Anger, 1965, unpublished; reproduced with permission.)

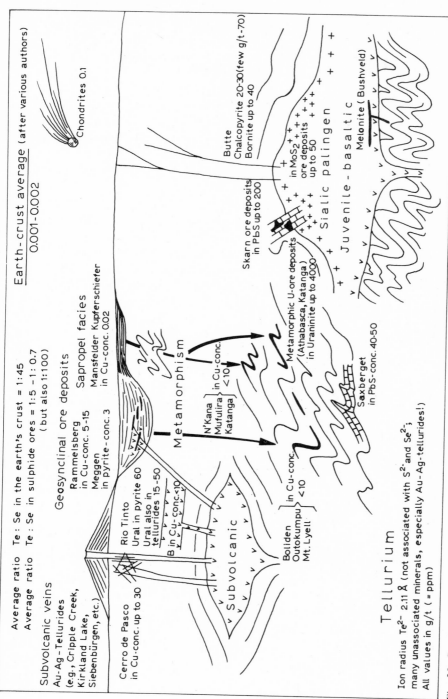

Fig. 30. Geochemical distribution of tellurium. (After J. Feiser, H. Borchert and G. Anger, 1965, unpublished; reproduced with permission.)

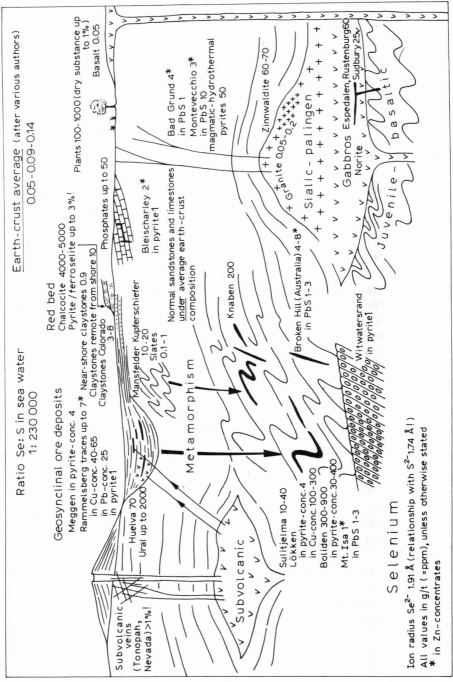

Fig. 31. Geochemical distribution of selenium. (After J. Feiser, H. Borchert and G. Anger, 1965, unpublished; reproduced with permission.)

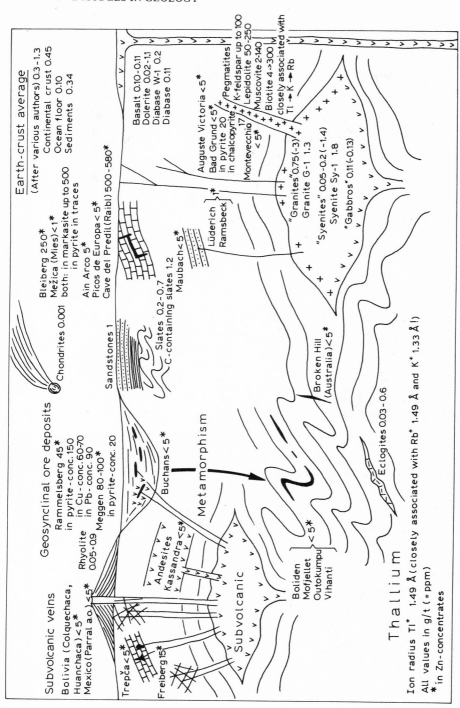

Fig. 32. Geochemical distribution of thallium. (After J. Feiser, H. Borchert and G. Anger, 1965, unpublished; reproduced with permission.)

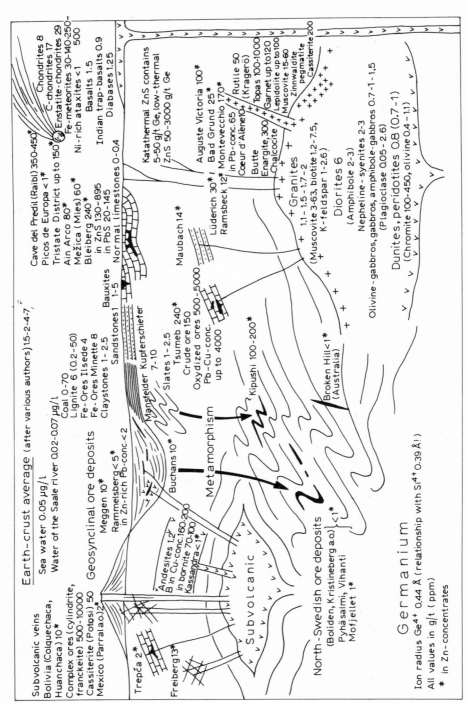

Fig. 33. Geochemical distribution of germanium. (After J. Feiser, H. Borchert, and G. Anger, 1965, unpublished; reproduced with permission.)

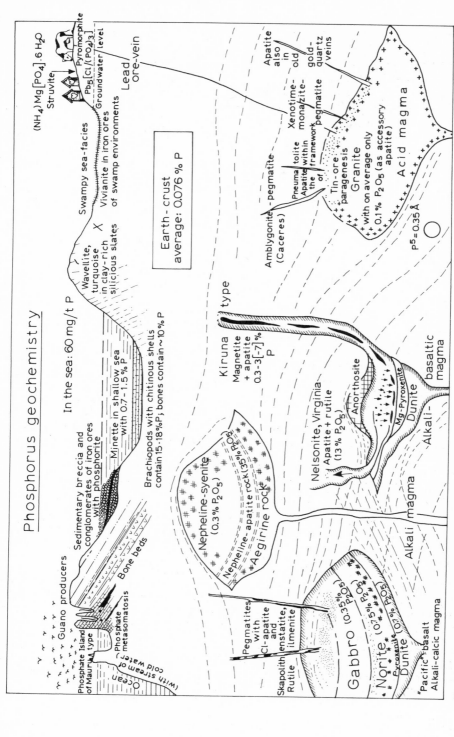

Fig. 34. Geochemical distribution of phosphate. (After H. Borchert, 1950, unpublished; reproduced with permission.)

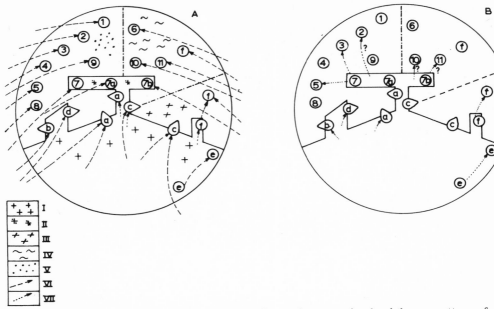

Fig. 35. Schematic representation of ore genesis according to the conventional and the new patterns of thought. (After Amstutz, 1964).

A. This figure shows the domination of the myth of epigenetic replacement and of the unknown depth ("deep seated sources"). Epigenetic introduction from the outside is an axiomatic condition for the formation of most ore deposits. This pattern corresponds essentially to the creationistic, pre-Darwinian beliefs in paleontology.

B. This figure pictures the pattern of ore genesis theories according to the new "petrographic" or integrated theory, according to which ore deposits normally formed at the same time and essentially within or very near the observed host rock. Just as man and animals in the evolution theory of paleontology, ore deposits are, in the new theory, considered a normal integral part of rock evolution.

I = igneous intrusive rocks (known!); II = igneous extrusive or subvolcanic rocks (known); III = metamorphic igneous rocks or migmatites; IV = metamorphic sedimentary rocks; V = sedimentary rocks (non-, or partly metamorphic); VI = introduction from the (unknown) outside assumed; VII = some migration probable, possible, or (?) questionable.

List of major types of ore deposits for which a syn-endo as well as an epi-exo origin[1,2] has been proposed. In sediments and volcanic rocks: 1 = Arkansas-type barite deposits; 2 = Mississippi Valley-type deposits (including the barite and fluorspar deposits in the same type of sediments); 3 = Rammelsberg and similar deposits; 4 = magnesite, rhodochrosite and siderite deposits of the Alps and elsewhere; 5 = Kupferschiefer and/or Red Bed copper deposits as well as various disseminated to massive copper–lead–zinc deposits, for example of the Kuroko type; 6 = Blind River, Witwatersrand and similar deposits; 7 = propylitic deposits of copper, gold, and other metals; 7a and 7b = deposits of sulfides, oxides and native elements (Cu, Ag, Au) in or near volcanic rocks (often with spilitic phases); 8 = Mina Ragra type vanadium deposits; 9 = Colorado Plateau or "sandstone type" uranium deposits; 10 = iron deposits of the Lake Superior type; 11 = Ducktown, Broken Hill, Outukumpu, Falun, and similar deposits in metamorphic belts.

In and adjacent to igneous rocks: a = porphyry copper deposits in and around intrusions (including the Climax molybdenum deposit); b = Granite Mt., Utah, deposits of magnetite, and similar deposits; c = tin deposits in and around intrusions; d = contact deposits, pipe deposits, perimagmatic vein deposits; e = chromite deposits; f = pegmatites.

[1] Endogenous = from within (the host rock or its own source); exogenous = from outside (the host rock or its own source).

[2] Syngenetic = at the same time as the host rock; epigenetic = at any later time than the host rock.

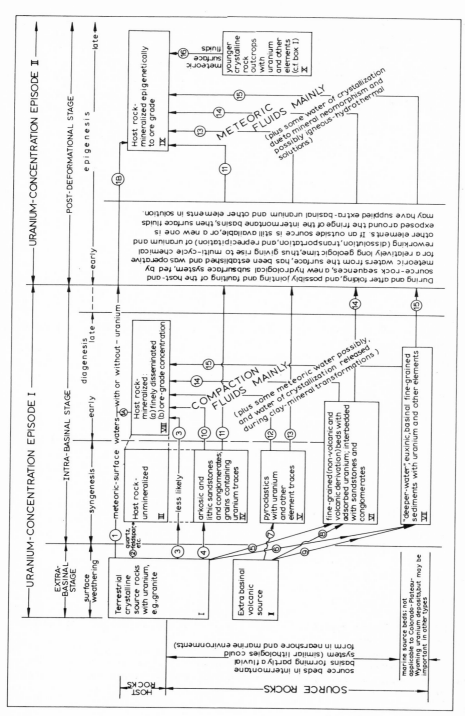

Fig. 36. Conceptual model of the origin of uranium in sandstones. (For details, see text, and Wolf, 1976, fig. 79.)

(7) The scheme in Fig. 36 is, by necessity, very generalized and incorporates at least two varieties of uranium deposits, namely, the Colorado Plateau—Wyoming (called the Western States-type in Rackley's chapter in Vol. 7 for reasons dicussed by him) and the Texas uranium types depicted on the left-hand side in the diagram. The former originated in a continental intermontane basin within the deposits of a fluvial—lacustrine complex, whereas the latter formed within sediments that accumulated under nearshore and marine conditions (Eargle and Weeks, 1971). Thus, the types of source and host rocks are controlled by the total geologic setting.

(8) Although syngenetic ore concentrations in general are known in sediments, the uranium deposits discussed here are of a "secondary" origin. Even very early diagenetic precipitates in sands are secondary in comparison to the detrital sand grains which accumulated prior to the introduction of the chemical precipitates. Late diagenetic products in sediments, as is well known now, show all the characteristics of *both* early-diagenetic and epigenetic processes (see for example Vine and Tourtelot, 1970). For this reason, it is often difficult, if not impossible, to distinguish between diagenetic material and that formed much later (usually called "epigenetic"). Nearly all uranium deposits in sandstones have epigenetic features, but the *precise time* of formation as related to the host sediments (i.e., whether or not early diagenetic and, therefore, nearly contemporaneous with the sand grains) is not discernable in many investigations, if textures, fabrics, local structures, and stratigraphic properties are considered.

(9) As a consequence of folding, jointing, faulting, and subsequent subaerial erosion, a new geological situation is established, which is different in most respects from those that existed under the original sedimentary milieu. A new surface and subsurface hydrologic system is initiated, with meteoric water playing the major role. There are several alternative geological settings that are significant in uranium genesis: (a) The tectonism resulted in folding and fracturing of the sediments that already contained some diagenetically-concentrated uranium. The meteoric subsurface fluids passing through the folded beds (lines 11, 13, 14 and 15 in Fig. 36) dissolved, transported and reprecipitated the uranium, forming "roll" structures (cf. Rackley in Vol. 7). Repeated, cyclic movement of the uranium eventually resulted in an economic ore deposit. (b) As a result of tectonism, intermontane basins are formed that were fringed by crystalline rock outcrops, such as granite. Chemical weathering may supply uranium to the surface meteoric water which, upon reaching the sedimentary rocks, penetrates the latter (box IX). The subsurface meteoric water may have picked up additional uranium from the arkosic sandstones and pyroclastics while passing through them. Hence, there is a transition to the conditions mentioned in case (a). Upon reaching a chemically reducing milieu, these meteoric waters precipitated the urianium, possibly in the form of "rolls". (c) If the mechanical erosion of the crystalline rocks (box X, Fig. 36) fringing the intramontane basin formed a new sedimentary complex, then these fine- to coarse-grained deposits would have been exposed to the diagenetic processes which were already mentioned under (1) through (8) above. All the discussions presented earlier, therefore, are applicable here again.

TABLE II

Four types of ore deposits with several sub-varieties (for details, see text)

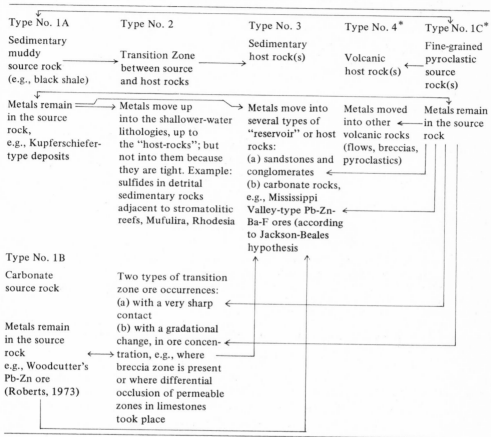

Type No. 1A	Type No. 2	Type No. 3	Type No. 4*	Type No. 1C*
Sedimentary muddy source rock (e.g., black shale)	Transition Zone between source and host rocks	Sedimentary host rock(s)	Volcanic host rock(s)	Fine-grained pyroclastic source rock(s)
Metals remain in the source rock, e.g., Kupferschiefer-type deposits	Metals move up into the shallower-water lithologies, up to the "host-rocks"; but not into them because they are tight. Example: sulfides in detrital sedimentary rocks adjacent to stromatolitic reefs, Mufulira, Rhodesia	Metals move into several types of "reservoir" or host rocks: (a) sandstones and conglomerates (b) carbonate rocks, e.g., Mississippi Valley-type Pb-Zn-Ba-F ores (according to Jackson-Beales hypothesis	Metals moved into other volcanic rocks (flows, breccias, pyroclastics)	Metals remain in the source rock
Type No. 1B Carbonate source rock	Two types of transition zone ore occurrences: (a) with a very sharp contact (b) with a gradational change, in ore concentration, e.g., where breccia zone is present or where differential occlusion of permeable zones in limestones took place			
Metals remain in the source rock e.g., Woodcutter's Pb-Zn ore (Roberts, 1973)				

* Note: Volcanic exhalative and hydrothermal metal supply source is not considered here.

Wolf (1976, fig. 52 and table VIII) has also modellized the sources and origins of copper in four types of sedimentary-volcanic host rocks, as illustrated here in Fig. 37 and Table II. Using this data as a basis for discussion, let us consider some genetic problems. When contrasting mineralization within fine-grained clastics with that occurring within carbonate host rocks, the following problem should be taken into account in future investigations of both the Mississippi Valley-type lead—zinc and the Rhodesian-type copper deposits. Those researchers adhering to the syngenetic—diagenetic origin have usually proposed that the clay-rich sediments may have been the *source* of the metal-bearing fluids (as well as for hydrocarbons in some instances), which moved into a suitable

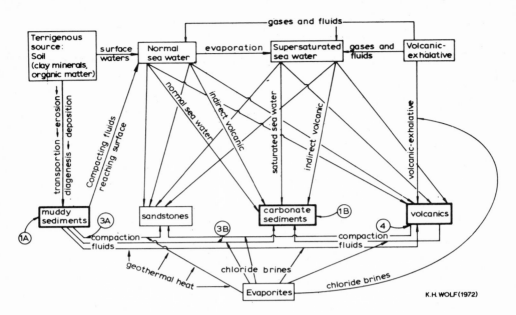

Fig. 37. Conceptual model showing sources for metal ions, i.e., terrigenous rocks, normal and saturated sea water, volcanic—exhalative emanations, and possibly evaporites, and chlorite brines derived
from evaporites. Four rock varieties may serve as hosts for the ore bodies, namely, mudstones or
shales, sandstones, carbonates, and volcanics. The types of copper ores given by the four numbers
(e.g., 3A and 3B) are discussed by Wolf, 1976. One should note that this model is simplified, as it does
not indicate, for example, that the detritus in sandstones (e.g., arkoses) and the limy and/or dolomitic
components in the carbonates can supply chemical elements for ore mineralization under favourable
conditions. (After Wolf, 1976, fig. 52.)

reservoir or *host* rock where precipitation of the ore minerals occurred. Considering that
ore mineralization has been located in both "source" and "host rock" lithologies, the
question should be posed as to a possible genetic relationship between them. An attempt
should also be made to establish a conceptual model composed of end-members and
several transitional examples in between as shown in Table II and Fig. 37. If the metal
originated in the black-shale, euxinic environment and was precipitated there (type No.
1A), then it would form a Kupferschiefer-type metal deposit. Similarly, if the shale was
the source of oil, but the oil remained there for various possible geologic reasons, the
result is an oil shale. If the metal-bearing fluids moved from the shaly environment to a
locality where there is a lithologic transition, e.g., an interbedded shale—siltstone and
carbonate facies, and where the carbonate is impermeable or does not offer the geochemical conditions required for precipitation, deposition of the ore minerals may occur within
the clastics around the carbonate unit (type No. 2, Table II). This may have occurred in
the case of some deposits in Rhodesia. If the carbonate unit is permeable and the geochemical milieu is conducive to precipitation, however, then the carbonate may become

the host rock for the ore minerals (type 3b, Fig. 37) (e.g., Mississippi Valley-type deposits). Roberts (1973) suggested that the carbonate sediments themselves were the precursors for the metal ions that upon remobilization became concentrated to form the Pb—Zn accumulations of the Woodcutter's mine in Australia (type 1B, Table II and Fig. 37).

It should be pointed out that in the above discussion no reference has been made to a volcanic source for the metals. In Table II it has been proposed that pyroclastic material, with its relatively high trace-element content, could be the source for the metal ions, as has been proposed for the Colorado Plateau uranium ores (which constitute one possible variety of type 3 ore among several others). If the metals remain in the volcanic source rock as concentrations, the type No. 1C ore would be formed, whereas if the metals move into another variety of volcanic host rock, possibly because the latter is more permeable and more porous, the ore would be of type No. 4 (Table II and Fig. 37). As the arrows indicate, there are other combinations of possible migration routes, namely, metals from the pyroclastics could be moved into sandstones or into carbonate host rocks. Although no volcanic—exhalative, hydrothermal source has been considered in Table II, it has been given a proper place in Fig. 37 discussed earlier.

The above-described, rather simple model in Fig. 37 has been used as a basis to prepare a more all-inclusive conceptual diagram combining the numerous factors and processes responsible for the origin of ores in sedimentary—volcanic piles (Fig. 38). This scheme comprises "boxes" in the upper part (I to IV), depicting the four major sources for the metals, and boxes listing the four groups of lithologies that commonly act as hosts for the ore bodies (A to D). One of the sources is extra-basinal, two are intra-basinal, and one is volcanic, which can be both extra- and intra-basinal as visualized in relationship to the area of sedimentation and ore mineralization. It is important to realize, as mentioned earlier, that any of the so-called "host" rocks can at the same time serve as a source of primary dispersed trace and minor metals which can be reworked and concentrated. The chloride brines derived from evaporites are depicted at the bottom of the model as another source (box E). The numerous genetic interrelationships, ranging from simple to complex, are illustrated by the arrows.

As illustrated by the brief explanations accompanying Figs. 36 to 38, geological discussions can be made more precise by employing diagrammatic models. Although they are idealized, they serve their purpose in compelling the researcher and explorationist to consider all possible genetic interpretations.

Tectonic and orogenic mechanisms, often related to the geosynclinal cycle — but more recently in connection with continental drift and plate tectonics — have been modellized, as one earlier example illustrated. Laznicka (1973) discussed the development of non-ferrous metal deposits in geological time during a geosynclinal phase. Although many of the ore deposits treated by him are not of the stratabound variety under consideration here, his information is summarized below to present the whole sequence of ores generated during one cycle — it would be meaningless to take the strata-bound ores out of their total metallotectonic context.

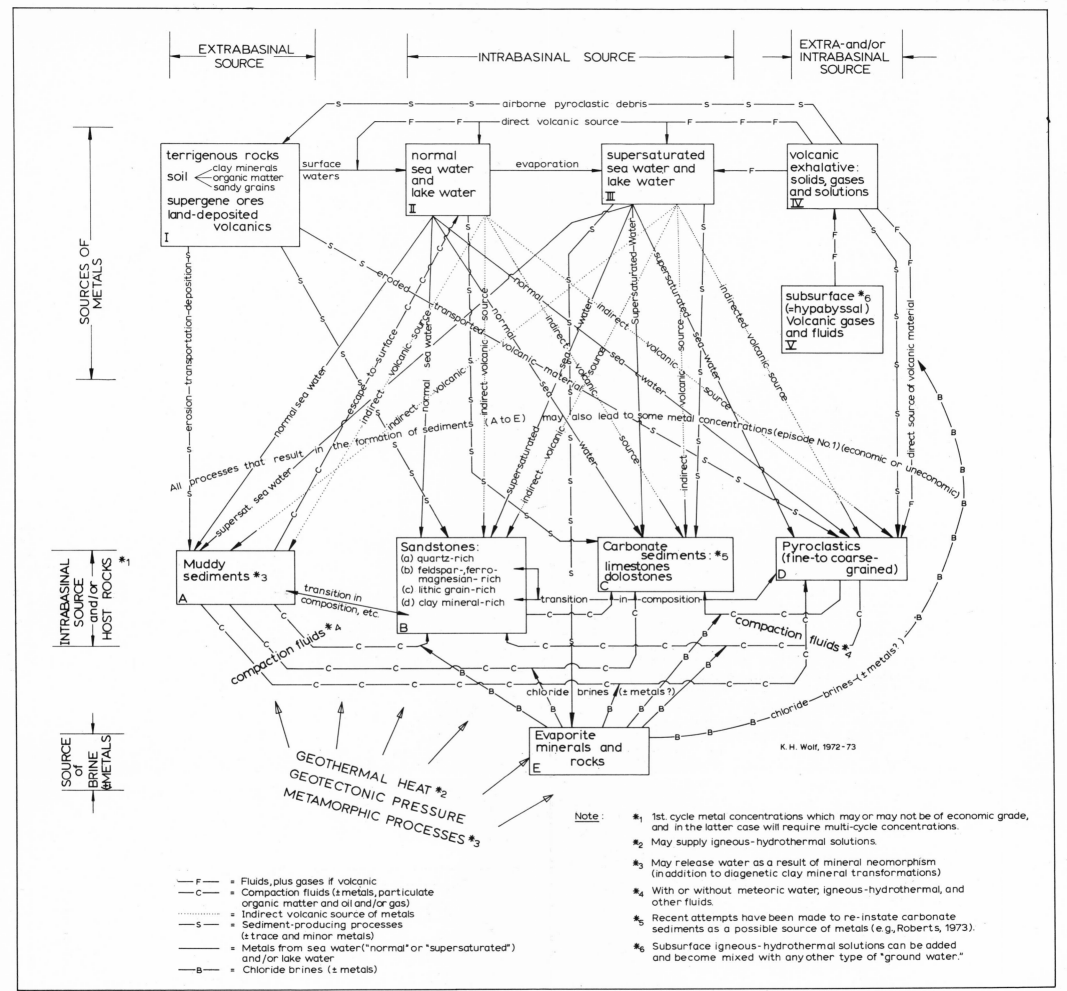

Fig. 38. Conceptual model expanded from Fig. 37 to include all possible sources of metal ions, modes of transportation, environments of precipitation, and types of host rocks in the origin of strata-bound ores in sediments and volcanics. (After Wolf, 1976, fig. 92.)

Laznicka (1973) found that in orogenic belts along the junctures of the ocean/continent type of lithospheric plates (i.e., geosynclinal orogenic belts of the older nomenclature), the metal deposits associated with the older rocks are genetically related to the rocks derived from subcrustal sources, such as the mantle, that were tapped by faults and plate subduction zones. Also, the absolutely oldest metal deposits on the averaged world scale of metal distribution are also the comparatively oldest within a modellized single tectonic cycle, as shown by Laznicka (Fig. 39). This appears to apply equally to mantle-derived lithologic suites of ultrabasic and basic rocks occasionally accompanying more acidic differentiated rocks situated on rifted platforms and Atlantic-type continental margins. As a consequence of limited or incomplete magmatic differentiation, however, and due to discontinuous outcrops, this established and expected succession of metallic deposits is rarely available in its entirety.

As illustrated in Fig. 39, the "primary" or earliest-formed metallic minerals, such as Cr and Ni derived from the upper mantle, accumulated early in the geotectonic evolution or any other processes that affect the crust. Should these deposits be exposed to the surface, they are quickly dispersed and, consequently, are not hypogenetically concentrated into ores. The late-formed "secondary" chemical elements (e.g., Sn, W, Hg, Sb), on the other hand, are concentrated under more tranquil geological conditions due to repeated episodes of deposition, hybrid magmatism, more complete magmatic differentiation, mobilization, and other processes. The metallic elements that undergo multi-cycle accumulations (e.g., Au, Ag, Cu, Zn, and Pb) take the intermediate position between the earlier-formed primary and the late-stage secondary metals, and are concentrated through-

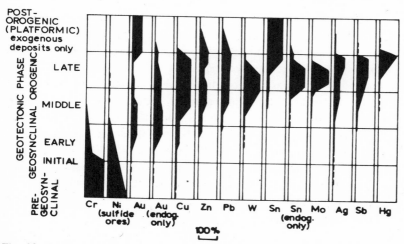

Fig. 39. Ore deposition in the geotectonic cycle. It is based on the geosynclinal theory and it is generally valid for ocean/continent-type junctures of lithosphere plates. The "pregeosynclinal phase" includes all ore-forming processes which took place before the moving lithosphere plate reached the zone of subduction (ocean trench). (After Laznicka, 1973, fig. 2; reproduced with permission of *Can. J. Earth Sci.*)

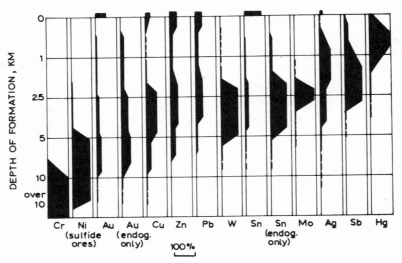

Fig. 40. Depth of ore deposition. The deposits formed in the geosynclinal and pregeosynclinal phases (some massive sulfides) are plotted in depth around 5 km, in which they presumably appeared in the middle development stages, in association with hypogene vein deposits. (After Laznicka, 1973, fig. 3; reproduced with permission of *Can. J. Earth Sci.*)

out the tectonic cycle, as illustrated by the distribution pattern in Fig. 39.

   Laznicka made a point (p. 21) of significance related to the above discussion, namely, that although geologically younger lithologic assemblages are generally more complex and diverse, these differences are not wholly the consequence of evolution in geologic time. The differences are predominantly the result of comparisons of rocks formed at different depths, under varying pressure-temperature conditions and, consequently in different facies zones. When these variables are taken into account, Laznicka found that his out-lined succession in Fig. 39, although being modellized, applies to deposits of Archaean, Paleozoic and Meso-Cenozoic ages. In effect then, the depth of ore formation, as given by the tentative diagrams for twelve non-ferrous metal accumulations in Fig. 40, as well as the depth of denudation (i.e., the depths to which surface erosion exposed the rocks) have to be considered concomitantly. Although new research results will change the concepts on ore genesis and, therefore, the depth of formation, the data by Laznicka is summarizing our present state of consensus.

   The preliminary diagram (Fig. 40) shows the likely association of ores as a function of depth of erosion, assuming that all twelve metal deposits originated penecontemporane-ously. This is, of course, too idealized so that the diagram is merely a model to be used as a guide — nevertheless, it is of practical use where at least a number of the ore types are associated. Taking a depth of 1 km of denudation, the following is expected: Hg > Ag,

Pb, Zn, Sb, Au > Cu, Sn; W

at 2.5 km: Mo, Sn. W > Sb, Ag, Pb, Zn, Cu, Au
at 5–10 km: Ni $\pm$ Cr > Au, Cu, Zn > Pb, Ag, Mo, Sn, W

Using published data, Laznicka calculated that at the average estimated post-Jurassic erosion rate of 1 m/10^5 years, the 1-km depth zone would be exposed in 100 million years and could expose Cretaceous subvolcanic rocks:

1 km	100 m.y.	Cretaceous subvolcanic deposits
2.5 km	250 m.y.	Permian mesothermal deposits
5 km	500 m.y.	Upper Cambrian–Lower Ordovician hypothermal deposits

The present rate of denudation, however, varies by a factor of up to 60 because of the mature landscape, e.g., from 1 m/$8 \cdot 10^4$ years in platform areas to 1 m/$1.32 \cdot 10^3$ years in high mountains like the Himalayas. As a consequence of these differences in erosion rates, the possibility exists that Pliocene subvolcanic deposits in a mountain belt at a depth of 1 km have the same opportunity to be near the present-day earth surface as a Mississippi Valley-type Pb–Zn ore that originated only at a depth of 300 to 500 m. As Laznicka pointed out (p. 24), mineralizations of stable platforms have a better chance to be preserved over a long geologic period than those in tectonically mobile belts.

Where "primary" metal accumulations (e.g., Cr and Ni) are associated along the same depth zone with "secondary" (e.g., Hg, Mo and Sb) metal accumulations, the former usually predate the latter and have been brought into juxtaposition by a number of emplacement mechanisms. Laznicka concluded his arguments (pp. 25, 26) by stating that as the result of denudation (due to uplift in many instances), the ore deposits that formed during the Precambrian in the upper 1–2.5 km depth level, have long been removed, so that Sb and Hg are absent in shield areas (with occasional minor exceptions). This argument is not completely factual in the case of Ag, for example, for this metal has a wider depth range. Also, our present state of nomenclature and classification (see Gabelman's Chapter 3, this Vol.) has not been sufficiently refined to sort out various important sub-types of the major groups of ores. This lack of information is not necessarily a reflection of an absence of methods required to establish these sub-types, but is believed to indicate the need of more super-detailed investigations by employing the techniques already to our disposal. From this viewpoint, it is significant that Laznicka recognized two types of porphyry-copper deposits:

(1) The Cerro Colorado-type (in Rio Tinto), formed during the upper-early and lower-middle phases in eugeosynclinal volcano–sedimentary sequences within andesites, albitophyres, Na-rich porphyries and syenites associated spatially with massive pyrite–chalcopyrite sulfides. This deposit has a low Mo-content and lacks zonally arranged Pb–Zn mineralization. As a result of downwarp after ore mineralization, this type of deposit was

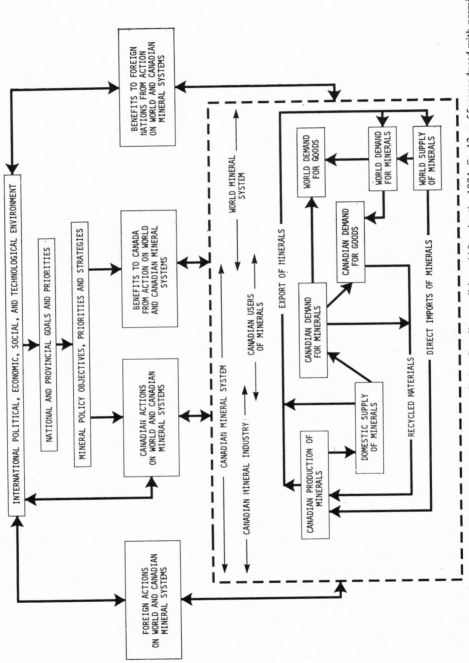

Fig. 41. Policy environment of the mineral system as applied to Canadian conditions. (After Austin, 1974, fig. 12, p. 55; reproduced with permission of Can. Inst. Min. Metall.)

preserved even since the Precambrian (e.g., McIntyre Cu-ore body, Timmins, Ontario).

(2) The porphyry-copper per se ores occur in calc-alkaline to hyper-alkaline sequences produced during the middle to late geosynclinal development phases. This type of ore culminated in the Late Mesozoic and Early Cenozoic (although members of their host rocks are present since the Precambrian).

A host of conceptual models designed by economists are available from the published literature, but it seems that the modelling of the complex variables as seen by economic geologists is of a more recent vintage — undoubtedly, much of the information has not been generally released in journals. Let us consider merely one simple example in Fig. 41 which applies to the Canadian situation (Austin, 1974) and depicts the mining industry as a system within a much larger world system encompassing the whole international political, economic, social and technological environment.

Numerous other examples could have been supplied, and the reader may wish to consult the papers by Robertson and Vandeever (1952) and Routhier (1967) for modellized mineral parageneses and metallogenesis, respectively. The technique of conceptualizing or modelling is merely one more technique employed during the several phases of the scientific method, which forms the foundation of all geological enquiries in both field and laboratory, and, consequently, is part of the research and exploration philosophy and architecture.

REFERENCES

Austin, J., 1974. Canadian mineral industry by the year 1999. *Can. Inst. Min. Metall., Bull.*, 1974: 49–56.

Amstutz, G.C. (Editor), 1964. *Sedimentology and Ore Genesis.* Elsevier, Amsterdam, 185 pp.

Amstutz, G.C., 1971. Introductory talk. *Soc. Min. Geol. Japan, Spec. Issue*, 3: 251.

Chorley, R.J., 1964. Geography and analogue theory. *Ann. Assoc. Am. Geogr.*, 54: 127–137.

Eargle, D.H. and Weeks, A.M.D., 1971. Geologic relations among uranium deposits, South Texas, Coastal Plain region, U.S.A. In: G.C. Amstutz and A.J. Bernard (Editors), *Ores in Sediments.* Springer, New York, N.Y., pp. 101–115.

Griffiths, J.C., 1967. *Scientific Method in Analysis of Sediments.* McGraw-Hill, New York, N.Y., 508 pp.

Hallberg, R.O., 1972. Sedimentary sulfide mineral formation — an energy circuit system approach. *Miner. Deposita*, 7: 189–201.

Kendall, M.G. and Buckland, W.R., 1960. *A Dictionary of Statistical Terms.* Oliver and Boyd, London, 590 pp.

King, H.F., 1973. Some antipodean thoughts about ore. *Econ. Geol.*, 68: 1369–1380.

Koch, L.E., 1949. Tetraktys — the system of the categories of natural science and its application to the geological sciences. *Aust. J. Sci.*, 11: 1–31.

Krumbein, W.C. and Graybill, F.A., 1965. *An Introduction to Statistical Models in Geology.* McGraw-Hill, New York, N.Y., 475 pp.

Laznicka, P., 1973. Development of nonferrous metal deposits in geological time. *Can. J. Earth Sci.*, 10: 18–25.

Lloyd, D., 1972. *The Idea of Law.* Penguin Books, Middlesex, 363 pp.

Moody, J.D., 1973. Petroleum exploration aspects of wrench-fault tectonics. *Am. Assoc. Petrol. Geol.*, 57: 449–476.

Northern Miner, 1975. State of geosciences – report calls for more exploration. *Northern Miner (Canada)*, Jan. 16, 1975.

Pettijohn, F.J., Potter, P.E. and Siever, R., 1972. *Sand and Sandstone.* Springer, New York, N.Y., 618 pp.

Potter, P.E. and Pettijohn, F.J., 1963. *Paleocurrents and Basin Analysis.* Springer, New York, N.Y., 296 pp.

Pretorius, D.A., 1966. Conceptual geological models in the exploration for gold mineralization in the Witwatersrand basin. *Econ. Geol. Res. Unit, Univ. Witwatersrand, Inf. Circ.,* 33: 29 pp.

Roberts, W.M.B., 1973. Dolomitization and the genesis of the Woodcutters Pb–Zn prospect, Northern Territory, Australia. *Miner. Deposita,* 8: 35–56.

Robertson, F.S. and Vandeever, P.L., 1952. A new diagrammatic scheme for paragenetic relations of the ore minerals. *Econ. Geol.,* 47: 101–105.

Routhier, P., 1967. Le modèle de la genèse. *Chron. Mines Rech. Min., Paris,* 363: 177–190.

Vine, J.D. and Tourtelot, E.B., 1970. Geochemistry of black shale deposits – a summary report. *Econ. Geol.,* 65: 253–272.

Whitten, G.H.T., 1964. Process-response models in geology. *Geol. Soc. Am.,* 75: 455–464.

Wolf, K.H., 1971. Textural and compositional transitional stages between various lithic grain types (with a comment on "Introductory Detrital Modes of Graywake and Arkose"). *J. Sediment. Petrol.,* 41: 328–332; Errata, 889.

Wolf, K.H., 1973a. Conceptual models, I. Examples in sedimentary petrology, environmental and stratigraphic reconstructions, and soil, reef, chemical and placer sedimentary ore deposits. *Sediment. Geol.,* 9: 153–194.

Wolf, K.H., 1973b. Conceptual models, II. Fluvial-alluvial, glacial, lacustrine, desert, and shorezone (beach-bar-dune-chenier) sedimentary environments. *Sediment. Geol.,* 9: 235–260.

Wolf, K.H., 1973c. Book review of "Carbonate Sediments and Their Diagenesis" by R.G.C. Bathurst. *Earth-Sci. Rev.,* 9: 152–157.

Wolf, K.H., 1976, Influence of compaction on ore genesis. In: G.V. Chilingar and K.H. Wolf (Editors), *Compaction of Coarse-Grained Sediments, 2.* Elsevier, Amsterdam, in press.

Chapter 3

CLASSIFICATIONS OF STRATA-BOUND ORE DEPOSITS

JOHN W. GABELMAN

INTRODUCTION

"Classifications of strata-bound ore deposits" was written originally as a section of the chapter[1]: "Strata-bound ore deposits and metallotectonics", to define descriptive and genetic process parameters for the consideration of the role of tectonics in mineralization processes which produce strata-bound deposits. The various classifications conceived and discussed were based on a critical review of all the mineralization processes which conceivably could produce such deposits. This review appears in Volume 4 as chapter 2: "Strata-binding mineralization processes".

All three chapters are sequentially dependent, although their introductory and illustrative material has not been repeated except for Table I. Thus it is recommended that the review of processes be read first and the analysis of tectonics and mineralization last.

GENERAL STATEMENT

The expression "single stratigraphic unit" in the *A.G.I. Glossary of Geology* (Gary et al., 1972) definition of "strata-bound" could signify anything from a single microstratum to a stratigraphic series, and includes them all. For example, a stratigraphic system, including facies from continental to marine and lithologies from sandstone to limestone, could contain veins that did not extend beyond the system boundaries. These would then be strata-bound deposits. The central Kentucky barite veins in the Lower Ordovician series (Plummer, 1971) and the base-metal veins in the San Juan tuff of southwestern Colorado (Burbank, 1947) are factual examples.

Close examination of even the most bound deposits reveals that few are truly and consistently conformable. Most exhibit some form of variation in shape, attitude, or position with respect to the layering. The nature of the variation is usually a form of transgression at some scale. All degrees of any of the types of variations are somewhere represented.

The type and degree of binding are highly indicative of the emplacement mechanism and genesis interpreted for a deposit or district. Therefore, an analysis of strata-binding,

---

[1] See Volume 4, Chapter 3.

in relation to tectonic features should properly begin with a review of the types of strata-binding, and the possible ways of classifying ore deposits on this basis. The classifications commonly used in the literature (Lindgren, 1933; Loughlin and Behre, 1933; Ridge, 1968) are not sufficiently useful in this regard because they do not address specifically the aspect of strata-binding. All of the eight classifications discussed here have been used to compare deposits, but generally not as inclusively as here. The extension of these classifications to all possible types of strata-bound deposits illustrates deficiencies in most of the classifications which might have seemed ideal for a shorter range of deposit types. The writer believes these deficiencies also indicate that one or a few mineralization processes cannot account for all the strata-bound deposits represented. Rather, a large variety of processes, some very different, appear to have been involved. By analyzing the various classifications in terms of all the possible processes, it is hoped that common denominators will emerge which can help explain the systematic geographic relations between deposits of such diverse origin.

A distinction should be made at the outset between "strata-bound" and "stratiform." "Stratiform" clearly refers to the stratified shape of a deposit without regard to its emplacement mechanism or genesis, although a sedimentational mechanism may be implied by the user. It is also by definition a strata-bound deposit. However, a strata-bound deposit need not be stratiform.

The external relations of mineral deposits to their matrices, their shape, and internal structures and textures offer permissive (Snyder, 1967), but seldom conclusive, evidence which allows them to be interpreted and classified in a variety of ways. That some of the interpretations are contradictory indicates that some critically weighted evidence was more permissive than restrictive.

PARAMETERS FOR CLASSIFICATION

The intent of the following classifications is to compare the different types of deposits in the most ways possible. Therefore, the writer has tried to conceive of the maximum number of most significant theoretical bases for comparison. For each basis a complete range of possible conditions was set forth. The important variable factor influencing the range is identified, and one or more example deposits fitting the conditions is cited.

Many more classifications can be conceived than those presented here. However, those presented are believed to include most of the significant possibilities in genesis, metal source, transport, emplacement mechanism, and compatibility with host rock.

No attempt has been made to determine the *correct* origin, process, or mechanism for any deposit or district cited as an example. In a few cases there is minimum argument over possible choices. However, for most the origin is still controversial. In numerous cases the same district has been cited in the literature as an example product of conflicting processes, e.g., the Mississippi Valley replacements, copper shales, or uraniferous sandstones.

TABLE I

Example strata-bound ore deposits (arranged by rock type and metal concentrations)

(I) SEDIMENTARY ROCKS

    (A) Whole-rock ores
       (1) Precambrian banded iron-formation:
           Lake Superior ores (Baley and James, 1973)
       (2) Phanerozoic banded iron-formation:
           Bathurst, New Brunswick (Boyle and Davies, 1973)
       (3) Oolitic iron ore:
           Birmingham, Ala. (Simpson and Gray, 1968)
           New Mexico (Kelley, 1951)
       (4) Bedded phosphate:
           Phosphoria Formation, western U.S. (McKelvey, 1959)
           Florida phosphates (Mansfield, 1943)
       (5) Bedded barite:
           Washington State, U.S.A. (Mills, 1971)
           Nevada (Shaw et al., 1969)
           Arkansas (Scull, 1958)
           Meggen, Germany (Amstutz et al., 1971)
       (6) Evaporites, salt:
           U.S.A. (Phalen, 1919)
       (7) Evaporites, anhydrite:
           U.S.A. (Withington and Jaster, 1960)
       (8) Evaporites, related to strata-bound metal deposits:
           U.S.A. (Renfro, 1974)

    (B) Argillaceous sedimentary rocks
       (1) Disseminated metals
          (a) modern deep-sea ferromanganese nodules:
              Mid-Atlantic ridge (Scott and Scott, 1974)
              World oceans (Morgenstein, 1972; Horn et al., 1972)
          (b) modern metalliferous flysch:
              Nazca plate—Peru trench (JOIDES Sci. Staff, 1974)
          (c) copper shales:
              White Pine, Michigan (Brown, 1971)
              Boleo, Santa Rosalia, Mexico (Wilson and Rocha Morena, 1955)
          (d) vanadium shales:
              Minas Ragra, Peru (Hewett, 1909; Colo. School Mines, 1961)
              Phosphoria Formation, western U.S.A. (McKelvey, 1949)
          (e) uranium shales:
              Chattanooga shale, eastern U.S.A. (Glover, 1959; Landis, 1962)
              Sweden (Armands, 1972)
          (f) baritic shales:
              Arkansas (Scull, 1958)
              Meggen, Germany (Amstutz et al., 1971)
       (2) Massive metal concentrations
          (a) Kuroko deposits:
              Japan (Sato, 1974; Horikoshi, 1969; Matsukuma and Horikoshi, 1970; Sakai and
              Matsubaya, 1974)
          (b) other massive base-metal sulfide deposits (possibly not different from Kuroko
             deposits):

TABLE I *(continued)*

          Cyprus (Constantinou and Govett, 1973)
          Canada (Sangster, 1972)
          Ramsbeck and Rammelsberg, Germany (Amstutz et al., 1971)
          Tasmania (Brathwaite, 1974)
          Skellefte belt (Boliden), Sweden (Grip, 1950; Du Rietz, 1953)
    (c) bedded barite:
          Arkansas (Scull, 1958)
          Meggen, Germany (Zimmermann, 1970)

(C) Arenaceous sedimentary rocks
  (1) Disseminated metals
    (a) littoral (beach) semi-concentrations:
          Nile delta (Overstreet, 1967)
          India (Overstreet, 1967)
          Tennessee (Hershey, 1969)
    (b) alluvial placers:
          gold, platinum, tin, etc., in most orogenic systems (Lindgren, 1933)
          uranium, Pakistan (Miller, 1963)
          uraniferous conglomerates (see below)
    (c) uraniferous sheet conglomerates:
          Witwatersrand, S.Afr. (Whiteside, 1970)
          Elliot Lake, Can. (Roscoe, 1969)
    (d) vanadium—uranium in sandstones:
          Colorado plateau (Gabelman, 1971)
          Argentina (Stipanicic, 1970)
    (e) copper in red beds:
          New Mexico (Finch, 1933; Woodward et al., 1974)
    (f) copper in sheet conglomerates:
          Michigan (Weege et al., 1972)
    (g) lead, barite, and fluorite in sandstone:
          Eastern Caledonides (Laisvall), Sweden (Grip, 1950, 1967)
    (h) manganese in sandstones:
          Um Bogma, Sinai (Mart and Sass, 1972)
  (2) Massive concentrations
      none identified

(D) Calcareous sedimentary rocks
  (1) Disseminated metals
      prominent on fringes of massive metal replacements
  (2) Massive metals
    (a) tactites:
          Sierra Nevada, Lone Pine, Calif. (Gray et al., 1968)
    (b) magnetite replacements:
          Cornwall, Pa. (Hickok, 1933)
          Iron Springs, Utah (Mackin, 1947, 1968)
    (c) mesothermal base-metal replacements:
          Leadville, Colo. (Emmons et al., 1927)
          Pioche, Nev. (Westgate and Knopf, 1932; Gemmil, 1968)
          Eureka, Nev. (Nolan and Hunt, 1968)
          Santa Eulalia, Mexico (Gonzales, 1956; Hewitt, 1968)

TABLE I *(continued)*

    (d) epithermal sulfosalt replacements:
         Naica, Mexico (Gonzales, 1956)
         Sierra del Cal, Durango, Mexico (Gabelman, unpublished)
         Eureka, Nev. (Nolan and Hunt, 1968)
         Tintic, Utah (Morris, 1968)
         Terlingua, Texas (Phillips, 1905; Turner, 1905)
    (e) Appalachian-type zinc replacements:
         Appalachian belt (Hoagland, 1971)
         East Tennessee (Maher, 1971; Hill et al., 1971)
    (f) Mississippi Valley-type base-metal replacements:
         southeast Missouri (Doe and Delevaux, 1972; Gerdemann and Meyers, 1972)
         English Pennines (Dunham, 1967)
         Alpine geosyncline, Europe (Maucher and Schneider, 1967)
         general (Heyl et al., 1974)
    (g) fluorite replacements:
         Illinois—Kentucky (Heyl and Brock, 1961; Brecke, 1965; Grogan and Bradbury, 1967)
    (h) barite replacements:
         east Tennessee (Fagan, 1969)
         Kentucky—Illinois (Brecke, 1965)
         Arkansas (Scull, 1958)
         Washington State (Mills, 1971)
         Meggen, Germany (Zimmermann, 1970)
    (i) manganese replacements:
         Gilman, Colo. (Gabelman, observation)
         Pioche, Nev. (Gabelman, observation)
         Pryor Mountains, Wyo.—Mont. (Gabelman, observation)

(E) Carbonaceous sedimentary rocks
  (1) Disseminated metals (as adsorptions or complexes)
    (a) uraniferous lignites:
         North and South Dakota (Densen et al., 1955)
         Spain (Arribas, 1974)

(F) Asphaltic sedimentary rocks
  (1) Disseminated metals (as adsorptions or complexes)
    (a) vanadiferous asphaltite:
         Minas Ragra, Peru (Hewett, 1909; Colo School Mines, 1961)
    (b) uraniferous asphaltite or humate:
         Grants, New Mex. (Granger, 1963)
         Witwatersrand (Davidson and Bowie, 1951)

(G) Volcanic sedimentary rocks (described under volcanic rocks where deposits in both types are similar)
  (1) Mafic volcanic material
    (a) native copper:
         Michigan (White, 1968)
    (b) sideritic tuffaceous sediments:
         Carpathian siderites (Krautner, 1970)

TABLE I *(continued)*

(II) IGNEOUS ROCKS

(A) Intrusives
- (1) Ultramafic and mafic
  - (a) disseminated metals:
    layered complexes (Cr, Fe, Ti, V, Pt): Bushveld (Willemse, 1969)
    differentiated complexes of thick massive layers:
    copper–nickel: Sudbury (Souch et al., 1969)
    platinum: Sudbury (Cabri and Gills, 1974)
    post-magmatic disseminations in conformable bodies:
    iron (magnetite): Sanford Lake, N.Y. (Gross, 1968)
- (2) Felsic and alkalic
  - (a) disseminated metals (conformable granitoid and aplitic bodies):
    uraniferous alaskites – Rossing, southwest Africa (Von Backström, 1970); northeast Brazil (Gabelman, observation)
    uraniferous syenitic complexes – Illimausaq, Greenland (Sörensen, 1970)
  - (b) disseminated metals (pegmatitic sills):
    uraniferous alkalic granite pegmatites – Bancroft, Ontario (Lang et al., 1962)

(B) Volcanics
- (1) Ultramafic and mafic
  - (a) disseminated metals:
    amygdaloidal copper – Michigan (Weege et al., 1972)
  - (b) massive metal concentrations:
    none identified
- (2) Felsic and alkalic
  - (a) disseminated metals:
    preferred metal enrichments (Krauskopf, 1967)
    uraniferous tuffs: Basin-range province (Coats, 1955)
    beryllium–fluorine–manganese–uranium tuffs: Aguachile, Mexico (McAnulty et al., 1963); Spor Mt., Utah (Shawe, 1968)
  - (b) massive metal concentrations
    none identified

(III) METAMORPHIC ROCKS

(A) Argillaceous sedimentary rock equivalents
- (1) Disseminated metals
  - (a) copper phyllites:
    Stekenjokk, Sweden (Zachrisson, 1971)
    Rhodesian copperbelt (Mendelsohn, 1961)
  - (b) base-metal phyllites:
    Bathurst, New Brunswick (Davis, 1972)
  - (c) vanadium–uranium–molybdenum–graphitic phyllites:
    Alabama (Jones, 1929)
    Korea (Gabelman, unpublished)
  - (d) gold phyllites and schists:
    Homestake, S. Dakota (Slaughter, 1968)
    Carolina slate belt (Worthington and Kiff, 1970)

TABLE I *(continued)*

        (e)   magnetitic iron-formation or schist:
              Lake Superior region (Marsden, 1968)
              Atlantic City, Wyo. (Pride and Hagner, 1972)
   (2) Massive metal concentrations
        (a)   copper phyllites and schists:
              Ducktown, Tenn. (Mauger, 1972)
        (b)   base-metal phyllites and schists:
              Newfoundland (Strong, 1974)
              Canada, general (Sangster, 1972)

(B) Siliceous sedimentary rock equivalents
   (1) Disseminated metals
        (a)   uraniferous conglomerates:
              Witwatersrand, S. Afr. (Whiteside, 1970)
              Elliot Lake, Ont. (Roscoe, 1969)
        (b)   uraniferous–thoriferous (monazite, zircon) quartzites:
              Goodrich quartzite, Mich. (Vickers, 1956)
              Santa Elena, Spain (Arribas, 1974)

(C) Calcareous sedimentary rock equivalents
   (1) Disseminated metals
        (a)   zinc replacements:
              Balmat-Edwards, N.Y. (Lea and Dill, 1968)
   (2) Massive metal concentrations
        (a)   skarns:
              Central Sweden (Magnusson, 1950)
              Bishop tungsten district, Calif. (Gray et al., 1968)
        (b)   zinc replacements or recrystallizations:
              Balmat-Edwards, N.Y. (Lea and Dill, 1968)
              Franklin, N.J. (Frondel and Baum, 1974)

(IV) SOILS

(A) Migratory concentrations
   (1) Bog iron concentrations:
        Mesabi Range, Minn. (Royce, 1942)
        Tennessee (Gordon, 1913; Burchard, 1927)
   (2) Supergene manganese concentrations:
        Spain (Pastor et al., 1956)
        various localities (Varentsov, 1964; Borchert, 1970)

(B) Residual concentrations:
   (1) Phosphate:
        Central Tennessee (Smith, 1924; Martin and Wilding, 1937)
   (2) Bauxite:
        Arkansas (Harder, 1937; Gordon and Tracey, 1952)
   (3) Nickel laterite:
        Cuba, Puerto Rico (Heidenreich, 1959)
        California (Page et al., 1972)

It has sufficed for this writer that the citation of a district qualifies it for inclusion under the process for which it was cited. This establishes the possibility that the process is basically valid, although not necessarily completely or correctly interpreted and applied. It also emphasizes the inadequacy of existing qualifying data. Because most of the examples are repeated in several classifications, all examples have been tabulated in a separate list (Table I), grouped according to host rock and internal texture, and including references.

CLASSIFICATION BY MAJOR CONTROLLING PROCESS

Deposits can be classified according to their basic formative process (Table II). Thus, phosphate beds result from marine sedimentation, coal from littoral or paludal sedimenta-

TABLE II

Classification by major geologic process

Process			Environment or facies	Example type of strata-bound deposit
main process	variety	phase		
Sedimentation	chemical precipitation	marine	deep	(1) manganese-iron nodules; (2) base-metalliferous shales
			shallow	(1) banded iron-fm.; (2) oolitic iron-fm.; (3) phosphate; (4) vanadiferous shale
			marginal	(1) vanidiferous shale; (2) uraniferous lignite; (3) manganiferous ls; (4) evaporites
		continental	lacustrine	(1) evaporites
	physical settling	marine	marginal	(1) Th–U–RE beach placers
		continental	fluvial	(1) Au placers; (2) U placers
Intrusion	magmatic		ultramafic to mafic	(1) layered ultramafic complexes
	late- or post-magmatic	deuteric or hydrothermal	mafic	(1) differentiated thick massive-layer complexes
	post-magmatic	pegmatitic	felsic	(1) pegmatite sills; (2) alaskite sills
	magmatic hydrothermal	hypothermal	deep	siliceous rock replacements by: (1) Fe, (2) Au, (3) Cu, (4) Th–U
		mesothermal	moderately deep	(1) limestone replacement mantos
		epithermal	shallow	(1) limestone replacements by Pb, Zn, F, Ba, (2) sandstone impregnations by Pb
		telethermal	near surficial	(1) Appalachian or Miss. Valley-type limestone replacements; (2) V–U sandstone impregn. (3) red-bed copper

TABLE II *(continued)*

Process			Environ-ment or facies	Example type of strata-bound deposit
main process	variety	phase		
Volcanism	extrusion, lava flow		subaerial	
	pyroclastic		subaerial	(1) uraniferous tuffs; (2) beryllium−manganese−fluorine tuffs
	hydrothermal		submarine	(1) exhalogenic massive sulfides
			subaerial	(1) uraniferous tuffs; (2) Be−Mn−F tuffs; (3) strata-restricted veins
Metamorphism	pyrometasom-atism	hydrothermal silication	carbonate rocks	tactites of Fe, W, Cu, Pb, Zn
	regional metamorphism	recrystal-lization-mobilization	silicate rocks	(1) banded Fe-fm; (2) Au or Cu phyllites
			carbonate rocks	(1) skarns of Fe, W, Cu, Pb, Zn
Tectonism	orogeny	hydrothermal		(1) limestone replacements of base metals; (2) dissemination of base metals; (3) impreg-nations
	taphrogeny	hydrothermal		many types of deposits arranged in belts enclosing lineaments (Front Range belt)
	epeirogeny	"hydrotepid"		(1) red-bed copper; (2) V−U sandstone impregnations
Surface destruction	dissolution−concentration	supergene enrichment	surficial	(1) V−U sandstone impregna-tions; (2) Cu in all rocks
		laterogene enrichment	shallow	(1) V−U sandstone impregna-tions; (2) Cu sandstone impregnations
	residual concentration		surficial	(1) bog Fe; (2) sed. Mn; (3) residual phosphate; (4) clay; (5) Au saprolite
	lateritization		surficial	(1) bauxite; (2) Ni laterite

tion, and gold placers from alluvial sedimentation. All are distinctive by their external relations, shape, and internal features. The major process is readily recognizable for some types of deposits, such as layered intrusive complexes, contact-metamorphic tactites or banded iron-formations. However, the parent process of others is only implied, for example the massive sulfides in massive flysch; or is inconclusive such as for the copper or uranium impregnations in sandstone. Recognition of the process may be indirect such as the interpretation of an orogenic or taphrogenic affiliation from regional zonal patterns as described below. The great disadvantage of this classification is that the uncertainties greatly out-number the certainties, and effectively render the classification too interpretive and, therefore, unsuitable.

CLASSIFICATION BY DIRECT EMPLACEMENT MECHANISM

Table III outlines a classification by direct emplacement mechanism, for which evidence is more direct. Often a combination of mechanisms is indicated as at Pryor Mt.,

TABLE III

Classification by direct emplacement mechanism

Mechanism		Environment	Facies	Example type of strata-bound deposit
principal	subtype			
Physical settling (sedimentation)		marine	deep shallow	Au–U conglomerates; diamond placers
		marginal marine	littoral	U conglomerates; heavy-mineral beach sands
		continental	alluvial	Au placers
Chemical precipitation	supersaturation	marine lagoonal, lacustrine		phosphates, sulfates evaporites
	temperature or pressure change	marine		volcanic metal exhalations
	pH or Eh change	marine		Fe–Mn nodules, metalliferous shales, massive volcanic metal exhalations, Fe-fm, U shales
		lagoonal or continental		uraniferous lignites
Chemical weathering	lateritization	continental	arid	enriched gossans in soil, limest., sandst.
			semi-arid	enriched gossans in soil, limest., sandst.
			humid	residual phosphates, bauxite, Ni laterite

TABLE III *(continued)*

Mechanism		Environment	Facies	Example type of strata-bound deposit
principal	subtype			
Organic secretion	anaerobic bacteria	marine or continental	shallow marine, lagoonal, lacustrine	sulfide in shales, coals, or carbonaceous sandstones
	complexing in liquid hydro-carbons	continental	deformed sedi-mentary basins	metalliferous oils or asphaltites
	syngenetic adsorption	marine		uraniferous organic shales
	epigenetic adsorption	continental		uraniferous lignites and coals
Metasomatism		sedimentary	calcareous	mesothermal limestone re-placements, Miss. Valley limestone replacements
			arenaceous and argillaceous	sandstone cement replacements
		intrusive		
		volcanic	flows	
			tuffs	U, Be, F, Mn tuffs
		metamorphic		hydrothermal skarns, tactites
Cavity filling			calcareous karst	base metal mantos; U, F, Mn mantos
	competent rock fracturing	orogenic or taphrogenic belts	volcanic rock complexes	competent strata-bound veins
			volcanic, sedi-mentary or meta-morphic inter-layered com-plexes	competent strata-bound joint stockworks
			sandstone porosity	red bed copper, V–U sandst. impregnations
Magmatic crystallization		mafic		thinly layered complexes, thick-layered differentiated complexes
		felsic		U granites and pegmatites, U, Be tuffs
Metamorphic recrystallization				Fe-formation, skarns

Wyoming (Hart, 1958) where in addition to filling limestone caves, rhodochrosite, fluorite, pyrite and chalcopyrite formed in the walls by metasomatic replacement. In carbonaceous sandstones the H_2S produced by bacteriogenic sulfate reduction is believed to have reduced and precipitated ore metals.

One difficulty with this classification is that the emplacement mechanism illustrates little about the source or migration mechanism of the elements involved. Where evidence suggests these factors several very different alternatives usually fit the facts, again rendering the classification too interpretive to be useful.

CLASSIFICATION BY HOST LITHOLOGY

Classifying strata-bound deposits according to the lithology of the host formation (Table IV) is the simplest and most-used method. I believe that this classification contains a subtle bias toward the syngenetic interpretation. Thus, the preference of copper for the Nonesuch shale at White Pine (Ensign et al., 1968; Brown, 1971), or of uranium for the Salt Wash sandstone in Colorado (Gabelman, 1971), outwardly suggests that the copper was indigenous to the shale and uranium to the sandstone. In some instances there is direct evidence to support this interpretation, e.g., chromium or nickel (Souch et al., 1969) in layered intrusives. However, for most types such as the limestone replacements, shale disseminations or sandstone impregnations, the association is permissive at best.

Indirect evidence locally may support such a conclusion. The logic that in a given area confinement of all deposits of a mineral to a single time-stratigraphic unit should indicate mineral cogenesis with the sediment, is a natural and valid working assumption in the absence of conclusive evidence, provided there is not equally or more conclusive evidence for another origin. However, where evidence is only permissive, the geologist is obligated to consider other mechanisms that could have produced the same selectivity. These include preferred formational permeability; and selective diagenetic, structural, or supergene ground preparation. In many strata-bound districts careful inspection reveals quite a stratigraphic "spread" to mineral deposit distribution in a variety of lithologies. The east Tennessee zinc district (Hoagland, 1971; Maher, 1971), Uravan vanadium—uranium belt (Gabelman, 1971), and Grants uranium belt (Granger, 1963; Gabelman, 1956, 1971) are examples. Reliance on the restriction of the preponderance of occurrences to the "favored" unit of the spread to support the conclusion of binding is seldom valid because of the variety of circumstances which will permit an epigenetic preference. There is also the possibility that significant deposits in neighboring beds simply have not yet been discovered.

Where evidence for syngenesis is compelling as for many Canadian massive sulfides (Sangster, 1972), Japanese Kuroko deposits (Horikoshi, 1969) or the Boleo (Mexico) copper shales (Wilson and Rocha Morena, 1955), the assumption that the metals and sediment were derived from the same source still may be unwarranted. This is illustrated

TABLE IV

Classification by host lithology

Major rock classification	Type and lithology		Environment	Example type of strata-bound deposit
Sedimentary	carbonates		shallow marine	Miss. Valley-type replacements; epithermal replac. and stockworks; Rocky Mt.-type mesothermal replac.
	carbonaceous rocks (coal, lignite)		marginal marine, continental	U-lignites
	shale (incl. carbonaceous sh.)		marine	metal disseminations in shale, massive sulfide deposits (Kuroko, base-metal), sea-floor nodules
	sandstone		marginal marine, continental	sedimentary Fe and Mn residua; red-bed copper reconcentrations; V−U reconcentrations; gold placers
	conglomerate		continental	U in channel conglom.; Au in sheet conglom.; U in sheet conglom., native Cu conglom. (Mich.)
Igneous	intrusive	felsic	plutonic	pegmatite sills; conformable U-Th granites; layered complexes (Stillwater)
		mafic intermediate mafic	hypabyssal	porphyry coppers differentiated complexes (Sudbury)
		ultramafic		layered complexes
	volcanic	felsic intermediate mafic	shallow	veins in most competent lavas Cu amygdaloidal basalts (Mich.)
		felsic intermediate	surficial	uraniferous tuffs; beryllium tuffs
Metamorphic	contact	silicated limestone		tactites
	regional	carbonates		skarn
		argillaceous		graphite phyllite and schist; V phyllites, U phyllites
		siliceous (sedimentary) siliceous (ign. and metamorphic)		Au−U conglomerates; U conglomerates
	hydrothermal	silicified limestones		Au cherts

by the drastic change in popular interpretation for these deposits from one of co-sediment-source deposition to that of sea-floor hypogene exhalation, within the same framework of evidence, once the possibility of exhalation was recognized.

CLASSIFICATION BY CHEMICAL REACTIVITY

A classification based on host-rock chemical reactivity highlights tendencies of strata-bound deposits to be diagenetically or epigenetically introduced or modified. Truly syngenetic strata-bound deposits should occur independently of host reactivity. Major rock types are listed with an accompanying estimate of porosity, permeability, chemical reactivity, and the relative abundance of strata-bound deposits within them (Table V). The Table illustrates at least three different classes of prominent strata-bound ore deposits. One in rocks of very low textural accessibility, but high competency and reactivity is represented by the abundant carbonate replacements. Since these are marine chemical precipitates which are areally extensive and uniform, their co-precipitated metals would be extremely dispersed with little opportunity for diagenetic or epigenetic reconcentration. Therefore, the easiest mineralization is extrinsically epigenetic. The second class is of very low access porosity, permeability, reactivity, and competency, and is represented by the metalliferous shales, including massive sulfides. In order to contain prominent concentrations, the metals are most simply interpreted as sedimentationally syngenetic. The third class represented by the sandstones is characterized by high porosity and permeability, average competency, and low reactivity, and allows epigenetic mineralization as the simplest process. This mineralization could involve intrinsic or extrinsic metals. However, the mineralization would not be prominent if it depended on the reactivity of the detrital sand grains. The third class is made prominent by the higher reactivity of an indigenous or introduced cement, and sandstone mineralization must be considered either a replacement of reactive cement, like the carbonates, or an open-space impregnation.

CLASSIFICATION BY SOURCE OF METALS

This classification (Table VI) has been used extensively in the literature although more in discussing specific deposits or processes, than for complete analysis. Outwardly it is simple and useful because of the few choices of ultimate source (intratelluric or surficial; mantle or crust). However, this utility is deceiving and the classification contains the same major deficiency as others by requiring assumptions of parent process and source for which evidence is not conclusive. Again, because the features of many strata-bound deposits are permissive to several conflicting sources and processes, the classification does not provide an adequate means of distinction.

TABLE V

Classification by host rock reactivity

Rock	Average textural porosity	Average textural permeability	Structural competency	Relative reactivity	Relative ore deposit abundance*	Example type of strata-bound deposit
Sedimentary						
limestone	extremely low	very high	very high	very high	B	Mississippi-Valley replacements
dolomite	low	low	very high	high	A	mesothermal dolomite replacements; tactites
shale	extremely low	extremely low	extremely low	low	D	Kuroko massive sulfides; uranium shales; copper shales
marl	very low	very low	very low	moderate	C	U marls (Argentina)
conglomerate	very high	very high	high	very low	D	Au–U conglomerates (Witwatersrand); U conglomerates (Elliot Lake)
sandstone	high	high	high	very low	E	red-bed copper; Colo. Plateau-U sandstones
Igneous crystalline						
ultramafic	low	very low	high	moderate to low	D	layered complexes (Bushveld)
mafic	low	very low	high	low to moderate	D	differentiated intrusive complexes (Sudbury)
intermediate	low	low	high	moderate	D	Silverton veins
felsic	low	low	high	moderate to high	C	pegmatite sills; uraniferous alaskite
Igneous pyroclastic						
mafic	high	high	very low	low	F	?
felsic	high	high	very low	moderate	C	uraniferous tuffs: beryllium tuffs
Metamorphic						
marble	low	low	high	very high	B	skarns
quartzite	extremely low	extremely low	very high	extremely low	F	uraniferous quartzite (fossil zircon placers)
schist	moderate to low	moderate to low	low	low	E	residual or rocks massive sulfides; Great Gossan lead
gneiss	very low	very low	high	moderate to low	E	banded Fe-formation

* Comparative scale range from A = most abundant to F = least abundant.

TABLE VI

Classification by source of metals

Metal source	Transporting and/or concentrating process	Example type of strata-bound deposit
Intratelluric mantle	layered ultramafic intrusion	Bushveld, Musk Ox, Skaergaard
	differentiated mafic intrusion	Sudbury complex
	submarine volcanism	Kuroko Cu., massive sulfides
	subaerial volcanism	uraniferous tuffs, strata-bound veins and stockworks
deep to intermadiate crust	regional metamorphism	skarns
	batholithization hydrothermal series	contact carbonate tactites
	hypabyssal intrusion hydrothermal series	mesothermal carbonate replacements contact carbonate tactites
	subaerial volcanism	uraniferous tuffs; beryllium tuffs
Epidermic shallow crust	reconcentration by orogenic heat	mesothermal carbonate replacements; Appalachian telethermal carbonate replacements
	reconcentration by taphrogenic heat	Colo.-Plateau V−U; Miss.-Valley limestone replacements
	lithogenic secretion	Colo.-Plateau V−U in sandstones; Cu in red beds
	diagenesis of new sediments (dewatering)	U in sandstones (Texas)
	new sedimentation	heavy minerals in beach sands; Au−U in sheet conglomerates
	destructive erosion and transport	alluvial placers
Surface sea water	selective chemical precipitation	phosphates; Fe-formations; U shales; evaporites

CLASSIFICATION BY SOURCE OF TRANSPORTING FLUIDS

The general sources outlined in Table VII are again few and simple, but the specific sources become increasingly obscure with greater geologic detail. Thus, the proportion of released rock fluids, meteoric water, and connate water in repeated geologic cycles is nearly impossible to determine. Is there truly juvenile water, or is basically meteoric or marine water recycled through orogenesis? We still do not know the depths to which meteoric ground water can circulate or its variation in saturation with depth. As the possible variations increase the classification seems to demonstrate a number of intercommunicating processes and types of fluid which might be involved in a larger fluid cycle—one governed by orogenesis or taphrogenesis. Therefore, the classification to be useful, again requires basic assumptions that are not warranted without additional information.

TABLE VII

Classification by source of transporting fluid

Source	Type or specific source	Description	Example type of strata-bound deposit
Earth's interior	mantle	mantle devolatilization (amagmatic)	deep-sea Mn nodules; massive sulfide deposits; metal disseminations in shales; taphrogenic hydrothermal deposits
	deep crust	subducted sea water (amagmatic)	volcanogenic hydrothermal deposits; submarine exhalation deposits
Magma	mantle magma deep crustal magma	residual fluids residual fluids	layered ultramafic complexes; differentiated mafic complexes, magmatic hydrothermal deposits; submarine exhalation deposits; (mesothermal carbonate replacements, V, U, Cu in sandstones and carbonates)
	middle crustal magma	residual magmatic fluids + connate + meteoric fluids	
	volcanic	magmatic + connate + meteoric or marine fluids	
Existing rocks	released by recrystallization	magmatic fluids and/or connate fluids and/or meteoric fluids	orogenic lithogenic hydrothermal deposits; taphrogenic hydrothermal deposits; (Appalachian and Miss.-Valley carbonate replacements)
Forming rocks	diagenetically released	marine, connate, and clay-released fluids	lithogenic–hydrothermal deposits; Gulf Coast U impregnations in sandstones
Oceans	connate	trapped in lithifying sediments	lithogenic–hydrotepid deposits (V, U, Cu, Pb, Zn in sandstones and carbonates)
	sea water	concentrated in isolated bodies	marine precipitates; (phosphate, metal shales, iron-formation, evaporites)
Atmosphere	meteoric water	involved in hydrologic cycle	lithogenic–hydrologic deposits (Cu, Pb, Zn, V, U, Mo in sandstone impregnations and carbonate replacements)
Mixtures	all above types		all epigenetic–hydrologic deposits are candidates for this category

It is true that modern isotopic chemistry has provided means to differentiate several types of water, but even this has not permitted resolution of the systematic relations between those types involved in the orogenic cycle. For example, connate water was once meteoric or marine, and sea and meteoric water are constantly exchanged through their own cycle. (For compaction fluids, see Wolf, 1976.)

CLASSIFICATION BY DIRECTION OF TRANSPORTING FLUIDS

This distinction has so frequently been the basis for discussions of ore deposits that perhaps more than any classification it has conditioned their interpretation and the exploration for others. Its original purpose was to distinguish between deposits formed by

TABLE VIII

Classification by direction of transporting fluids

Fluid direction	Origin of fluid	Migration process	Example type of strata-bound deposit*
Hypogene	devolatilization of mantle	pneumatolysis	carbonatite sills
	magma generation in mantle	magmation	layered intrusive complexes
	magma generation in crust	magmatic differentiation	normal hydrothermal series
	metamorphic devolatilization of crust	regional metamorphism	magnetitic schists; graphitic schists; skarns
	orogenic devolatilization of crust	compressive folding, thrusting and magmation	normal hydrothermal series (Mississippi-Valley deposits)
	dewatering of sediments by burial	diagenesis	uraniferous sandstones (Texas)
	mixtures		
Laterogene	mobilized connate water	lateral secretion, and lithogenesis	uraniferous sandstones (Texas)
	metamorphic dewatering of rocks	regional metamorphism and lithogenesis	uraniferous slates and schists (Korea)
	orogenic dewatering of rocks	orogeny and lithogenesis	orogenic hydrothermal deposits; lithogenic hydrologic deposits (Miss.-Valley type); Cu, V, U in sandstones
	meteoric water	lithogenesis	lithogenic hydrologic deposits (Miss.-Valley); Cu, V, U in sandstones
	magmatic water	magmatic differentiation	telethermal deposits; Cu, V, U in sandstones
	volcanic water	magmatic differentiation; meteoric hydrothermalism	telethermal deposits; Cu, V, U in sandstones; Miss.-Valley replacements
	mixtures		
Supergene	meteoric water	meteoric lithogenesis	Fe, Mn, Cu, V, U, Mo in sandstones; Miss.-Valley replacements
	volcanic water	magmatic differentiation; meteoric hydrothermalism	U-tuffs; CaF_2 is replaced; Be-tuffs
	mixtures		

* The example deposits have been attributed by some geologists to the corresponding fluids, but the relation in most cases has not been demonstrated conclusively.

juvenile fluids carrying metals from depth, and descending meteoric waters redepositing metals leached from the surface. The former were originally assumed to be entirely magmatic until the possibilities of orogenic or metamorphic fluids were better recognized. This sharp distinction still lives in the minds of many geologists who are reluctant to admit genetic relations between deposits representative of each. An example is the well known argument over the hypogene or supergene origin of Mississippi Valley replacements or Colorado Plateau vanadium—uranium impregnations. The deposits are considered to be necessarily of one type or the other because the hypogene and supergene concepts are thought to be mutually exclusive. This argument is also used to deny an association of meteoric secretionary processes with tectonic processes.

However, most serious students of ore deposits have recognized that such clear distinctions are not so simple. Hydrothermal can no longer imply magmatic because most of the water was isotopically shown to be meteoric, even where heated and rising (White, 1957a, b, 1974). A variety of other fluid sources is recognized as indicated in Table VII. The direction of movement locally may not be consistent or regionally representative. The attempt to expand the classification (Table VIII) to include the laterogene deposits formed by deep-connate or released rock water concentrating deep metals, immediately encounters the problems of variable metal sources, water types, water sources, local movement direction and even direction reversals with time. Creation of the classification is theoretically possible, but assignment of example deposits to specific categories is nearly impossible without assumptions for which evidence is not adequate. The writer again concludes that the persistent use of such a classification as this worsens the genetic arguments rather than resolving them.

CLASSIFICATION BY RELATIVE AGE OF DEPOSIT AND HOST

As with the classifications by process, host lithology, origin, and direction of fluid movement, that by relative age requires assumptions based on permissive evidence, which cloud the issue. This classification (Table IX) is perhaps older than the others and earlier in the century was the center of the controversy over the origin of uranium and copper deposits in sandstones. At that time both the classification and the arguments were simpler: syngenetic vs. epigenetic. The increasing recognition of evidence for epigenesis was a main reason for expansion of the classification to accommodate concepts of diagenetic mineralization. Most of these additions were made by those most interested in keeping the age of mineralization as nearly as possible contemporaneous with that of sedimentation. Thus terms like penecontemporaneous were introduced.

When analyzed as completely as possible, this classification again emphasizes the large variety of subprocesses and conditions which conceivably could cause the formation of the example deposits. It also emphasizes the uncertainty of any relative age for any deposit chosen. To place a deposit in any category mostly requires an assumption beyond

TABLE IX

Classification by relative age of deposit and host rock

Relative age	Dominant process		Type of concentrating fluid		Example type of strata-bound deposit
Syngenetic	physical settling	dissemination	sea water, turbidity currents		uraniferous shales; (Witwatersrand ?); Au, Th, U
		concentration	sea water		beach placers (Egypt)
			meteoric water	fluvial	Valley placers (Elliot Lake)
	chemical precipitates	supersaturation evaporation	sea water	marine	phosphates
		selective extraction		lacustrine	oil shale
		hypogene exhalation	hypogene exhalation fluid		massive sulfides
		organic secretion	marine	lacustrine	biogenic sulfides
		adsorption	marine	lacustrine	Chatanooga shale; Swedish U shale
		overwhelming supply			oil shale; U tuffs
Diagenetic (penecon-temporaneous)	dehydration replacement recrystallization		sea water, connate water		U in sandstones (Texas) dolomitization phosphates
Epigenetic	replacement		juvenile (amagmatic)		massive mantos, lenses or pods; disseminations in any rock type
	pore impregnation		magmatic metamorphic connate	all are possible	low-grade mantos, lenses or pods in clastic rocks; altered massive rocks, or porous layered volcanic or metamorphic rocks
	cavity filling		marine		limestone karst fillings
	fracture filling		meteoric		stockworks in selected strata
	fault and unconformity filling		mixtures		veins in selected strata

the limit of supporting evidence. There is no way to recognize one process as most probable; therefore, the safest assumption is the multiplicity of parent processes.

COMMON DENOMINATOR OF CLASSIFICATIONS

One important benefit from the analysis of classifications is the recognition that strata-binding can result from a large variety of geologic processes (almost the entire range), and that the character of deposits commonly cannot definitely identify one process, but is permissive to several. Thus, the conclusion is reached that most of these processes could be considered responsible at some place, and that some other means is required to determine which.

Such a conclusion not only is believed valid, but also is very useful. Both a strata-bound and stratiform are geometric descriptive terms and should not be used to connate a particular genesis. However, in their geometric sense they are valuable.

CLASSIFICATION ADOPTED – GEOMETRIC AND STRATIGRAPHIC CONFORMITY

Description

A classification based on mineral-deposit geometry and the extent of geometric conformity within strata (Table X) was chosen for this paper to maximize empiricism in deducing and comparing mineralization processes, and to assist definition of the relations of mineral districts to tectonic features. This classification is considered most practical because it allows comparison of multiple features. Although the most empirical of the classifications reviewed, it is also the most suggestive of emplacement mechanism and time, which in turn indicate something about genesis. Yet it minimizes genetic assumptions.

The degree of conformity of geometry of a deposit to the geometry of sedimentational features within confining strata is determined, and the various types and degrees of conformity are compared. The scale of comparison extends from microscopic to megascopic, and the range of comparison includes the chemistry, mineralogy, texture, structure, and history of both the ore and its surroundings. As some strata-binding is possible in all mineralization processes, the range of genesis extends from magmatic to sedimentational.

Inferences

A number of subtle relations not revealed in other classifications become apparent in the one adopted here.

Many districts by themselves, or in areas noted for bedding control of ore distribution, also contain mineralogically similar and apparently contemporaneous transgressive veins or pipes. These include the English Pennines (Dunham, 1967), Ramsbeck, Western Germany (Amstutz et al., 1971), Leadville (Emmons et al., 1927), Gilman, Colorado (Radabaugh et al., 1968); and the East Tennessee (Fagan, 1969; Hoagland, 1971; Maher, 1971; Laurence, 1971), Kentucky–Illinois (Brecke, 1965; Heyl and Brock, 1961), and Central Kentucky (Plummer, 1971) districts in the Appalachian Basin, as well as other districts in the central United States (Heyl, 1968). Ore could be syngenetic and epigenetic

TABLE X

Classification by degree of stratigraphic confinement

		Conformable magmatic units					Sedimentary or subdivided metamorphic units						
		intrusive		volcanic (incl. tuffs)									
Type of rock occupancy by ore minerals / filling of open spaces		massive	layered	group of units	whole single unit	partial single unit	System	Series	Group	Formation	Member	Bed	Stratum
Faults	Large faults	St. Lawrence, CaF$_2$		Huancavelica, Peru	Silverton San Juan tuff, BM								
Faults	Minor faults	Front Range, gran. pegmat. sills		Huancavelica, Pb, Ag, As	Silverton cald., BM, Chih., U			Kentucky Ba	Kentucky Ba	Ramsbeck Pennines Pb, Ba, F			
Joints	Selective joints	Chihuahua, U			Sierra del Lobo, Mex., U	Chihuahua, U				Todilto ls, N. Mex., U			
Joints	All joints local	porphyry Cu											
porosity	Partial cavity impregnation	N.E. Brazil, U					Rhodesian Cu belt	Rhodesian Cu belt	Rhodesian Cu belt	Rhodesian	V–U ss Cu ss	V–U, Cu ss	
porosity	Complete cavity impreg-								Terlingua, Hg	Terlingua, Hg		Pennines	

TABLE X *(continued)*

Process	Mode										
Replacement	Partial local	porphyry Cu	Bancroft U, Rossing, U, NE Brazil, U	Spor Mt., Be	Rhodesian copper belt	Rhodesian copper belt	Rhodesian copper belt	Rhodesian copper belt	Todilto U, Balmat (Meggen), Terlingua	Miss. Valley, E. Tenn.	Peru tactite, Colo. tactite, Miss. Valley, Monte Tamara, skarn
Replacement	Complete local	Sudbury, Cu, Ni							Meggen Pb, Ba, I	Gilman, Miss. Valley skarn	Miss. Valley Ba, F.; Gilman. skarn, Pennines
Replacement	Partial extensive									skarn	skarn
Syngenetic crystallization	Partial local	Illimausaq. U, Sudbury		Kuroko Cu					Meggen, Pb, Ba, F	Rammelsberg	Ba
Syngenetic crystallization	Complete local	Conway gran. U	U-tuff	Kuroko Cu					Fe fm	Fe fm	Fe fm, oil shale
Syngenetic crystallization	Partial extensive	Bushveld complex							Phosphoria fm rand	Witwatersrand	
Deposition	Complete extensive								Fe fm	Oil shale, Chatt. U. shale; Phosphoria fm	Lodev U., Ba, Fe fm; Chattanooga shale, Fe fm

Type of rock occupancy by ore minerals

(BM = base metals)

TABLE XI

Correlation of diverse gradients (temperature-paragenesis, emplacement mechanism, genetic process)

Metal zone	Emplacement mechanism
(1) Chromium−magnetite replacement disseminations	massive siliceous rock replacement
(2) Magnetite massive and disseminated replacements	
(3) Tin replacement disseminations, pegmatites and veins	
(4) Tungsten replacement disseminations	
(5) Gold replacement disseminations and veins	
(6) Molybdenum replacement disseminations, skarns and veins	
(7) Copper replacement disseminations and veins, massive replacements	
(8) Copper-zinc skarns, intrusive stock-related veins, or massive exhalations	
(9) Base-matel intrusive stock-related veins, tactites	
(10) Zinc−lead mesothermal replacement mantos and veins	
(11) Lead−silver−sulfosalt epithermal replacement mantos and veins	
(12) Sulfosalt−fluorite−Hg volcanic-related veins and carbonate replacements	
(13) Gold volcanic-related veins	
(14) Appalachian-type telethermal base-metal carbonate replacement mantos	
(15) Mississippi Valley-type telethermal Zn or Pb replacement mantos	
(16) Ba−F telethermal carbonate replacement mantos	
(17) Cu, V, or U impregnations in sandstone	
(18) Metalliferous fumaroles and hot springs	
(19) Uraniferous lignites	
Uraniferous shales	
Phosphates	
Evaporites	

Emplacement mechanism columns (rotated labels): metamorphic carbonate silication (skarn); siliceous rock disseminated replacement; reactive rock vein filling; non-reactive rock vein filling

at the same place with a history of remobilizations. It is possible also that beneath a certain stratigraphic level representing sea floor at one time, ore might occupy transgressive feeders which issued from the floor and supplied syngenetic deposits above it. Ramdohr advanced this theory for Ramsberg in 1953. Since then Amstutz (1962), Horikoshi (1969), Constantinou and Govett (1973), Sakai and Matsubaya (1974), and many others have emphasized submarine exhalation. It has become a standard interpretation for the Kuroko-type and other massive sulfide deposits in volcanic−sedimentary eugeosynclinal facies. It may be possible for successively later exhalative feeders to be superimposed on earlier syngenetic sulfide layers to create what, after orogenic deformation, may appear to be an intimate mixture of deposit geometries. These might then easily be interpreted as necessarily epigenetic (since the conformable portions could be explained by selective

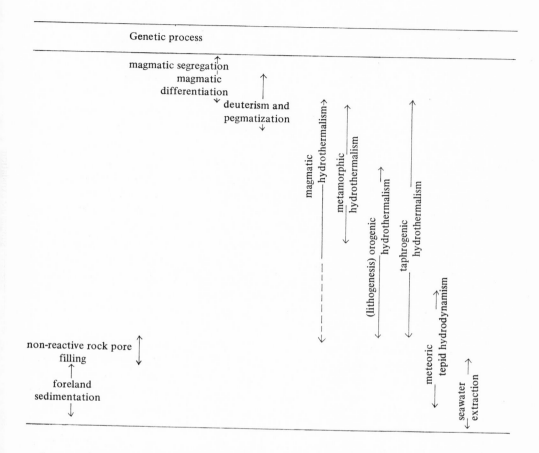

Genetic process

magmatic segregation
magmatic differentiation
deuterism and pegmatization
magmatic hydrothermalism
metamorphic hydrothermalism
(lithogenesis) orogenic hydrothermalism
taphrogenic hydrothermalism
meteoric tepid hydrodynamism
seawater extraction
non-reactive rock pore filling
foreland sedimentation

replacement). However, it should be possible to differentiate these circumstances on the basis of diagenetic or epigenetic textures or structures, because deposits created or modified diagenetically cannot be considered purely sedimentational or syngenetic.

These alternatives are apropos of uranium impregnations in sandstone. Thought to be truly syngenetic by many geologists for years, deeper mining disclosed convincing evidence that in their present, even unoxidized form, they are epigenetic. The interpretation favored by syngeneticists then shifted to one of diagenetic or epigenetic concentration by laterogenesis of originally syngenetic alluvial disseminations. It was difficult, if not impossible, throughout both controversies, to produce clear evidence for original syngenesis without also providing good evidence for epigenesis. Although the epigenesis may have been a remobilization, the possible syngenesis was masked by the later process. (Editor's note: see the related diagrammatic conceptual model in Wolf's Chapter 2.)

Several authors have described lateral zoning within limestone or shale replacement strata-bound districts, and considered it evidence for epigenesis. These include the Kentucky—Illinois fluorspar district (Heyl and Brock, 1961; Brecke, 1965), White Pine Brown, 1971, 1974) the English Pennines (Dunham, 1967), and Monte Tamara, Sardinia (Zuffardi, 1967). Others (Govett and Whitehead, 1974) find a sedimentational explanation for the zoning.

A striking feature brought out by Table X is that completely and extensive stratiform and strata-bound deposits are of only a few mineralogical types and are very distinctive. These include the phosphates, iron-formations, uraniferous black shales, and oil shales. Here, significant, though possibly disseminated, quantities of economic-type elements are deposited directly as primary sediment. Distinctively in each case the original distribution of these elements and the conditions of sedimentation are extensively uniform throughout very large areas, as would be expected of a sedimentary constituent either transported long distances to large basins of quiet deposition, or secreted uniformly from the basin water. These are widely recognized without argument as truly syngenetic deposits, but they are also relatively unique in comparison with all the others.

The outstanding common denominator from all the districts in Table X, exclusive of the extensive whole-rock ore beds, is the local irregular discontinuity of ore minerals within orebodies. Other significant denominators are the irregularities of orebody shapes, and the marked tendency of orebodies to cluster in small areas within the much larger area of a district. Similarly, districts occupy a minute portion of the uniform regional area of a host bed.

The degree of stratigraphic confinement is indicative. In the extensive whole-rock ore beds, the ore-bearing intervals constitute a major portion of the reference stratigraphic unit such as phosphates or iron-formation, and very thin individual ore strata may continue for lengths astronomical orders of magnitude greater than their thickness. In all the other strata-bound deposits, however, ore strata occupy only a small percentage of the reference stratigraphic unit, but the in-bed dimensions of orebodies are still large compared to their thickness. The degree of stratigraphic confinement and conformity decreases as observation is focused on continually smaller details within deposits. Usually a smaller portion of the rock matrix contains ore minerals and these may be erratically shaped and distributed. Sulfides in the Sudbury norite irruptive and sphalerite in dolomitized limestone in east Tennessee are examples. Therefore, in close detail within deposits, few are truly bound by thin strata. Bedding streaking, common in the metalliferous shales, would be the closest approach to such binding, but the streaks are highly discontinuous.

Interestingly, deposits bound within stratigraphic intervals above formation rank are rare. This may be because of the wide variety of lithologies, substantial thickness, and long time represented by sequences of group or greater size.

REFERENCES

Amstutz, G.C., 1962. Massive sulfide deposits: origin and genesis. *Can. Min. J.*, 83 (12): 42–48.

Amstutz, G.C., Zimmermann, R.A. and Schot, E.H., 1971. The Devonian mineral belt of western Germany (the mines of Meggen, Ramsbeck and Rammelsberg) In: *Sedimentology of Parts of Central Europe, Guidebook 8th. Int. Sedimentol. Congr.*, pp. 254–272.

Armands, G., 1972. Geochemical studies of uranium, molybdenum and vanadium in a Swedish alum shale. *Acta Univ. Stockh., Stockh. Contrib. Geol.*, 27 (1): 148 pp.

Arribas, A., 1974. Carateres geologicos de los yacimientos Espanoles de uranio: (Discurso pronunciado en la solemne apertura del Curso Academico 1974–75, Univ. Salamanca.)

Baley, R.W. and James, H.L., 1973. Precambrian iron-formations of the United States. *Econ. Geol.*, 68: 934–959.

Borchert, H., 1970. On the ore deposition and geochemistry of manganese. *Miner. Deposita*, 5: 300–314.

Boyle, R.W. and Davies, J.L., 1973. Banded iron-formations. *Geochim. Cosmochim. Acta*, 37: 1389.

Brecke, E.A., 1965. Origin of the Illinois–Kentucky fluorspar deposits. *Econ. Geol.*, 49: 891–902.

Brathwaite, R.L., 1974. Geology and origin of the Rosebery ore deposit. Tasmania. *Econ. Geol.*, 69: 1086–1101.

Brown, A.C., 1971. Zoning in the White Pine copper deposit, Ontonagon County, Michigan. *Econ. Geol.*, 66: 574–582.

Brown, A.C., 1974. An epigenetic origin of stratiform Cd–Pb–Zn sulfides in the lower Nonesuch shale, White Pine, Michigan. *Econ. Geol.*, 69: 271–274.

Burbank, W.S. (Editor), 1947. *Mineral Resources of Colorado, 2*. Colo. Miner. Resour. Board, Denver, Colo.

Burchard, E.F., 1927. The brown ores of west-middle Tennessee. *U.S. Geol. Surv., Bull.*, 795: 53–112.

Cabri, L.J. and Gilles, J.H., 1974. Mineralogical investigations of the plantinum-group elements in the Sudbury area deposits – a preliminary report (abstr.). *Geol. Assoc. Can./Mineral. Assoc., Can., Ann. Meet., St. Johns, Nfld., May 19–22, 1974*, p. 16.

Coats, R.R., 1955. Uranium and certain other trace elements in felsic volcanic rocks of Cenozoic age in western United States. In: L.R. Page, H.E. Stocking and H.B. Smith (Editors), *Contributions to the Geology of Uranium and Thorium by the U.S. Geological Survey and Atomic Energy Commission for the United Nations International Conference on the Peaceful Uses of Atomic Energy, Geneva, Switzerland, 1955* – U.S. Geol. Surv., Prof. Pap., 300: 75–78.

Colo. School Mines Res. Found. (D.C. Seidel), 1961. *Colorado Vanadium, a Composite Study*. State Colo., Metal Min. Board, Denver, Colo., 155 pp.

Constantinou, G. and Govett, G.J.S., 1973. Geology, geochemistry, and genesis of Cyprus sulfide deposits. *Econ. Geol.*, 68: 843–858.

Davidson, C.F. and Bowie, S.H.U., 1951. On thucholite and related hydrocarbon–uraninite complexes, with a note on the origin of the Witwatersrand gold ores. *Gr. Br. Geol. Surv. Bull.*, 3: 1–18.

Davis, G.H., 1972. Deformational history of the Caribou strata-bound sulfide deposit, Bathurst, New Brunswick, Canada, *Econ. Geol.*, 67: 634–655.

Densen, N.M., Bachman, G.P. and Zeller, H.D., 1959. Uranium-bearing lignite in northwestern South Dakota and adjacent states. In: *Uranium in Coal in the Western United States–U.S. Geol. Surv., Bull.*, 1055: 11–57.

Doe, B.R. and Delevaux, M.H., 1972. Source of lead in southeast Missouri galena ores. *Econ. Geol.*, 67: 409–425.

Dunham, K.C., 1967. Veins, flats, and pipes in the Carboniferous of the English Pennines. In: J.S. Brown (Editor), *Genesis of Stratiform Lead–Zinc–Barite–Fluorite Deposits–Econ. Geol., Monogr.*, 3: 201–207.

Du Rietz, T., 1953. Geology and ores of the Kristineberg deposit, Vesterbotten, Sweden. *Sver. Geol. Unders., Arsb.*, 45: 90 pp.

Emmons, S.F., Irving, J.D. and Loughlin, G.F., 1927. Geology- and ore deposits of the Leadville mining district, Colorado. *U.S. Geol. Surv., Prof. Pap.*, 148: 360 pp.

Ensign, C.O. Jr., White, W.S., Wright, J.C., Patrick, J.L., Leone, R.J., Hathaway, D.J., Trammel, J.W., Fritts, J.J. and Wright, T.L., 1968. Copper deposits in the Nonesuch shale, White Pine, Michigan. In: J.D. Ridge (Editor), *Ore Deposits in the United States–1933/1967*. Am. Inst. Min. Eng., Graton/Sales Vol., pp. 460–488.

Fagan, J.M., 1969. Geology of the Lost Creek barite mine. In: W.D. Hardeman (Editor), *Papers on the Stratigraphy and Mine Geology of the Kingsport and Mascot Formations (Lower Ordovician) of East Tennessee–Tenn. Div. Geol., Rep. Invest.*, 23: 40–44.

Finch, J.W., 1933. Sedimentary copper deposits of the western states. In: J.W. Finch et al. (Editors), *Ore Deposits of the Western States*. Am. Inst. Min. Eng., Lindgren Vol., pp. 481–487.

Frondel, C. and Baum, J.L., 1974. Structure and mineralogy of the Franklin zinc–iron–manganese deposit, New Jersey. *Econ. Geol.*, 69: 157–180.

Gabelman, J.W., 1956. Structural control of uranium deposits in the Zuni–Mt. Taylor region, northwest New Mexico (abstr.). *Econ. Geol.*, 51: 114.

Gabelman, J.W., 1971. Sedimentology and uranium prospecting. *Sediment. Geol.*, 6: 145–186.

Gary, M., McAfee, P. Jr. and Wolf, C., 1972. *Glossary of Geology*. Am. Geol. Inst., Washington, D.C., 805 pp.

Gemmil, P., 1968. The geology of the ore deposits of the Pioche district, Nevada. In: J.D. Ridge (Editor), *Ore Deposits in the United States–1933/1967*. Am. Inst. Min. Eng., Graton/Sales Vol., pp. 1128–1147.

Gerdemann, P.E. and Meyers, P.E., 1972. Relationship of carbonate facies patterns to ore distribution and to ore genesis in the southeast Missouri lead district. *Econ. Geol.*, 67: 426–433.

Glover, L., 1959. Stratigraphy and uranium content of the Chattanooga shale in northeastern Alabama, northwestern Georgia and eastern Tennessee. *U.S. Geol. Surv., Bull.*, 1087-E: 133–168.

Gonzales Reyna, J., 1956. Los yacimientos de antimonio de Mexico. In: J. Gonzales R. (Editor), *Riqueza Minera y Yacimientos Minerales de Mexico. Int. Geol. Congr., 20th, Mex.*, pp. 203–213.

Gordon, C.H., 1913. Types of iron ore deposits in Tennessee. *Tenn. Geol. Surv., Res. Tenn.*, 3: 84–95.

Gordon, M. Jr. and Tracey, J.I. Jr., 1952. Origin of the Arkansas bauxite deposits. In: *Problems of Clay and Laterite Genesis*. Am. Inst. Min. Eng., pp. 12–34.

Govett, G.J.S. and Whitehead, R.E.S., 1974. Origin of metal zoning in stratiform sulfides: a hypothesis. *Econ. Geol.*, 69: 551–556.

Granger, H.C., 1963. Mineralogy. In: V.C. Kelley (Editor), *Geology and Technology of the Grants Uranium Region – State Bur. Mines Miner. Tech., New Mex. Inst. Min. Technol., Mem.*, 15: 21–37.

Gray, R.F., Hoffman, J.J., Bagan, R.J. and McKinley, H.L., 1968. Bishop tungsten district, California. In: J.D. Ridge (Editor), *Ore Deposits of the United States–1933/1967*. Am. Inst. Min. Eng., pp. 1531–1554.

Grip, E., 1950. Lead and zinc deposits in northern Sweden. In: *Geology, Paragenesis, and Reserves of Lead and Zinc–Int. Geol. Congr., 18th*, 8: 362–370.

Grip, E., 1967. On the genesis of the lead ores of the eastern border of the Caledonides in Scandinavia. In: J.S. Bram (Editor), *Genesis of Stratiform Lead–Zinc–Barite–Fluorite Deposits – Econ. Geol., Monogr.*, 3: 208–218.

Grogan, R.M. and Bradbury, J.C., 1967. Origin of the stratiform fluorite deposits of southern Illinois. In: J.S. Bram (Editor), *Genesis of Stratiform Lead–Zinc–Barite–Fluorite Deposits (Mississippi Valley-type Deposits – Econ. Geol., Monogr.*, 3: 40–51.

Gross, S.O., 1968. Titaniferous ores of the Sanford Lake District, New York. In: J.D. Ridge (Editor), *Ore Deposits in the United States–1933/1967*. Am. Inst. Min. Eng. Graton/Sales Vol., pp. 140–154.

Harder, E.C., 1937. Bauxite. In: S.H. Dolbear and O. Bowles (Editors), *Industrial Minerals and Rocks*. Am. Inst. Min. Eng., pp. 111–128.

Hart, O.M., 1958. Uranium deposits in the Pryor—Big Horn Mountains, Carbon County, Montana, and Big Horn County, Wyoming. *U. N. Surv. Raw Mater. Resour., Int. Conf. Peaceful Uses At. Energy, 2nd, Geneva, Sept. 1958*, pp. 523—526.

Heidenreich, W.L., 1959. Nickel—cobalt—iron-bearing deposits in Puerto Rico. *U.S. Bur. Mines*, Rep. Ing., 5532: 68 pp.

Hershey, R.E., 1969. Tennessee. In: Southern Interstate Nuclear Board, *Uranium in the Southern United States* — U.S. At. Energy Comm., WASH—1128; pp. 91—97.

Hewett, F.F., 1909. Vanadium deposits of Peru. *AIME Trans., Min. Eng.*, 40: 274—299.

Hewitt, W.P., 1968. Geology and mineralization of the main mineral zone of the Santa Eulalia district, Chihuahua, Mexico. *Soc. Min. Eng., AIME Trans.*, 241: 228—260.

Heyl, A.V., 1968. Minor epigenetic, diagenetic, and syngenetic sulfide, fluorite, and barite occurrences in the central United States. *Econ. Geol.*, 63: 585—594.

Heyl, A.V. and Brock, M.R., 1961. Structural framework of the Illinois—Kentucky mining district and its relation to mineral deposits. *U.S. Geol. Surv. Prof. Pap.*, 424D: 3—6.

Heyl, A.V., Landis, G.P. and Zartman, R.E., 1974. Isotopic evidence for the origin of Mississippi Valley-type mineral deposits: a review. *Econ. Geol.*, 69: 992—1006.

Hickok, J.H., 1933. The iron ore deposits at Cornwall, Pennsylvania. *Econ. Geol.*, 28: 193—255.

Hill, W.T., McCormick, J.E. and Wedow, H. Jr., 1971. Problems on the origin of ore deposits in the Lower Ordovician formations of east Tennessee. *Econ. Geol.*, 66: 799—804.

Hoagland, A.D., 1971. Appalachian strata-bound deposits: their essential features, genesis, and the exploration problem. *Econ. Geol.*, 66: 805—810.

Horikoshi, E., 1969. Volcanic activity related to the formation of the Kuroko-type deposits in the Kosaka district, Japan. *Miner. Deposita*, 4: 321—345.

Horn, D.R., Horn, B.N. and Delach, M.N., 1972. Distribution of ferromanganese deposits in the world ocean. In: D.R. Horn (Editor), *Ferromanganese Deposits on the Ocean Floor*. Off. Int. Decade Ocean Expl., Natl. Sci. Found., Washington, D.C. pp. 9—17.

JOIDES Scientific Staff, 1974. Leg 34, oceanic basalt and the Nazca plate. *Geotimes*, 19: 20—24.

Jones, W.B., 1929. Summary report on graphite in Alabama. *Ala. Geol. Surv., Circ.*, 9: p. 27.

Kelley, V.C., 1951. Oolite iron deposits of New Mexico. *Am. Assoc. Pet. Geol. Bull.*, 35: 2199—2228.

Krauskopf, K.B., 1967. Source rocks for metal-bearing fluids. In: H.L. Barnes (Editor), *Geochemistry of Hydrothermal Ore Deposits*. Holt, Rinehart and Winston, New York, N.Y., pp. 1—33.

Krautner, H.G., 1970. Die hercynische Geosynklinal-erzbildung in der rumänischen Karpaten und ihre Beziehungen zu der hercynischen Metallogenese Mittel-europas. *Miner. Deposita*, 5: 323—344.

Lambert, I.B. and Sato, T., 1974. The Kuroko and associated deposits of Japan: a review of their features and metallogenesis. *Econ. Geol.* 69: 1215—1236.

Landis, E.R., 1962. Uranium and other trace elements in Devonian and Mississippian black shales in the central mid-continent area. *U.S. Geol. Surv. Bull.*, 1107-E: 287—336.

Lang, A.H., Griffin, J.W. and Steacy, H.R., 1962. Canadian deposits of uranium and thorium. *Can. Geol. Surv., Econ. Geol. Sect.*, 16: 324 pp.

Laurence, R.A., 1971. Evolution of thought on ore controls in Tennessee. *Econ. Geol.*, 66: 696—700.

Lea, E.R. and Dill, D.B. Jr., 1967. Zinc deposits of the Balmat—Edwards district, New York. In: J.D. Ridge (Editor), *Ore Deposits of the United States—1933/1967*. Am. Inst. Min. Eng., Graton/Sales Vol., pp. 20—48.

Lindgren, W., 1933. *Mineral Deposits*. McGraw-Hill, New York, N.Y., 929 pp.

Loughlin, G.F. and Behre, C.H. Jr., 1933. Classification of ore deposits. In: *Ore Deposits of the Western States*. Am. Inst. Min. Eng., Lindgren Vol., pp. 17—55.

Mackin, J.H., 1947. Some structural features of the intrusions of the Iron Springs district. *Utah Geol. Soc., Guidebook Geol. Utah*, 2: 62 pp.

Mackin, J.H., 1968. Iron ore deposits of the Iron Springs district, Southwestern Utah. In: J.D. Ridge (Editor), *Ore Deposits of the United States—1933/1967*. Am. Inst. Min. Eng., pp. 992—1019.

Magnusson, N.H., 1950. Lead and zinc deposits of central Sweden. In: *Geology, Paragenesis and Reserves of Lead and Zinc—Int. Geol. Congr., 18th, Gr. Br.*, 7: 371—379.

Maher, S.W., 1971. Regional distribution of mineral deposits beneath the pre-Middle Ordovician unconformity in the southern Appalachians. *Econ. Geol.*, 66: 744—747.

Mansfield, G.R., 1943. Phosphate resources of Florida. *U.S. Geol. Surv., Bull.*, 934: 82 pp.

Marsden, R.W., 1968. Geology of the iron ores of the Lake Superior region in the United States. In: J.D. Ridge (Editor), *Ore Deposits in the United States—1933/1967.* Am. Inst. Min. Eng., Graton/ Sales Vol., pp. 489—506.

Mart, J. and Sass, E., 1972. Geology and origin of the manganese ore of Um Bogma, Sinai. *Econ. Geol.*, 67: 145—155.

Martin, H.S. and Wilding, J., 1937. Phosphate rock. In: S.H. Dolbear and O. Bowles (Editors),*Industrial Minerals and Rocks.* Am. Inst. Min. Eng., pp. 543—570.

Matsukuma, T. and Horikoshi, E., 1970. Kuroko deposits in Japan, a review. In: T. Tatsumi (Editor), *Volcanism and Ore Genesis.* Univ. Tokyo Press, Tokyo, pp. 153—179.

Maucher, A. and Schneider, H.J., 1967. The Alpine lead—zinc ores. In: J.S. Brown (Editor), *Genesis of Stratiform Lead—Zinc—Barite—Fluorite Deposits (Mississippi Valley-type Deposits)- Econ. Geol., Monogr.*, 3: 71—89.

Mauger, R.L., 1972. A sulfur-isotope study of the Ducktown, Tennessee district, U.S.A. *Econ. Geol.*, 67: 497—510.

McAnulty, W.N., Sewell, C.K., Athinson, D.R. and Rasberry, J.M., 1963. Aguachile beryllium-bearing fluorspar district, Coahuila, Mexico. *Geol. Soc. Am., Bull.*, 74: 735—743.

McKelvey, V.E., 1949. Geological studies of the western phosphate field. In: *Symposium on Western Phosphate Mining — AIME Min. Branch. Trans.*, 184: 270—279.

Mendelsohn, F., 1961. Ore genesis—summary of the evidence. In: F. Mendelsohn (Editor), *Geology of the Northern Rhodesian Copperbelt.* McDonald, London, 523 pp.

Miller, J.M., 1963. Uraninite-bearing placer deposits in the Indus alluvium near Hazro. Pakistan. *Geol. Surv. Gr. Br., At. Energy Div., Rep.*, 254: 9 pp.

Mills, J.W., 1971. Bedded barite deposits of Stevens County, Washington. *Econ. Geol.*, 66: 1157—1163.

Morgenstein, M., 1972. Manganese accretion at the sediment—water interface at 400 to 2400 meters depth, Hawaiian archipelago. In: D.R. Horn (Editor), *Ferromanganese Deposits on the Ocean Floor.* Off. Int. Decade Ocean Expl., Natl. Sci. Found., Washington, D.C., pp. 131—138.

Morris, H.T., 1968. The main Tintic mining district, Utah. In: J.D. Ridge (Editor), *Ore Deposits of the United States—1933/1967.* Am. Inst. Min. Eng., Graton/Sales Vol., pp. 1043—1073.

Nolan, T.B. and Hunt, R.N., 1968. The Eureka mining district, Nevada. In: J.D. Ridge (Editor),*Ore Deposits of the United States—1933/1967.* Am. Inst. Min. Eng., Graton/Sales Vol., pp. 966—991.

Overstreet, W., 1967. The geologic occurrence of monazite. *U.S. Geol. Surv., Prof. Pap.*, 530: 327 pp.

Page, N.V., Riley, L.B. and Haffty, J., 1972. Vertical and lateral variation of platinum, palladium and rhodium in the Stillwater Complex. *Econ. Geol.*, 67: 915—923.

Pastor, M., Doetsche, J., Lizaur, J. and De la Concha, S., 1956. Criaderos de manganeso de España. In: J. Gonzales Reyna (Editor), *Symposium Sobre Yacimientos de Manganeso—Int. Geol. Congr., 20th, Mex.*, 4/5: 25—50.

Phalen, W.C., 1919. Salt resources of the United States. *U.S. Geol. Surv., Bull.*, 669: 284 pp.

Phillips, W.B., 1905. The quicksilver deposits of Brewster County, Texas. *Econ. Geol.*, 1: 155—162.

Plummer, L.N., 1971. Barite deposition in central Kentucky. *Econ. Geol.*, 66: 252—258.

Pride, D.E. and Hagner, A.F., 1972. Geochemistry and origin of the Precambrian iron formation near Atlantic City, Fremont County, Wyoming. *Econ. Geol.*, 67: 329—338.

Radabaugh, R.E., Merchant, J.S. and Brown, J.M., 1968. Geology and ore deposits of the Gilman (Red Cliff—Battle Mountain) district, Eagle County, Colorado. In: J.D. Ridge (Editor), *Ore Deposits of the United States—1933/1967.* Am. Inst. Min. Eng., Graton/Sales Vol., pp. 641—680.

Renfro, A.R., 1974. Genesis of evaporite-associated stratiform metalliferous deposits—a Sabkha process. *Econ. Geol.*, 69: 33—45.

Ridge, J.D. (Editor), 1968. Changes and developments in concepts of ore genesis. In: J.D. Ridge (Editor), *Ore Deposits of the United States—1933/1967.* Am. Inst. Min. Eng., Graton/Sales Vol., pp. 1713—1834.

Roscoe, S., 1969. Huronian rocks and uraniferous conglomerates in the Canadian Shield. *Geol. Surv. Can., Pap.*, 68-40: 205 pp.

Royce, S., 1942. Lake Superior iron deposits. In: W.H. Newhouse (Editor), *Ore Deposits as Related to Structural Features*. Princeton Univ. Press, Princeton, N.J., pp. 54–62.

Sakai, H. and Matsubaya, O., 1974. Isotopic geochemistry of the thermal waters of Japan and its bearing on the Kuroko-ore solutions. *Econ. Geol.*, 69: 974–991.

Sangster, D.F., 1972. Precambrian volcanogenic massive sulfide deposits in Canada: a review. *Geol. Surv. Can., Pap.*, 72-22: 44 pp.

Sato, T., 1974. Origin of the green tuff metal province of Japan (abstr.). *Geol. Assoc. Can./Mineral. Assoc. Can., Ann. Meet., St. Johns, Nfld., May 19–22, 1974*, p. 91.

Scott, M.R. and Scott, R.B., 1974. Rapidly accumulating manganese deposit from the median valley of the Mid-Atlantic Ridge. *Am. Geophys. Union, Geophys. Res. Lett.*, 1: 355–358.

Scull, B.J., 1958. Origin and occurrence of barite in Arkansas. *Ark. Geol. Cons. Comm., Inf. Circ.*, 18: 101 pp.

Shawe, D.R., 1968. Geology of the Spor Mountain beryllium district, Utah. In: J.D. Ridge (Editor), *Ore Deposits in the United States–1933/1967*. Am. Inst. Min. Eng., Graton/Sales Vol., pp. 1148–1161.

Shawe, D.R., Poole, F.G. and Brobst, D.A., 1969. Newly discovered bedded barite deposits in East Northumberland Canyon, Nye County, Nevada. *Econ. Geol.*, 64: 245–254.

Simpson, T.A. and Gray, T.R., 1968. The Birmingham red-ore district. In: J.D. Ridge (Editor), *Ore Deposits of the United States – 1933/1967*. Am. Inst. Min. Eng., Graton/Sales Vol., pp. 187–206.

Slaughter, A.L., 1968. The Homestake mine. In: J.D. Ridge (Editor), *Ore Deposits in the United States–1933/1967*. Am. Inst. Min. Eng., Graton/Sales Vol., pp. 1436–1459.

Smith, R.W., 1924. Geology and utilization of Tennessee phosphate rock. *AIME, Trans.*, 74: 127–146.

Snyder, F.G., 1967. Criteria for origin of stratiform orebodies with application to southeast Missouri. In: J.S. Brown (Editor), *Genesis of Stratiform Lead–Zinc–Barite–Fluorite Deposits–Econ. Geol., Monogr.*, 3: 1–13.

Sörensen, H., 1970. Occurrence of uranium in alkaline igneous rocks. In: *Uranium Exploration Geology*. Int. At. Energy Agency, Vienna, pp. 161–168.

Souch, B.E., Podolsky, T. and Geological Staff, the Internat. Nickel Company of Canada, 1969. The sulfide ores of Sudbury: their particular relationship to a distinctive inclusion-bearing facies of the Nickel Irruptive. In: H.D.B. Wilson (Editor), *Magmatic Ore Deposits–Econ. Geol., Monogr.*, 4: 252–261.

Stipanicic, P.N., 1970. Conceptos geostructurales sobre la distribuicion de los yacimientos uraniferos con control sedimentario en la Argentina y posible aplicacion de los mismos en el resto de Sudamerica. In: *Uranium Exploration Geology*. Int. At. Energy Agency, Vienna, pp. 205–218.

Strong, D.F., 1974. Mineral occurrences of the Morten's Harbour area. In: D.F. Strong (Editor), *Plate Tectonic Setting of Newfoundland Mineral Occurrences–Guidebook NATO Adv. Stud., Inst. Metallogeny Plate Tectonics. Mem. Univ., St. Johns, Nfld., May, 1974*, pp. 62–71.

Turner, H.W., 1905. The Terlingua quicksilver deposits. *Econ. Geol.*, 1: 265–281.

Varentsov, I.M., 1964. *Sedimentary Manganese Ores*. Elsevier, Amsterdam, 119 pp.

Vickers, R.C., 1956. Geology and monazite content of the Goodrich quartzite, Palmer area, Marquette County, Michigan. *U.S. Geol. Surv., Bull.*, 1030-F.

Von Backström, J.W., 1970. THe Rossing uranium deposit near Swakopmund, Southwest Africa, a preliminary report. In: *Uranium Exploration Geology*. Int. At. Energy Agency, Vienna, pp. 143–148.

Weege, R.J., Pollock, J.P. and the Calumet Division Geological Staff, 1972. The geology of two new mines in the native copper district of Michigan. *Econ. Geol.*, 67: 622–633.

Westgate, L.G. and Knopf, A., 1932. Geology and ore deposits of the Pioche district. *U.S. Geol. Surv., Prof. Pap.*, 171: 79 pp.

White, D.E., 1957a. Thermal springs and epithermal deposits. *Econ. Geol., 50th Anniv. Vol.*, 1: 99–154.

White, D.E., 1957b. Magmatic, connate, and metamorphic waters. *Geol. Soc. Am. Bull.*, 68: 1659–1682.

White, D.E., 1974. Diverse origins of hydrothermal ore fluids. *Econ. Geol.*, 69: 954–973.

White, W.S., 1968. The native copper deposits of northern Michigan. In: J.D. Ridge (Editor), *Ore Deposits in the United States—1933/1967.* Am. Inst. Min. Eng., Graton/Sales Vol., pp. 303–325.

Whiteside, H.C.M., 1970. Uraniferous Precambrian conglomerates of South Africa. In: *Uranium Exploration Geology.* Int. At. Energy Agency, Vienna, pp. 49–73.

Willemse, J., 1969. The uraniferous magnetic iron ore of the Bushveld igneous complex. In: H.D.B. Wilson (Editor), *Magmatic Ore Deposits – Econ. Geol., Monogr.,* 4: 187–208.

Wilson, I.V. and Rocha Morena, V.S., 1955. Geology and mineral deposits of the Boleo copper district, Baja California, Mexico. *U.S. Geol. Surv., Prof. Pap.,* 273: 134 pp.

Withington, C.F. and Jaster, H., 1960. Selected annotated bibliography of gypsum and anhydrite in the United States. *U.S. Geol. Surv., Bull.,* 1105: 126 pp.

Wolf, K.H., 1976. Ore genesis influenced by compaction. In: G.V. Chilingar and K.H. Wolf (Editors), *Compaction of Coarse-Grained Sediments, 2.* Elsevier, Amsterdam, in press.

Woodward, L.A., Kaufman, W.H., Schumacher, O.L. and Talbott, L.W., 1974. Strata-bound copper deposits in Triassic sandstone of Sieira Nacimiento, New Mexico. *Econ. Geol.,* 69: 108–120.

Worthington, J.E. and Kiff, I.T., 1970. A suggested volcanogenic origin for certain gold deposits in the slate belt of the North Carolina Piedmont. *Econ. Geol.,* 65: 529–537.

Yates, R.G. and Thompson, G.A., 1959. Geology and quicksilver deposits of the Terlingua district, Texas. *U.S. Geol. Surv., Prof. Pap.,* 312: 114 pp.

Zachrisson, E., 1971. The structural setting of the Stekenjokk orebodies, entral Swedish Caledonides. *Econ. Geol.,* 66: 641–652.

Zimmermann, R.A., 1970. Sedimentary features in the Meggen barite–pyrite–sphalerite deposit and a comparison with the Arkansas barite deposits. *Neues Jahrb. Mineral. Abh.,* 113: 179–214.

Zuffardi, P., 1967. The genesis of stratiform deposits of lead–zinc and barite in Sardinia. In: J.S. Brown (Editor), *Genesis of Stratiform Lead–Zinc–Barite–Fluorite Deposits–Econ. Geol., Monogr.,* 3: 178–191.

Chapter 4

SOME TRANSITIONAL TYPES OF MINERAL DEPOSITS IN VOLCANIC AND SEDIMENTARY ROCKS

PAUL GILMOUR

INTRODUCTION

Yu.A. Bilibin published his masterly "Metallogenic Provinces and Metallogenic Epochs" in 1955, but several years passed before English-language translations became generally available (Bilibin, 1968). In the meantime, one or two adaptations of Bilibin's ideas were applied to other areas — notably the Canadian segment of the Appalachian belt (McCartney and Potter, 1962) — and these provided an introduction to his work for those lacking access to the original. During the period which elapsed between the Russian publication of Bilibin's work and McCartney and Potter's adaptation, the present writer independently proposed that classifications of mineral deposits should be based on the criteria Bilibin employed, and briefly illustrated the principles with reference to deposits of copper and widely associated metals (Gilmour, 1962). Revised and expanded versions of this scheme were presented on subsequent occasions (Gilmour, 1966 and 1971).

No matter how elegantly contrived, however, no classification of natural objects or phenomena is capable of accommodating all of their attributes and the classification which forms the basis of Bilibin's study was no exception. In particular, it fails to depict the fact that some mineral deposits seem to possess the characteristics of (or appear to be transitional between) more than one major group. The scheme proposed by the present writer suffers from the same defect, of course.

Judging by published articles and conversations, most geologists seem to regard major classes of mineral deposits (such as "Mississippi Valley" lead–zinc deposits or "porphyry" copper–molybdenum deposits) as groups of examples which should — and to varying degrees do — conform to fairly simple stereotypes or models. Perhaps it would be more realistic to regard major classes of deposits as groups made up of examples, some of which only approximate the stereotype and may exhibit characteristics of other important classes. This practice would be analogous, say, to the treatment by palaentologists of genera and species which are defined statistically, rather than by comparison with the morphology of a single genotype or holotype (e.g., the Carboniferous, non-marine lamellibranchs).

In the descriptions of deposits which follow (as distinct from the accompanying comments on genesis), the writer has made a deliberate effort to employ descriptive, rather than genetic, terms. Unfortunately, the terminology of the earth sciences is so permeated with genetic implications that it is impossible to achieve this goal consistently. Since the threefold classification of rocks is based on their inferred mode of origin, it is not even possible to speak or write about rock types without embracing genetic concepts. Were it otherwise, H.H. Read (1943) would not have needed to note that there are "granites and granites", and debates about the nature of the so-called "porphyries" found in many mining camps would never have arisen. It is not possible to use the simple word "basalt" without implying, "a rock (*believed to have*) consolidated from a melt". Genetic and inferential words like "hypogene", "supergene", "epigenetic" and "syngenetic" are bandied about without qualification or explanation by most mining geologists as if they are descriptive terms which refer to physical properties susceptible to precise measurement and objective application (comparable to size, weight or shape), rather than words involving inferences indirectly derived from observation. It has become fashionable recently to speak and write of "volcanogenic" deposits, meaning the strata-bound massive pyritic sulphides found in eugeosynclinal rocks which are believed to have originated through the agency of volcanic processes. It is conceivable, even if improbable, that volcanism has nothing whatsoever to do with the genesis of this class of deposits. Terms like "volcanogenic" should be confined to the identification of genetic theories and actualistic terms, such as "Mississippi-Valley type" and "massive, pyritic sulphides", should be employed to characterize groups of mineral deposits.

A CLASSIFICATION OF DEPOSITS OF COPPER AND ASSOCIATED METALS

For the purposes of the present discussion concentrations of copper minerals, along with commonly associated metals, have been divided into two principal groups, depending on the nature of the rocks and structures with which they are associated. These two groups comprise deposits associated with orogenic belts and cratonic regions. It stands to reason that occurrences associated with a mobile belt, which has subsequently been incorporated into a shield area, might be said to be spatially related to both. The critical factor from the standpoint of the present classification is the role of the host rocks in the evolution of the craton, viz., whether or not they formed part of the "parental" mobile belt. If this is not clear to some readers, perhaps the discussion which follows will help.

In a paper which has become a modern "classic", Dewey and Bird (1970) proposed that mountain-building occurs in two basic ways, each of which gives rise to a characteristic type of orogenic belt, identified by them as "cordilleran" and "collision" (the use of non-equivalent terms, one of them genetic, the other empirical, seems unfortunate to the writer). These two types are more readily distinguished by the products of associated plutonic and tectonic activity than the nature of the layered rocks. Since Dewey and Bird's cordilleran type closely resembles the classical model of an orogenic zone, it was used as the basis for the outline presented here, but a similar scheme could easily be devised for deformed belts of the so-called collision type.

Deposits in orogenic belts

Fig. 1 shows an idealized cross-section of a mobile belt of cordilleran type on which the principal types of deposits of copper and associated metals have been superimposed according to the place of the host rocks in the orogen. Before proceeding to review the various classes of mineralization recognized a few general comments might be in order.

With few exceptions, the nomenclature used in relation to Fig. 1 closely follows

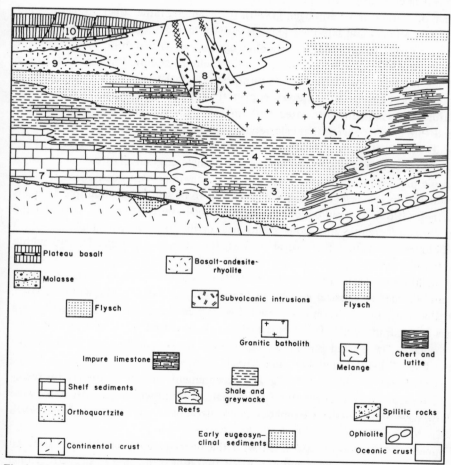

Fig. 1. Idealized cross-section of a "cordilleran" mountain belt (modified after Dewey and Bird, 1970). The numbers refer to the location of deposits of copper and associated metals relative to typical host rocks.

Dewey and Bird. These authors employed the term "ophiolite" in reference to the early basic and ultra-basic igneous activity associated with mobile belts and, for brevity, the expression has also been used here. Formerly believed to represent predominantly conformable intrusions (of the so-called "Alpine" type), these rocks are now increasingly regarded as flows. Dewey and Bird made scant reference to the overlying volcanic group of rocks, but they are important in the present context, since they constitute the soda-rich "spilite" association which is, practically by definition, characteristic of the eugeosynclinal suite. Irvine and Barager (1971) noted that greenstone belts of Precambrian age contain both calc-alkaline and tholeitic rocks, whereas volcanic rocks of the modern circum-Pacific belt are typically calc-alkaline. For this reason the present writer prefers to employ terms like "spilite–keratophyre" and "basalt–andesite–rhyolite" associations, in the manner of Turner and Verhoogen (1960), for example, rather than "tholeitic" and "calc-alkaline". Middle- and late-stage igneous activity includes "granitic" batholiths and volcanic rocks of the basalt–andesite–rhyolite association. The plutons, or sub-adjacent intrusions, related to this volcanic activity – the "minor" granitic intrusions of Turner and Verhoogen (op. cit. pp. 331–339) – form the principal hosts of another important class of deposits, namely the porphyry copper (molybdenum) deposits. So far as the nomenclature for sediments associated with mobile belts is concerned, Fig. 1 follows Dewey and Bird closely, except that calcareous rocks, commonly "impure", are shown to occur locally in sites other than the shelf zone proper.

Needless to say, it is impossible to represent in a single diagram all of the complexities of a dynamic process in which deformation is at least as important as sedimentation and igneous activity. In this respect, there is a great deal to be said for the multiple diagrams of McCartney (1965). It would probably be more accurate, for example, to depict the late-stage "minor" granitic intrusions mentioned in the previous paragraph associated with shelf limestones rather than, or as well as, the impure calcareous deposits within accumulations of flysch.

Some might object that current theories of plate tectonics render obsolete, or at least obsolescent, the use of classical geosynclinal terms (Sillitoe, 1973). The writer believes, however, that plate theory has not superseded geosynclinal models: it has merely placed them in a larger framework. This seems tacitly acknowledged by proponents of plate tectonics, such as Dewey and Bird (and even Sillitoe), who extensively employ terms like "eugeosyncline", "shelf", "ophiolite" and "flysch".

Perhaps the best way to explain Fig. 1 is to comment on each of the groups recognized, in a sequence which reflects the approximate order in which the host rocks were believed to have formed (the examples chosen are mostly, but not all, from North America).

(1) Copper–nickel deposits associated with the predominantly stratiform basic and ultra-mafic (ophiolitic) intrusions and extrusions found near the base of a eugeosynclinal assemblage (lower, right-hand portion of Fig. 1). Examples of this type of mineralization include numerous occurrences in Canada, such as those of Marbridge in Quebec and

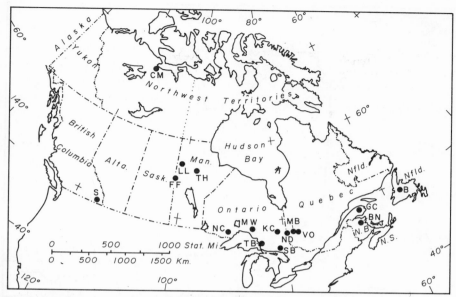

Fig. 2. Index map of the localities in Canada to which reference is made: B = Buchans mine; BN = Bathurst–Newcastle district; CM = Copper mine; FF = Flin Flon area; GC = Gaspe Copper mine; KC = Kidd Creek mine; LL = Lynn Lake area; MB = Marbridge mine; MW = Manitouwadge area; NC = North Coldstream mine; ND = Noranda district; TB = Tribag mine; TH = Thomson mine; S = Sullivan mine; SB = Sudbury district; VO = (town of) Val d'Or.

perhaps Thomson in Manitoba (Fig. 2), as well as the deposits near Kalgoorlie in Western Australia (Fig. 3). Mineralization of this type has certain features in common with copper–nickel deposits found in major, layered intrusions, such as the Sudbury "nickel irruptive" (see group 11, below), although they are distinguishable by details in the chemistry, mineralogy and petrology of the mineralization as well as the contrasted structural settings.

(2), (3), (4) and (5) Copper–zinc–lead mineralization associated with accumulations of eugeosynclinal volcanic and sedimentary rocks (lower right and centre, Fig. 1). Since it is proposed to enlarge on these occurrences below, no more need be said here.

(6) Lead–zinc (copper) concentrations in shelf limestone and shale (lower centre-left in Fig. 1). Together with deposits found in limestone and shale forming cratonic cover (group 13, below), deposits of this type comprise the important class commonly known as Mississippi-Valley deposits (Fig. 4). Generally regarded as lead–zinc deposits, some districts contain significant amounts of copper (Heyl et al., 1955) and the type may have affiliations with the "pyrometasomatic" deposits associated with porphyry copper mineralization (group 8). Transitional types found where shelf limestone passed laterally into cratonic cover are liable to exist. Host rocks deposited in a shelf zone are more likely to

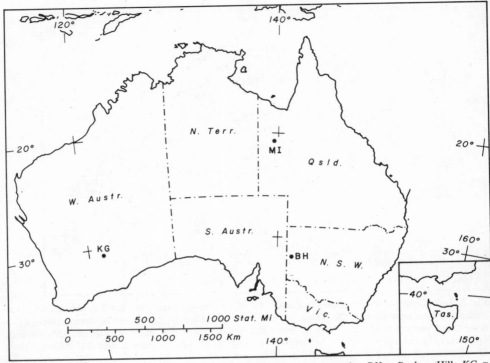

Fig. 3. Index map of localities in Australia to which reference is made: *BH* = Broken Hill; *KG* = Kalgoorlie; *MI* = Mount Isa.

be deformed than cratonic cover sediments (examples in the Rocky-Mountain zone compared with the Mississippi-Valley region itself – Fig. 4).

(7) Copper (zinc) deposits in basal, or marginal, clastic sediments of a shelf zone, with or without minor volcanic constituents (lower left, Fig. 1). The famous Kupferschiefer[1] of Mansfeld in the G.D.R.[2] (Fig. 5) might represent an example of this group. It should be noted parenthetically that the copper of the shale gives way both vertically and laterally to lead and zinc mineralization in calcareous sediments. Along with examples of groups 9 and 14, below, deposits of this variety are known (a little inappropriately, some have suggested) as "red bed" copper deposits.

(8) Copper-molybdenum (gold) deposits in late-orogenic, "minor" granitic intrusions and in the enclosing and overlying sedimentary and volcanic rocks (upper centre, Fig. 1). Examples of this type form the important group known as "porphyry" copper deposits.

---

[1] See Chapter 7 in Vol. 6 by Jung and Knitzschke for a complete treatment of the Kupferschiefer deposit.
[2] G.D.R. = German Democratic Republic.

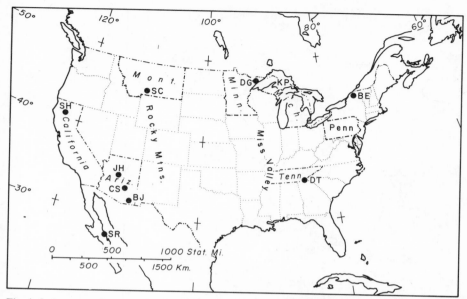

Fig. 4. Index map of localities in the U.S. and part of Mexico to which reference is made: *BE* = Balmat and Edwards district, N.Y.; *BJ* = Bisbee and Johnson districts, Ariz.; *CS* = Christmas and Superior districts, Ariz.; *Dg* = Duluth gabbro, Minn.; *DT* = Ducktown, Tenn.; *JH* = Jerome–Humbol area, Ariz.; *KP* = Keeweenaw Peninsula, Mich.; *SC* = Stillwater complex, Mont.; *SH* = Shasta district, Calif.; *SR* = Santa Rosalia (Boleo district), Baja California (Mexico).

The plutons with which porphyry copper deposits are associated appear to have been related to basalt–andesite–rhyolite ("calc-alkaline") volcanism. Formerly regarded as being restricted to rocks of Mesozoic or younger age, at least one Palaeozoic example is being exploited (Gaspe Copper in Quebec) and Precambrian occurrences may also exist (e.g., in the Bourlamaque "batholith", situated near Val d'Or, Quebec). The Tribag deposit in Ontario may also have porphyry affiliations. The spatially associated "bedded replacement" mineralization found in some porphyry copper districts (e.g., the tactite mineralization at Gaspe, Quebec, and Bisbee, Johnson Camp, Christmas and Superior, all in southern Arizona) bears some tantalizing similarities to types 5, 6 and 7 above.

(9) Copper (uranium–vanadium) mineralization in molasse-type sediments (upper left, Fig. 1). Volcanic debris is commonly present. Examples of this class include the copper mineralization in the Pennsylvanian sediments of Nova Scotia and New Brunswick as well as Pennsylvania itself. Deposits near Santa Rosalia in Baja California may also belong to this type (J.F. Wilson, 1955). As noted in relation to group 7, above, examples of this type have been included among "red bed" deposits.

(10) Copper deposits (commonly native) in post-orogenic flood, or plateau, basalts and the associated terrestrial sediments (upper left, Fig. 1). For economic reasons the best-known occurrence is the widespread mineralization in the Keeweenaw Peninsula of Michi-

Fig. 5. Index map of localities in Europe to which reference is made: *CS* = Central Sweden; *K* = Kupferschiefer (Mansfeld district); *M* = Meggen; *R* = Rammelsberg; *RT* = Rio Tinto (Huelva) district; *T* = Tynagh mine.

gan, but "geologic" examples, not presently commercial, include the copper mineralization of the Coppermine River series in the Northwest Territories of Canada and in the Triassic flood basalts of Nova Scotia.

Deposits in cratonic rocks

Concentrations of copper and associated metals found in cratonic regions may be organized in a similar manner and Fig. 6 shows the most important occurrences of this

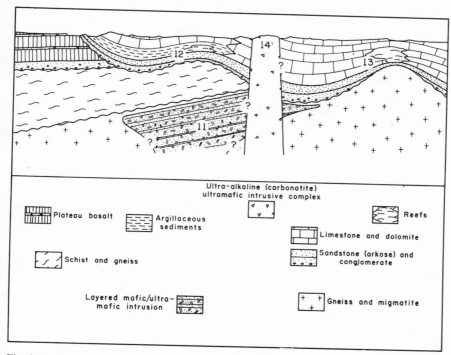

Fig. 6. Idealized cross-section of a cratonic region. The numbers refer to the approximate location of the occurrences of copper and associated metals described in the text.

general type superimposed upon a cross-section of an idealized cratonic block. As before, the best way of explaining this diagram may be to briefly review each of the groups individually in a sequence derived from the inferred relative ages of the host rocks:

(11) Copper—nickel (platinum-group metals, etc.) in layered basic complexes (lower left of Fig. 6). Examples of this group include the mineralization in the Sudbury "nickel irruptive", Ontario; the Duluth gabbro in Minnesota; and the Stillwater complex of Montana. The Bushveld complex of South Africa might also be placed here (Fig. 7). A number of authors have suggested that the Sudbury irruptive might have been localized by the explosive impact of a meteorite (Dietz, 1964 and Gross and Sijpkens, 1965). If correct, this interpretation would seem to exclude the inclusion of Sudbury as an example in a tectonic classification. On the other hand, layered mafic complexes, like the Stillwater complex, the Duluth gabbro or Sudbury irruptive, are commonly found at or near major structural contacts, suggesting a grand-scale tectonic, rather than meteoric, localization. The writer also believes that a case can be made for regarding the Bushveld complex as a metamorphosed accumulation of sediments containing a diversity of placer deposits. These points illustrate, if nothing else, the difficulty of trying to divorce genetic considerations from geological arguments!

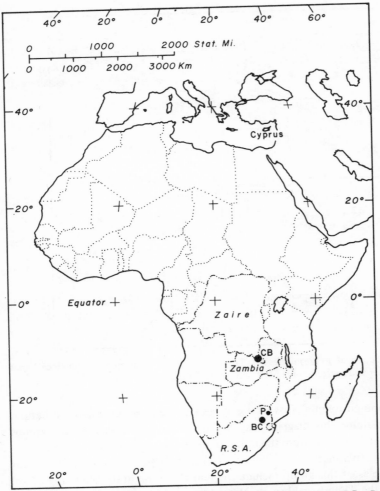

Fig. 7. Index map of localities in Africa to which reference is made: *BC* = Bushveldt complex; *CB* = Copperbelt (Zaire and Zambia); *P* = Palabora.

(12) Stratiform copper (cobalt–uranium–vanadium) mineralization localized at or near the base of a veneer of cratonic cover, that is to say, sediments deposited in a cratonic basin, a depression apparently formed by downwarping with little or no subsequent compression (De Sitter, 1956). The example best known to "western" geologists is the copper belt of Zaire–Zambia (Fig. 7). As previously noted, occurrences of this variety, along with groups 7 and 9, may be included in the broad group loosely known as "red bed" copper deposits.

(13) Lead–zinc (copper) deposits found principally in calcareous and dolomitic cra-

tonic cover (Fig. 2). These occurrences constitute the clearest examples of the class of Mississippi Valley-type deposits, but they grade into occurrences of type 6, above.

(14) Copper (iron, niobium, etc.) associated with ultra-alkaline, carbonatite complexes. Palabora in the Republic of South Africa provides the best known commercial representative, while numerous carbonatites in other areas provide non-commercial examples (Heinrichs, 1966).

EXAMPLES OF TRANSITIONAL SERIES OF MINERAL DEPOSITS

General statement

It has been noted already that the tectonic settings represented by Figs. 1 and 6 could be related in one of two ways. Mountain-building typically occurs in a narrow welt on the margin of a stable plate or craton. Thus, Figs. 1 and 6 might be said to depict two contrasted tectonic units which developed more or less contemporaneously. This could be represented roughly by placing the two diagrams side by side, with Fig. 1 to the right of Fig. 6, the clastic and calcareous rocks of the intervening shelf zone (and, presumably, any metalliferous deposits they contain) changing, in effect, laterally into cratonic cover. Alternatively, the rocks comprising an orogenic zone (shown in Fig. 1) may become "welded" together to form a part of a mature craton (Fig. 6). In this case, the two figures might be said to depict essentially the same rocks at different stages in development, it then being more difficult to envisage transitional rock types (and included metallic mineral deposits) being laid down in the temporal and developmental interval separating the stages represented by the two diagrams, an interval presumably characterized by uplift and denudation.

Within an orogeny, gradational or transitional variations in lithology (and, presumably, associated mineralization) might take several forms. Hutchinson et al. (1971) have related differences in base-metal mineralization found in eugeosynclinal rocks in variations in the depositional environments of the hosts and the present writer has drawn attention to gradational variations in both the composition of the non-ferrous components of eugeosynclinal mineralization and the lithology of the host rocks (Gilmour, 1971). Largely lateral variations within the group of Mississippi Valley-type lead–zinc deposits have been noted by numerous authors, notably Callahan (1967). Several writers have suggested that the characteristics of porphyry copper mineralization might vary widely over a fairly wide vertical range and they related the different appearance of various deposits to differences in the levels exposed by erosion (Lowell and Guilbert, 1970 and James, 1971). Comparison of Figs. 1 and 6 will reveal a number of other ways in which gradational changes might occur. It is proposed to confine the following discussion of transitional deposits to massive pyritic sulphides found in eugeosynclinal volcanic and sedimentary rocks and the adjacent portions of the accompanying shelf, or plate margin (groups 2, 3, 4 and perhaps 5, above).

Strata-bound, massive, pyritic sulphide deposits include some of the most famous and historically important copper occurrences exploited by Mediterranean and Atlantic cultures, such as the deposits of Cyprus and the Huelva district in Spain, the pyritic deposits of central Sweden and of Rammelsberg and Meggen in Germany. They also include some of the economically and technically important ore deposits exploited in recent years, as, for example, the Broken-Hill lode in New South Wales, the Sullivan deposit in British Columbia and the United Verde and Ducktown ore bodies located, respectively, in Arizona and Tennessee; and one of the most spectacular mineral discoveries of the last decade or more, namely, the Kidd Creek deposit near Timmins, Ontario[1].

The student of mineral deposits encountering descriptions of massive pyritic sulphides for the first time might reasonably ask what various representatives of this large and seemingly diverse group of deposits have in common. Although the Horne deposit in the Noranda district of Quebec has a good deal in common with the United Verde ore body located at Jerome in north-central Arizona and the Brunswick Mining and Smelting No. 6 and 12 ore bodies in New Brunswick with those at Rammelsberg, the Horne and Rammelsberg deposits do not appear to share many features. In an effort to create order out of the apparent diversity of massive sulphides, various authors have emphasized different features. Thus, Hutchinson (1973) has appealed to the age and tectonic setting of host rocks, whereas E. Horikoshi (written personal communication, 1973) relies on consideration of the structural setting, expressed in terms of plate tectonics.

The present writer has suggested that it might be helpful to arrange massive sulphides in a linear series ranging from concentrations of copper sulphides associated with rhyolitic pyroclastics, through copper—zinc (lead) sulphides in mixed volcanic and sedimentary rocks, to lead—zinc (copper) minerals in predominantly clastic sediments (Table I). Because of the criteria chosen as the basis for this scheme, the nature of the host rocks of massive sulphides, emphasis has, perforce, been placed on this feature at the expense of details concerning the mineralogical composition, form, and so on, of the actual deposits.

"Since" as J.G. Wargo (1960) succinctly put it, "nature makes the rocks and man the classifications", it would be very surprising indeed if all examples of massive sulphide mineralization could be fitted neatly into the proposed series. Although helpful as a first approximation, extended use has suggested that a series with only two end-members is not capable of differentiating what appear to be quite real and fundamental differences between some examples of massive sulphides. Consideration of numerous examples suggests that at least four important groups of host rocks should be accommodated in any empirical classification of massive sulphides, namely, spilitic lavas and associated pyro-

---

[1] The Broken Hill deposits have been described in Chapter 6 of Vol. 4.

TABLE I

A proposed linear (two end-member) arrangement of massive sulphide deposits based on observable features

EXAMPLES – with approximate position in series (left to right): Mount Isa mine, Queensland; Sullivan mine, Brit. Columbia; Meggen deposit, G.F.R.; Iron King mine, Ariz.; Buchans mine, Newfoundland; Rammelsberg, G.F.R.; B.M.&S. No.12, N.B.; East Shasta district, Calif.; Caribou mine, N.B.; Manitouwadge district, Ont.; Wedge mine, New Brunswick; Kidd Creek mine, Arizona; United Verde mine, Arizona; Ducktown district, Tenn.; West Shasta district, Calif.; Mattagami district, Quebec; Kuroko deposits, Honshu; Mavrouvouni, etc. Cyprus; Flin Flon distr. Manitoba; Horne mine, Noranda, Que.

	Group 1	Group 2	Group 3
CHARACTERISTIC HOST ROCKS	Shale and siltstone (may be dolomitic and may contain evidence of volcanic activity during deposition)	Rhyolitic crystal tuff (1) and flows, chert, andesitic flows and pyroclastic rocks, iron formation (2) and shale.	Predominantly andesitic and basaltic flows and pyroclastic deposits, chert and (3) rhyolite.
TYPICAL ORIGINAL STRUCTURES	Normal depositional sedimentary structures. (Local sinter accumulations?)	Various combinations of sedimentary and volcanic structures.	Constructional lenses and domes of pyroclastic debris, alteration pipes, intrusions.
PRINCIPAL SULFIDES	Pyrite / Sphalerite / Galena / (Chalcopyrite)	Pyrite / Sphalerite / Galena / (Chalcopyrite)	Pyrite / Pyrrhotite / Chalcopyrite / (Sphalerite)
PRINCIPAL RECOVERABLE METALS	Silver / Lead / Zinc	Gold / Silver / (Iron) Zinc / Lead / (Copper)	Gold / Copper / (Zinc)
COMMONLY ASSOCIATED MINERALS	Cassiterite / Tourmaline Gypsum and barite (4)	Quartz and iron oxides, notably haematite and magnetite	Quartz / Sericite

Special Notes:

1. Also known as arkose, quartz feldspar porphyry, leptite, quartz-feldspar gneiss, etc.
2. Not included among host rocks of Mesozoic or younger age.
3. Absent from some examples of massive sulfide host rocks of Mesozoic or Tertiary age.
4. Typically notable by absence from host rocks of Precambrian age.
5. Minor in host rocks of Mesozoic or Tertiary age (cf. Horikoshi, 1969, p. 323).

After Gilmour, 1966.

clastics; chert (including iron formation); limestone and dolomite; and greywacke, siltstone and shale. This is shown graphically in Fig. 8a.

In the attempt to organize massive sulphides which employed a linear series, the writer emphasized that copper–gold (zinc–silver) mineralization tends to be associated with volcanic rocks and structures, whereas zinc–lead–silver mineralization tends to be associated with sedimentary rocks and structures. Even if a system composed of four host-rock components is adopted the same general distinctions tend to be valid. Copper–gold (zinc–silver) mineralization is associated with spilitic lavas and pyroclastics (as well as spatially and probably genetically related sediments like chert and iron formation). With notable exceptions (some of which will be discussed below) pyritic deposits containing significant amounts of sphalerite, galena and silver are typically found in calcareous and greywacke-suite sediments. In practice, most variations in the host rocks of massive sulphide mineralization can be accommodated in two of the four possible "faces" of the four-component tetrahedron and, by referring to only three of the possible four end-members, they can be depicted in two dimensions (Fig. 8b and c). These two diagrams may be regarded as equivalent to the lower right-hand portion of Fig. 1.

Examples of massive sulphide deposits in both of the principal three-component fields

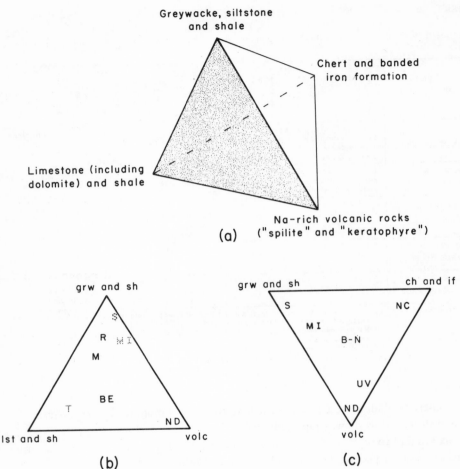

Fig. 8. (a). The principal host rocks of massive pyritic sulphide deposits arranged in a four-component (tetrahedral) system. (b) and (c): examples of massive sulphide deposits superimposed on two faces of the tetrahedral system of (a). *BE* = Balmat and Edwards district, N.Y.; *B—N* = Bathurst—Newcastle district, N.B.; *M* = Meggen deposit, German Federal Republic; *MI* = Mount Isa deposit, Qld.; *NC* = North Coldstream mine, Ont.; *ND* = Noranda district, Que.; *R* = Rammelsberg deposit, G.F.R.; *S* = Sullivan mine, B.C.; *T* = Tynagh deposit, Eire; *UV* = United Verde deposit, Ariz.

of host rocks will be reviewed in order to illustrate both the range of massive sulphide deposits and the transitional nature of the relationships of the "end-members".

Deposits associated with volcanic rocks, chert and greywacke

The above terms have been abbreviated for use in the sub-title, but, of course, they imply in full, "spilitic volcanic rocks", "chert and banded iron formation" and "greywacke, along with associated siltstone and shale". In other words, this section deals with the three-component system shown in Fig. 8c. When they refer to "strata-bound, massive pyritic sulphides", most geologists seem to have in mind deposits associated with eugeo-synclinal volcanic rocks, or their metamorphic equivalents, so it is proposed to begin by

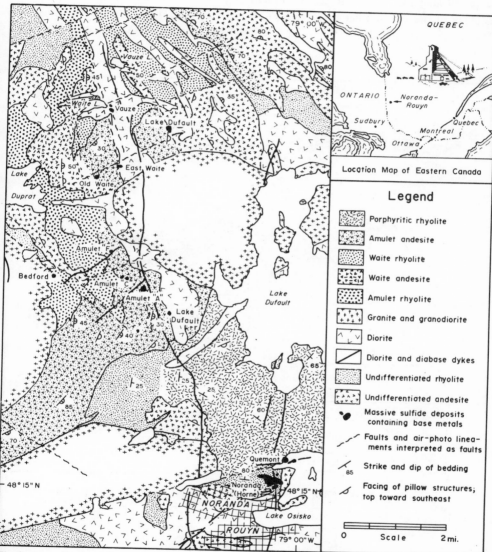

Fig. 9. Generalized geologic map of the Noranda district, Quebec, showing the locations of the princi-pal massive sulphide deposits, after Gilmour, 1965. (See fig. 3 in Chapter 5 of Vol. 6 by Sangster and Scott for a geologic map of the Abitibi Orogenic Belt and their fig. 7 for an additional map on the Noranda district.)

reviewing them. However, it is worth emphasizing that this subgroup is by no means the be-all or end-all of massive sulphides. Many of the largest and most famous examples do not belong to this subgroup at all.

The best examples known to the writer of massive sulphides associated exclusively with volcanic rocks occur in the Noranda district of northwestern Quebec. The general geology of the area has been reviewed so many times (e.g., M.E. Wilson, 1941 and 1948; Dresser and Dennis, 1949; Gilmour, 1965 and Spence, 1967) that no useful purpose

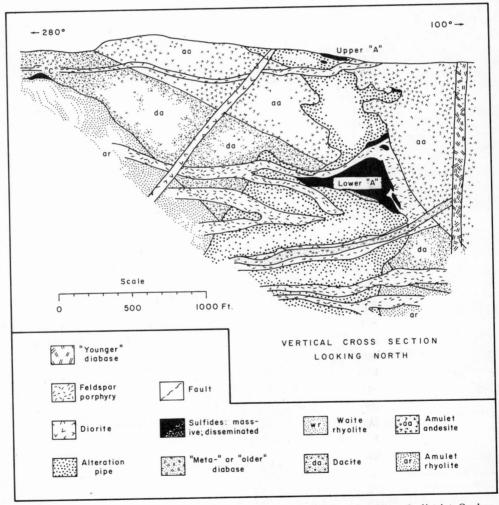

Fig. 10a. Cross-section of the Amulet "A" ore bodies, Waite-Amulet mine, Noranda district, Quebec. (After Suffel, 1948, and Price, 1953.)

would be served by repeating them here, although Fig. 9 serves both to summarize the salient geologic features and as an index map. In order to understand the geologic setting and form of the occurrences in the Noranda and other districts, a number of terms which have been used should be re-evaluated, however. To cite one example: the Horne deposit has been described as "pipe-like", yet, if the sulphide system represented by the numer-

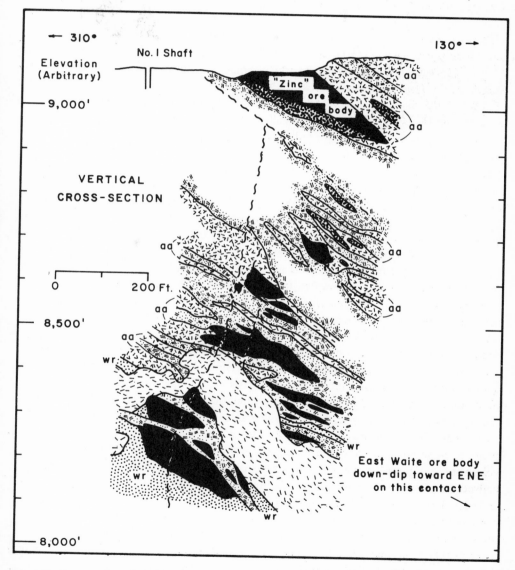

Fig. 10b. Cross-section of the "Old Waite" deposit, Waite-Amulet mine, Noranda district, Quebec. Explanation on Fig. 10a. (After Price and Bancroft, 1948, and Dresser and Denis, 1949; with permission of Can. Inst. Min. Metall.) (See figs. 8 and 10 in Chapter 5 of Vol. 6 by Sangster and Scott for additional information.)

ous ore bodies is viewed as a whole, it is seen to consist of a series of lenticular masses which are essentially conformable to the layering in the enclosing acid pyroclastic rocks. Since there are some structures in the district which all geologists have agreed are indeed "pipe-like", the term should obviously be employed judiciously.

Deposits associated with "alteration pipes"

Several ore bodies, or groups of ore bodies, scattered over an area which measures about one square mile (or, roughly, 2.5 km^2) were exploited at the Waite Amulet mine, Quebec. Most of these are associated with volcanic rocks and structures exclusively, some with "pipes" of intensely altered rocks.

The setting of the Amulet "A" deposit is best displayed by means of a vertical cross-section (Fig. 10a). The Amulet rhyolite differs greatly in appearance from place to place, ranging from very fine-grained welded tuff, or lava, to auto-brecciated lava or coarse-grained, welded ash flows. With certain qualifications (Descarreaux, 1973) rhyolitic and andesitic rocks are readily distinguishable in the Noranda district, but in the Amulet area intermediate varieties certainly exist. Dacite flows(?) separate the rhyolite from the overlying Amulet andesite in the vicinity of the upper and lower "A". A layer of laminated chert, which ranges in thickness from 2 or 3 inches (5—7 cm) to 30 feet (9 m), has generally been regarded as the uppermost unit of the rhyolite and dacite. The andesites are generally pillowed and tend to be easy to identify.

The layered volcanic succession was pierced by an "alteration pipe", a pipe-like mass of minerals believed to have resulted from the metasomatic replacement of the original rock, which not only cut the rhyolite and dacite, but, greatly reduced in diameter, penetrated the overlying andesite as well. The alteration pipe in the area of the "A" ore bodies consists of chloride, biotite cordierite, and anthophylite, although Rosen-Spence (1969) has argued convincingly that the last three minerals resulted from thermal metamorphism of chlorite during the emplacement of the Dufault granite. Since the composition of the chlorite in the pipe resembles that of andesite more closely than that of rhyolite, the greater development of the alteration pipe in the latter as compared to the former suggests a complicated history. Perhaps the rhyolite and dacite were subjected to the passage of a large volume of fluids over a long period, as, for example, during a phase of fumarolic activity which occurred toward the end of the rhyolitic (and dacitic) volcanism and prior to the eruption of the andesite. Subsequent gaseous activity, reactivated after the commencement of the next cycle, which began with the eruption of andesite flows, may have been relatively weak. Perhaps some of the silica or laminated chert which marks the top of the rhyolite and dacite may have been derived from the intermediate and siliceous rocks which were converted to chlorite.

Both the upper and lower "A" ore bodies lie within the alteration pipe, the former at the top of the combined rhyolite and dacite and the latter apparently well within the Amulet andesite. It has, however, been suggested (R.C.J. Edwards, oral personal communication, 1960) that the upper "A" might lie on the same horizon as the Waite rhyolite to the north (which represents a break in the eruption of andesite).

The Old Waite displays even better the stacking of massive sulphide lenses within an alteration pipe (Fig. 10b).

The mineralization exploited at the Old Waite illustrates another feature of massive sulphides which will be noted at several points below, namely, a stratigraphic zoning of base metals in which zinc and lead sulphides stratigraphically overly those of copper, regardless of the present attitude of the host rocks (see also Gilmour, 1965 and Gilmour and Still, 1968). Indeed, this zoning is present to such a degree at the Old Waite that the highest sulphide lens preserved was known as the "Zinc" ore body. To slip once again from description to the realm of interpretation: the fact that this stratigraphic zoning of metals exists even in deposits which are overturned (like Rammelsberg or the Iron King in north-central Arizona) strongly suggests that the metals in these deposits were localized, at the very least, before deformation and metamorphism took place.

Deposits in alteration pipes and interbedded with layered volcanic rocks

The Vauze deposit in Quebec presents, in close juxtaposition, examples both of sulphide minerals confined to a discordant alteration pipe (the B-2 ore body) and of a conformable lens of massive sulphides localized at the top of a series of rhyolitic flows and/or pyroclastic deposits (B-1) (Fig. 11a).

The original form of the Waite rhyolite in the Vauze area has been partially obscured by a diorite intrusion, but geological maps (both of surface and underground), stratum contours of contacts, and isopachs of the units all combine to indicate that a rhyolite dome, made up at least partly of pyroclastic material, existed at the site of the mine. The B-1 ore body lay on the southeast flank of this dome, whereas the alteration pipe with which the B-2 is associated lies near the axis. If any direct connection between the two ever existed, it was destroyed by the emplacement of the irregular mass of diorite.

A sill of "older diabase" (in the nomenclature of the Noranda district) is chilled against the massive chlorite of the alteration pipe. The diabase itself is devoid of a significant amount of sulphides. The formation of the alteration pipe and, presumably, the deposition of the magnetite, sphalerite, chalcopyrite and bornite it contains must have occurred while the volcanism was in progress.

Another beautiful example from the Noranda district of base-metal sulphides associated with an alteration pipe has been well-documented by Boldy (1968, see especially his figs. 8 and 10).

The stratigraphic zoning of base metals displayed by the separate lenses exploited at the Old Waite mine was well developed in the single ore lens of the B-1 ore body at Vauze. From hangingwall to footwall three intergradational zones were identified. A hangingwall zone, which ranged in thickness from a few inches to a few feet, consisted of very fine-grained, well-banded sulphides alternating with chert. This was succeeded down-

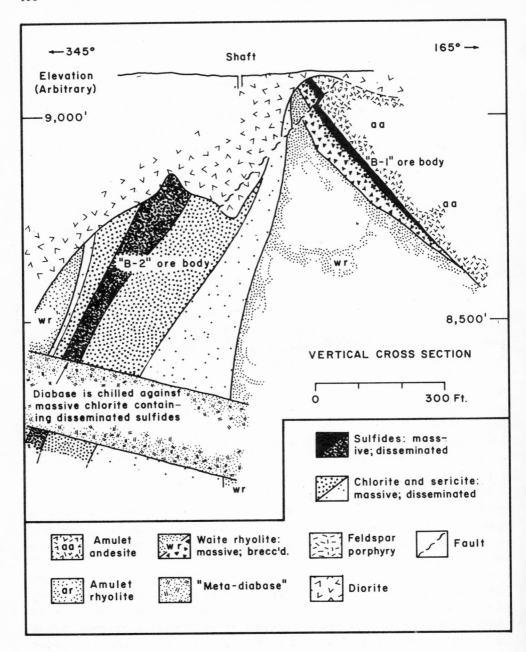

Fig. 11a. Cross-section of mineralization in a discordant pipe and conformable lens, Vauze mine, Noranda district, Quebec. (After Gilmour, 1965.)

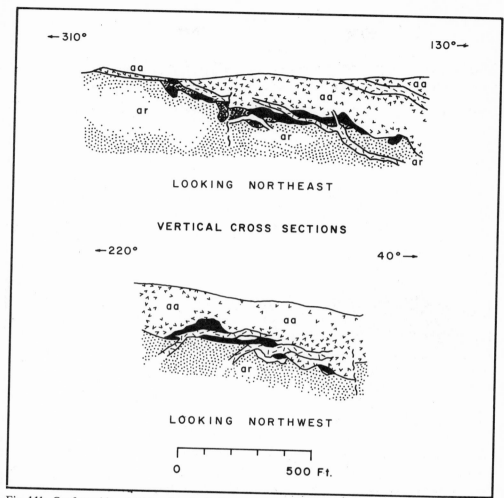

Fig. 11b. Conformable sulphides, Amulet "C" ore bodies, Waite-Amulet mine. Explanation on Fig. 11a. (After Suffel, 1948; with permission Can Inst. Min. Metall.)

ward by a zone a few feet thick of more or less homogeneous, or massive, fine-grained sulphides. The massive horizon gave way toward the footwall through a gradational contact to a zone consisting of scattered, coarse-grained mineralization and fragmental(?) rhyolite in which the proportion of sulphides progressively decreased downward so that the footwall zone graded into the underlying weakly mineralized or barren rhyolite breccia. The sulphide laminae in the banded hangingwall zone consisted principally of pyrite and sphalerite, whereas pyrite, pyrrhotite and chalcopyrite formed the bulk of the coarsely disseminated footwall mineralization. Pyrite predominated in the massive inter-mediate zone with lesser amounts of sphalerite and chalcopyrite in approximately equal proportions. Thus zinc tended to be concentrated toward the hangingwall and copper toward the footwall.

Other examples in the Noranda district of conformable deposits localized at the top of piles of rhyolitic flows and pyroclastics, include the "Bluff", "F" and "C" orebodies in the southern and central parts of the Waite Amulet property and the East Waite toward the north (Fig. 11b). The "C" orebodies in particular bear an uncanny resemblance to some of the "black ore" deposits of Honshu (cf. Horikoshi, 1969, fig. 16).

Bedded massive sulphides

Although traditionally described as a pipe-like deposit, the Horne mine provides the most impressive example in the Noranda district of a bedded or conformable deposit (M.E. Wilson, 1941; Price, 1948, and Dresser and Dennis, 1949).

The sulphide mineralization consists of two groups of bodies, one of which is made up of relatively "thick" lenticular masses and the other of thin layers (Fig. 12). The lenticular masses provided the bulk of the ore produced to date. In some level plans, the lenticular bodies almost appear to be discordant (because the length, measured along strike, and width, measured across strike, are almost equal), but in vertical cross-section the longest dimension (the height) is seen to be parallel to the bedding in the enclosing rhyolite pyroclastics. The principal sulphides in the lenticular bodies are pyrite, pyrrhotite and chalcopyrite. Over the greater part of the life of the mine the recovered grade produced from the lenticular bodies averaged between 2 and 2.5% copper and around 0.2 ounces per ton of gold.

The thin, tabular "No. 5 Zone" first attracted particular attention when the largest known body in the group was first discovered in the lower portions of the mine (around the 21st Level or so), but, in fact, No. 5 Zone had previously been identified as high as the 13th Level (Dresser and Dennis, 1949, fig. 60) and perusal of plans and sections indicates that the mineralization may be traced all the way up to the 1st Level (Dresser and Dennis, op. cit. and Price, 1953). The No. 5 Zone consists principally of pyrite accompanied by minor amounts of sphalerite and galena.

The concensus of opinion of geologists who have studied the Horne mine is that the volcanic host rocks become younger, or face, toward the north. Thus, the distribution of base metals in the cupriferous lenses and the No. 5 Zone repeats the stratigraphic zoning of metals noted above in the review of the Old Waite and Vauze deposits. While on the subject of the stratigraphic zoning of sulphide mineralization, it is worth noting that R.C.J. Edwards (1960, oral personal communication) pointed out a considerable time ago that the zoning exhibited by a single body (e.g., the Vauze B-1 ore body) or two or more lenses (e.g., the Old Waite and Horne mines) is also present on the scale of the Noranda district as a whole. The ore deposits in the western part of the district were mined principally for their copper content (although many contained an equivalent percentage of less-valuable zinc), whereas the deposits found in the stratigraphically higher forma-

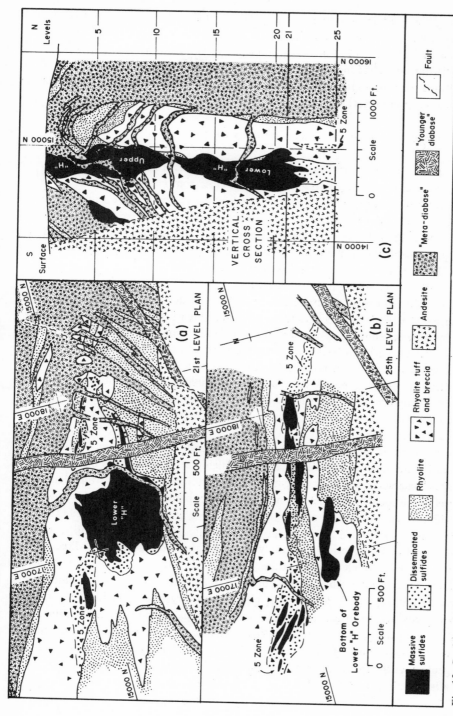

Fig. 12. Conformable massive sulphide deposits in a volcanic pile, Horne mine, Noranda district. (After Price, 1948 and 1953; with permission Can. Inst. Min. Metall.)

tions toward the east and northeast consist(ed) principally of pyrite and sphalerite with subordinate amounts of copper minerals.

Before leaving the subject of the transitional nature of occurrences as illustrated by the Noranda district, reference should be made to an important contribution which compared the transitional relationships of the igneous rocks of the Noranda district with that of the surrounding metasediments (Hutchinson et al., 1971).

Deposits associated with volcanic and sedimentary rocks

A further step in the transitional series being investigated (Fig. 9c) is illustrated by the principal deposits in the Jerome area in north-central Arizona (Figs. 4 and 13).

The Precambrian rocks which serve as the host rocks for massive sulphides in the Jerome—Humboldt—Mayer district are known as the Yavapai series (Anderson and Creasy, 1958, p. 9ff.). Those authors proposed that the Yavapai series be sub-divided into two groups, namely, the Ash Creek and Alder groups. This terminology is in use today, although the distinction between the sub-divisions is perhaps not quite so clear-cut as Anderson and Creasey appear to have believed.

Rocks of the Ash Creek group are exposed only near Jerome in the massif of Mingus Mountain, or the Black Hills, where they are readily identifiable as flows and ash flows overlain by a thick sequence of tuffaceous chert, "lean" iron formation and mafic flows (Fig. 13). The sequence has been cut by a variety of intrusions which range in composition from gabbro to granite (rhyolite). For the most part, these rocks are neither foliated nor significantly metamorphosed. Although the resemblance is partly obscured and complicated by Phanerozoic cover, basin-and-range(?) faulting and so on, the Ash Creek rocks clearly comprised a portion of a volcanic centre comparable to that exposed in the Noranda district.

Representatives of the Alder group, on the other hand, exposed around Humboldt and Mayer, are for the most part thinly layered and highly schistose and, in many instances, their original character has been obscured by metamorphism. A list of the identifiable rock types in the Alder group includes massive, vaguely pillowed, or otherwise flow-structured andesites; undeformed to weakly foliated volcanic agglomerates; very lean (i.e., low iron-oxide content) banded iron formation (originally identified as rhyolitic tuff by Anderson and Creasy, an error commonly made by those unfamiliar with Precambrian rocks) and phyllites which almost certainly originated as shales. The bulk of the Alder group, however, is made up of a variety of chloritic, sericitic and quartzose schists many of which contain grains (porphyroblasts, phenocrysts, crystal fragments?) of quartz or quartz and feldspar. Individual units in the Alder group pinch and swell and exhibit facies changes along, as well as across, strike so that formations tend to be traceable in a general way only. The appearance of a typical assemblage is provided by Fig. 14 which shows the surface geology at the Iron King mine located one mile west of Humboldt.

The relationship between the Ash Creek and Alder groups is also obscure, although the

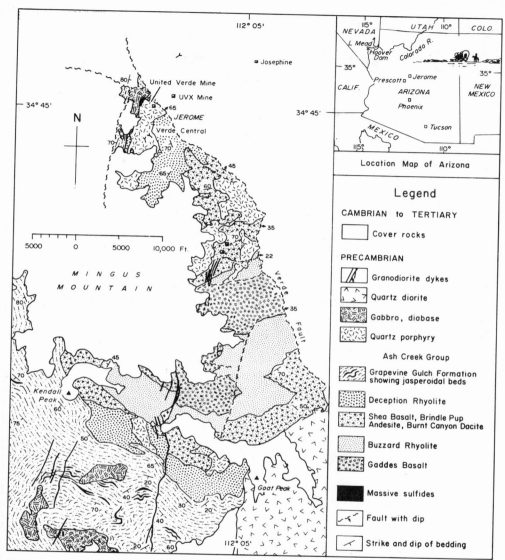

Fig. 13. Generalized geologic map of the Jerome district, north-central Arizona, showing the "style" of the typically thick-bedded succession of meta-volcanic rocks. The convolutions in the lithological contacts are largely a consequence of low dips and strong topographic relief (e.g., the base of the Phanerozoic cover which is nearly horizontal). (After Anderson and Creasey, 1958; with permission U.S. Geol. Surv.)

present writer believes that the latter originated as the marginal and/or late volcano—sedimentary facies of the former.

Because it contains the most important massive sulphides found in the Jerome—Hum-

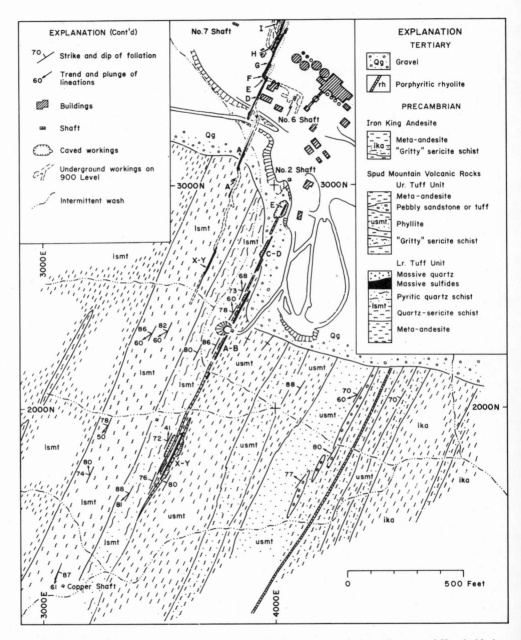

Fig. 14. Map of the surface geology at the Iron King mine located one mile west of Humboldt in north-central Arizona, showing a fairly typical section of the Alder group of the Yavapai series. (After Gilmour and Still, 1968; with permission Am. Inst. Min. Metall.)

boldt—Mayer district to date, the Ash Creek group is economically the more important of the two subdivisions of the Yavapai series. Unfortunately, a digression on the nature of the host rocks is required before proceeding to describe the mineral deposits, because, although the origin of most lithologies of the Ash Creek group has been established beyond reasonable doubt, the nature of some has been the subject of a good deal of speculation and debate.

The principal problem involves units which have been regarded as intrusions, although the writer believes that, either in whole or in part, they were originally extrusive. Such units include the "quartz porphyry" (contained among the predominantly flow-rocks forming the two principal eruptive sequences in the Jerome area) and some at least of the "gabbro" and "intrusive dacite" found in the Grapevine Gulch Formation, the thick bedded sequence which overlies the flows (and ash flows).

The writer and others argued some time ago that the "quartz porphyry" is actually a series of quartz and feldspar crystal tuffs and agglomerates (Gilmour, 1966). Evidence cited at that time in support of this interpretation included the "porphyritic" texture of fragmental facies of rhyolite, and the presence of the "purple porphyry", an iron-rich (haematite) facies at the contact between "normal porphyry" and the overlying tuffaceous, cherty sediments of the Grapevine Gulch Formation. After examining key outcrops Horikoshi proposed that the quartz porphyry is actually a flow, locally auto-brecciated (E. Horikoshi, written personal communication, 1973). Flow or (crystal-lithic) tuff, the essential point, so far as the present writer is concerned, is that the quartz porphyry was extrusive and interbedded with stratified rocks, rather than intrusive.

The writer also proposed that some of the gabbro in the stratified Grapevine Gulch sequence, and specifically that overlying the massive sulphides at the United Verde mine, is probably coarse-grained (recrystallized?) basalt. The evidence for this suggestion was acknowledged to be very weak and amounted to the conformable relations with the enclosing sediments and analogy with other areas. Again, Horikoshi examined critical outcrops and reported that he had seen pillows in the "gabbro" exposed along Hull Canyon, west of Jerome. He sent sketches and photos of these structures to this writer (E. Horikoshi, written personal communication, 1973).

In the meantime, the writer has mapped in detail some of the dacite described by Anderson and Creasey as being intrusive into the Grapevine Gulch rocks. The abundance of lithic fragments (making up 5—15% of the rock in most outcrops) in the dacite and the inclusion in the immediately overlying thin-bedded sediments of feldspar grains, like those in the dacite, indicate that the rock is a crystal (lithic) tuff.

These re-interpretations of the nature of the "porphyry" and "gabbro" were categorically rejected by the acknowledged experts on the area (e.g., S.C. Creasey, oral personal communication, 1965, and C.A. Anderson, written personal communication, 1967), although the latter subsequently accepted the extrusive nature of the "porphyry" (Anderson and Nash, 1972).

To repeat, it is unfortunate that space must be devoted to questions of this type, but it is not possible to write about mineral deposits and the rocks in which they occur, until they have been resolved.

The most important deposit discovered in the district to date was exploited at the United Verde mine, located just to the north of the town of Jerome. When the mine closed in 1953 almost 25 million tons averaging 7.0% copper, 2 ounces per ton of silver and 0.05 ounces per ton of gold had been produced. As in the case of the country rocks, the sulphides were also described in misleading terms. Anderson and Creasy (op. cit., p. 105 ff.) wrote of a "pipe", localized in the crest of "a small north-northwestward plunging anticline".

The pipe-like form is actually a function of the conformable mode of occurrence of the sulphides and of the steep dip of the country rocks, while the pipe-like aspect is exaggerated to some extent by the fact that mining failed to exploit the pyrite–sphalerite mineralization which forms a tabular or lenticular body, but concentrated instead on the scattered footwall chalcopyrite (and pyrite) which has a shorter strike-length than the principal massive sulphide body (Fig. 15).

The "anticlinal" plan-form may likewise be more apparent than real and may reflect a tabular deposit (of pyrite, sphalerite and minor chalcopyrite) which toward the north originally turned downward into a discordant "root" of chalcopyrite and chlorite which now projects, almost horizontally, in a southeasterly direction. Masses of chlorite, shearing, and other features which also strike in a southeasterly direction may also reflect this once-vertical root.

The immediate host rock for the massive sulphide mineralization is the "porphyry" facies of rhyolite. Lenses of jasperoidal (iron-rich) silica, much of which is apparently laminated, occurs in the same horizon. Small conformable (sills or flows?) and discordant (feeder dykes?) bodies of andesite are found in the same general horizon. Overlying the immediate host for the massive sulphides, and locally in contact with the upper portion of the sulphide mass, are a number of lenses of cherty tuff (Grapevine Gulch Formation). These are succeeded by the "gabbro" or recrystallized basalt(?) mentioned above. Interestingly enough, Anderson once acknowledged that he had not observed a significant amount of either sulphides or evidence of hydrothermal alteration (i.e., quartz or chlorite) in the gabbro (C.A. Anderson, oral personal communication, 1967).

The occurrence of silica in the horizon containing the massive sulphides is reminiscent of the banded chert noted above in connection with deposits in the Noranda district. Study of a large number of massive sulphide districts from around the world in rocks of different ages indicates that the mineralized horizon(s) is (are) commonly, or typically, marked by the presence of exceptional amounts of silica, iron and manganese as well as varying amounts of the non-ferrous base metals, copper, zinc and lead, of course. The mineralogical forms which these constituents have assumed varies from place to place. In the Noranda district, silica is present as thinly laminated chert and the bulk of the iron present occurs in sulphides. At Jerome, the bodies of jasperoidal quartz — silica, haematite and magnetite — play the same role. The "purple porphyry" of haematitic tuffaceous

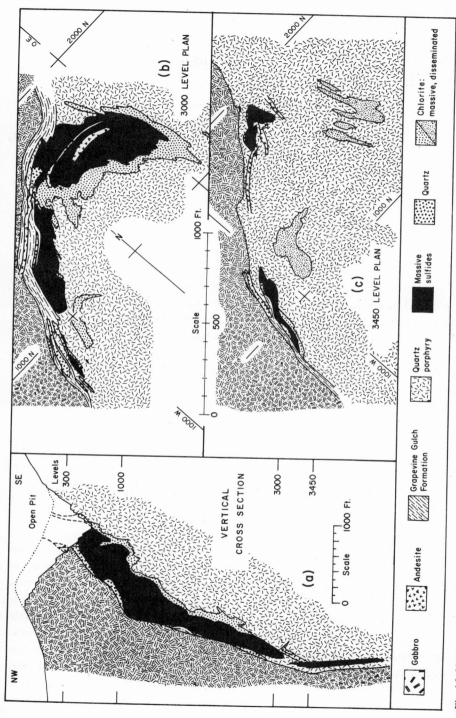

Fig. 15. Vertical cross-section and level plans of the United Verde mine, Jerome, Arizona. (After Anderson and Creasey, 1958.) (Cf. fig. 17 in Chapter 5 of Vol. 6 by Sangster and Scott.)

chert, exposed in Hull Canyon at the contact between the "porphyry" and overlying cherty sediments, is simply an expression of this very widespread phenomenon. The Miocene rocks in the "black ore" district of Honshu, Japan, contain the analogies of this characteristic lithology in the form of the "tetsusekiei", or ferruginous chert beds (Horikoshi, 1969).

In terms of geometry and stratigraphic zoning of metals, the United Verde sulphide mass resembles very closely, say, the Delbridge and Vauze deposits in the Noranda district. Unlike the Noranda deposits, however, which are bounded above and below by lavas and pyroclastics, the host rocks for the sulphides at the United Verde are succeeded by a thick accumulation of tuffaceous cherts and mafic flows(?), which appear to have been deposited during the waning stages of a period of volcanic activity.

The unfortunate debate about the origin of the "porphyry" in the Jerome area had its counterpart in other massive-sulphide districts. In some of these, geologic thinking actually retrogressed under the juggernaut of recent conventionality. In the West-Shasta district in California, for example (Fig. 4), Diller (1906) described the Balaklala rhyolite as a series of flows and tuffs. Graton (1910) subsequently concluded that the rhyolite is intrusive and identified it as alaskite porphyry. Careful work by Kinkel and his co-workers confirmed Diller's interpretation (Kinkel, et al., 1956). Interestingly enough, Kinkel noted that mine geologists in the district continued to characterize the Balaklala rhyolite as "tuff", even after the publication of Graton's report of 1910. The "porphyries" of the Rio-Tinto district, Spain, (Fig. 5), were likewise recognized as tuffs containing crystals and crystal fragments (Collins, 1922). Twelve years later this interpretation succumbed to the conventional wisdom of the day when D. Williams (1934) concluded that the porphyries are intrusive rocks. Opinion went full-circle in 1962 when Williams and Kinkel both reverted to Collin's view. Though perhaps not quite so well documented, the evolution of thought regarding the "porphyritic" rocks in the Bathurst—Newcastle district in New Brunswick (Fig. 2) followed much the same course, cf. Alcock (1941), Skinner (1956), and McAllister (1960), the latter presenting the views of F.A. Moss transmitted to him in 1957. Another example of this characteristic and significant lithology is provided by the leptites of central Sweden. While no single criterion is capable of identifying this class of rocks unequivocally, the chemical composition has been useful, especially in regard to metamorphosed examples (Table II).

Several other massive sulphide deposits occurred in the Ash Creek group of the Yavapai series, notably the United Verde Extension, subject of much speculation as to its relationship with the United Verde (Norman, Anderson and Creasey in Anderson and Creasey, 1958, p. 145—149). A rather large number of massive sulphides were found in the Alder group, although most of these were very small. The largest, the Iron King mine produced just over 5 million tons. More important however, from the standpoint of the present discussion, than their relative sizes, the Iron King provides a further gradation in the characteristics being reviewed. The lithology, textures, variations in composition, and so on, indicate that, although some units may have been directly derived from volcanism,

TABLE II

Chemical composition of representative quartz–feldspar rocks from massive sulphide districts compared to typical quartz keratophyres

	1	2	3	4	5	6
SiO_2	75.31	75.00	71.28	75.70	75.10	75.04
TiO_2	0.23	0.23	0.20	0.16	0.22	0.10
Al_2O_3	12.56	12.32	14.93	11.02	12.84	13.39
Fe_2O_3	0.91	1.65	0.37	0.46	0.70	1.61
FeO	2.10	2.39	1.52	3.05	1.36	0.37
MnO	0.04	0.10	0.02	0.00	0.04	0.05
MgO	1.30	0.93	0.64	0.70	0.30	0.18
CaO	0.54	1.29	2.30	1.56	0.32	0.40
Na_2O	4.25	3.20	4.57	3.66	5.12	6.36
K_2O	1.12	1.74	2.90	1.46	2.39	0.83
H_2O	1.59	0.43	0.43	0.59	1.22	1.31
CO_2	0.12	1.47	0.95	0.48	0.03	0.10
P_2O_5	0.07	0.06	0.08	0.00	0.03	0.08
BaO	–	–	–	–	0.04	0.00
Rest	–	2.89	0.02	–	0.05	–
					0.09	
Total	100.14	103.70	100.21	98.84	99.82	99.82

1 = Average of two porphyritic rhyolites (crystal tuffs), West Shasta, California (Kinkel et al., 1956, p. 23).

2 = Average of six quartz porphyries, Jerome district, Arizona (Anderson and Creasey, 1958, p.35).
3 = Average of two gneisses from Manitouwade, Ontario (Pye, 1957, p. 18).
4 = Average of five quartz-eye gneisses, Chisel Lake, Manitoba, (H. Williams, 1966, p. 15).
5 = Quartz keratophyre, Great King Island, New Zealand (Bartrum, 1936, p. 417).
6 = Quartz keratophyre, eastern Oregen (Gilluly, 1935, p. 235).

most represent volcanic material which was probably re-worked and subsequently deposited in a sedimentary environment, i.e., a significant but unknown distance from the centre (or centres) of volcanism (Gilmour and Still, 1968).

Thus the siliceous, sericitic and chloritic schists in the area of the Iron King mine appear to have been interbedded rhyolitic and andesitic tuffs (Fig. 16). Thin beds, a metre or so thick, which contain grains of feldspar and quartz a few millimetres in diameter and were formerly believed to represent sills of porphyritic rhyolite, were found in drill holes to have gradational contacts with the enclosing schists. These are now believed to be thin layers of crystal tuff (although, of course, such conclusions must always be qualified where metamorphic rocks are involved.)

The horizon which contains the massive sulphides includes a good deal of silica both in the form of siliceous sericite schists and as lenses of massive quartz (siliceous "Footwall" series of ore lenses and, ghastly term, "north-end quartz noses", see Fig. 15). It has been suggested that some or all of the silica was hydrothermally introduced (Creasey,

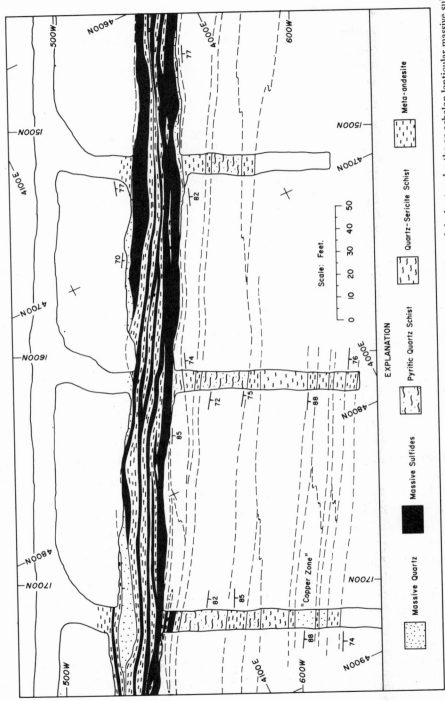

Fig. 16. Plan of a portion of the 22nd Level of the Iron King mine, showing the general lithology of the host rocks, the en-echelon lenticular massive sulphide layers and abundant silica (in quartz schist and "north-end quartz noses"). (After Gilmour and Still, 1968; with permission Am. Inst. Min. Metall.)

in Anderson and Creasey, 1958, p. 155—169), but the presence of pebbles of silica and jasper in the stratigraphically overlying rocks suggests that some at least of this silica was introduced before sedimentation and lithification ended. Again, it is tempting to speculate about the possible role of submarine fumaroles.

The Iron King is one of the relatively few examples known to the writer of a massive sulphide deposit in rocks which, on the basis of independent evidence, are believed to be overturned and in which the zoning of metals reverses that found in normal examples (see also section on Rammelsberg, p. 151, and Hutchinson, 1973, p. 1224). This observation seems to suggest that the sulphides were probably localized, at the very least, before the rocks were deformed (and metamorphosed?).

The Bathurst—Newcastle district

Previous sections have dealt with a particular setting, such as deposits associated with alteration pipes, but study of the Bathurst—Newcastle district in the Canadian province of New Brunswick (Figs. 2 and 17) is instructive, because the numerous deposits found within the area illustrate not only the type of gradations being examined here, but also host rocks which include a larger proportion of greywacke-suite sediments than those associated with the deposits already reviewed. The massive sulphides also contain a larger proportion of lead and silver (relative to copper and gold) than all but one of the foregoing examples.

It was recognized fairly early in the history of exploration of the district that most, if not all, of the important sulphide deposits are associated with rhyolite and quartz porphyry (e.g. Holyk, 1956). Strictly speaking "quartz porphyry" is a purely descriptive term, but, to most geologists, it also has implications of "intrusion". First identified by Alcock (1941), ideas about the nature of the porphyry in the Bathurst Camp underwent the same evolution as in West Shasta and elsewhere. Thus, Skinner (1956) regarded the quartz porphyry bodies as sills (although at the same time Holyk, op cit., described them as flows). The concensus of opinion today probably regards the rock type as a crystal (lithic) tuff which, locally, may have been eroded and redeposited as a sediment. It stands to reason that intrusive phases (feeder-dykes and plugs, etc.) may also exist (e.g., the porphyry plug in which the Nigadoo deposit is localized?).

The overall structure of the Bathurst—Newcastle district does not appear to have been satisfactorily explained and, consequently, the inter-relationships of mineralized areas are obscure. The district has been described both as a basin and as a dome, but certain observations suggest that the structure is a good deal more complicated than these terms imply. In the eastern third or so of the northern part of the district — around the Brunswick Mining and Smelting deposits — the fold axes of major and minor structures plunge at 60° or 65° toward the southwest. In the central area (between, say, Rocky Turn and the Wedge mine) axes are practically vertical. Finally, toward the west, in the Caribou area, fold axes seem to plunge steeply northeast. The writer believes that these

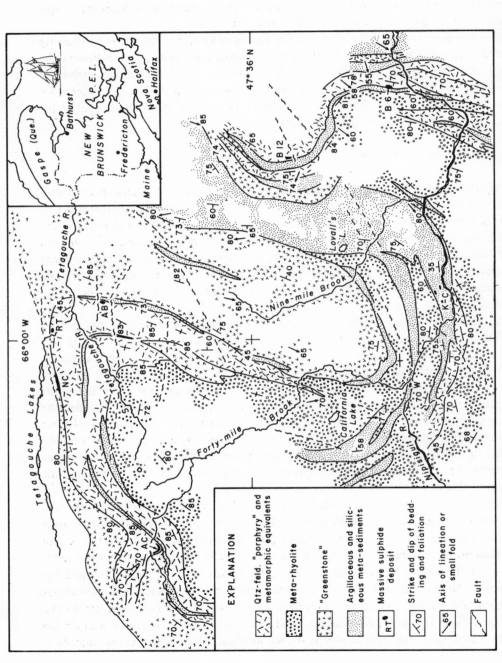

Fig. 17. Generalized geological map of the north-central portion of the Bathurst-Newcastle district in New Brunswick. Abbreviations for the names of massive sulphide deposits signify: *AB* = Armstrong Brook; *AC* = Anaconda Caribou deposit; *B6* = Brunswick Mining and Smelting No. 6 mine; *B12* = massive sulphide deposits signify: *AB* = Armstrong Brook; *AC* = Anaconda Caribou deposit; *B6* = Brunswick Mining and Smelting No. 6 mine; *B12* =

and other observations imply grand-scale *en echelon* folding and, possibly, local overturning of fold axes (Campbell, 1958; Mendelsohn, 1959).

Two of the earliest-found, largest, and certainly best known of the sulphide masses are

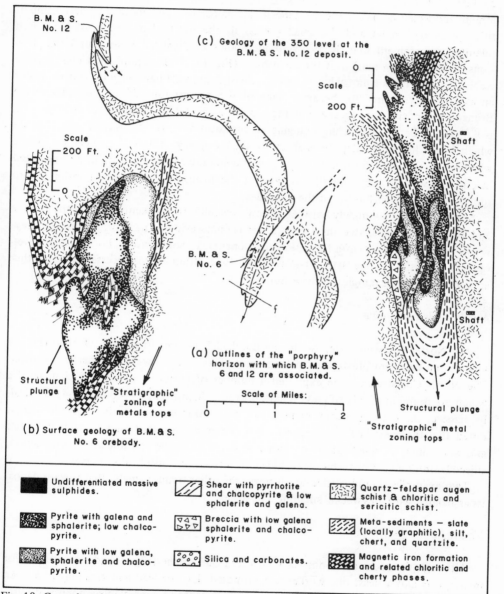

Fig. 18. General geologic maps of the Brunswick Mining and Smelting No. 6 and 12 deposits in the Bathurst district, N.B. (After Lea and Rancourt, 1958; with permission Can. Inst. Min. Metall.)

the No. 6 and 12 deposits of Brunswick Mining and Smelting (Lea and Rancourt, 1958; Stanton, 1959 and Boyle and Davies, 1964).

Although there has been a good deal of uncertainty about the form of the eastern contact of the broad band of quartz—feldspar crystal tuff with which the Brunswick Mining and Smelting No. 6 and 12 deposits are associated (and about the relationship of this "porphyry" band and the smaller mass to the east), there seems to have been unanimous agreement that the two sulphide bodies are localized in structural complications on the western margin of the main band (Fig. 18). This requires that one of these associated structures is not what it seems: one of them must have been overturned about an axis roughly normal to the axial plane of the present fold. *If* the sulphides in the Brunswick deposits were localized at the top of the crystal horizon prior to deformation, No. 6, which lies in a syncline plunging to the southwest, is right-way-up and the fold which hosts the No. 12 deposit is an antiform. Previously, it must have been a syncline which probably plunged toward the northwest quadrant. *If* the stratigraphic zoning noted in connection with other districts obtains at the Brunswick, this interpretation is also supported by the distribution of the sulphides.

Somewhere approximately between the two overall settings exemplified by the Jerome and Bathurst districts, two divergent trends become evident, namely, toward the grey-wacke—shale and chert—iron formation components. To some extent, both lines of development are visible in the Bathurst—Newcastle alone, although it is necessary to cite other areas for definitive examples of these trends.

North Coldstream mine

The North Coldstream deposit in the Burchell Lake area in western Ontario (Fig. 2) provides an excellent illustration of the evolutionary trend toward the chert association. The North Coldstream mine exploited a number of lenses of pyrite and chalcopyrite which are found in a body of grey to dark brown, aphanitic chert (Giblin, 1964). Typically massive, the chert is locally banded and brecciation was said to be common. Giblin noted that the chert had been variously interpreted as an inter-flow sediment, quartzite and a product of silicification of either basic or siliceous volcanic rocks. Nowadays, siliceous sinter would probably be included on this list of possible alternatives.

Two examples should suffice to illustrate the other trend, namely, that toward deposits associated with greywacke, siltstone and shale.

Mount Isa

Although the rocks at Mount Isa in northern Queensland (Fig. 3) had been studied for many years and at least one worker had proposed that the lead and zinc may have been contributed by volcanism (Stanton, 1962), the presence of appreciable amounts of pyroclastic material in the host siltstone and shale went unrecognized until fairly recently

(Bennett, 1965). A far more satisfactory explanation it seems to the writer, especially in light of the recognition of pyroclastic debris in the enclosing sediments, would be to regard the silica—dolomite as sinter deposits. It may be noted that, if valid, this suggestion is also in keeping with the sequence preserved at Rammelsberg in which silica and copper mineralization are both believed to have been localized early in the depositional sequence.

In considering the second example, it is necessary, in order to put the deposit in perspective, to revert to re-interpretation to a greater degree than hitherto.

Sullivan, B.C.[1]

The Sullivan ore body at Kimberly, B.C. (Fig. 2), is found at the contact between the Lower and Middle Members of the Aldridge Formation of Precambrian age (Freeze, 1966). Both members are made up of argillite, siltstone and sandstone, with argillaceous material predominating in the lower unit and arenaceous in the upper. The lower member is also characterized by thick lenses of intraformational conglomerate in its upper horizons. Freeze concluded that all of these rocks are turbidites (op. cit., pp. 266 and 267).

Igneous rocks in the Aldridge Formation are restricted to sills and dykes of quartz diabase, named Moyie intrusions, in the lower member. The only other igneous rocks known to occur in the area, the Purcell volcanics, are thin basalt flows found in the Siyeh Formation, some 21,000 ft. (about 6400 m) stratigraphically above the Sullivan ore horizon. Potassium-argon dating has obtained two ages for the Moyie intrusions, namely, 1500 and 1100 million years. The younger date is believed to correspond to the volcanic rocks found in the Siyeh, and Freeze argued that the older (i.e., 1500 million) age also post-dates the deposition of the topmost beds in the Lower Aldridge (Freeze, op cit., p. 267). The evidence does not appear very convincing to the present writer.

The Sullivan ore body is an extraordinary tabular lens of massive and banded sulphides which measures over 3000 ft. (about 900 m) in lenght and breadth, and, roughly, 150—200 ft. (45—60 m) in thickness. In vertical cross-sections the sulphides exhibit some, though not all, of the features of the stratigraphic zoning previously noted; pyrrhotite on the footwall is, in general, succeeded upward by lead—zinc mineralization which is, in turn, overlain by discontinuous lenticles of pyrite.

Unlike most massive sulphide deposits, which of course are typically devoid of significant alteration, a considerable amount of evidence of hydrothermal alteration is present, but, in common with those examples which do exhibit alteration phenomena, the bulk of the alteration underlies the ore zone. The most widespread, as well as the most spectacular, evidence of alteration is a large mass of tourmalinized rocks underlying the ore zone. The tourmaline mass is, roughly speaking, mushroom-shaped, with a "head" which is slightly smaller (in plan-view) than the sulphide lens and a "stem" which tapers downward. The mushroom is not quite concentric with respect to the centre of the ore body:

---

[1] See also the chapters by Sangster/Scott (Vol. 6) and Thompson/Panteleyev (Vol. 5).

it seems displayed about 500–100 ft. (175–350 m) toward the south or southwest. Chloritic alteration is relatively minor by comparison, being confined to masses immediately above or below the sulphides. The only type of alteration reported to reach horizons significantly above the ore zone was identified as albitization. Unlike the tourmalinization, albite alteration seems centred over the thickest portion of the sulphide mass.

Freeze discussed the origin of the Sullivan ore body at some length and concluded that it is of conventional epigenetic, hydrothermal origin. He specifically excluded a volcano–sedimentary genetic theory for a number of reasons which might be summarized as follows (Freeze, op cit., pp. 277 and 278):

(1) Study of the host rocks indicates that they were formed in a shallow to moderately deep marine environment within a basin which was well-supplied with clastic sediments, so that, according to Freeze, chemical precipitates would have been "diluted" by incoming mud. In addition, Freeze noted, turbidity currents appear to have been effective in homogenizing and respreading sediment on the basin floor.

There are several difficulties about these arguments, of course. If a hiatus in normal clastic sedimentation occurred, as the absence of quartz diabase intrusions in the Middle Aldridge and lithological break between the lower and middle members might be said to imply, chemical precipitates deposited during this period might not have been "diluted". The second is that Freeze seems to confuse the environments in which turbidities are believed to *originate* with those in which they *terminate*. Judging by the evidence marshalled by Freeze, the Sullivan host rocks were deposited *by* turbidity currents at some distance from the source area. This implies moderately deep water and largely precludes "homogenizing and respreading".

(2) If thermal springs were responsible for the introduction of sulphides, Freeze went on, "surely other minerals... commonly deposited around such springs would also be present. And surely the rocks through which these mineral-charged fluids had passed would bear their imprint...".

The first point presupposes that thermal springs all have the same composition, a debatable point at best. The possibility remains that some products of fumarolic activity, such as gypsum, anhydrite, barite, etc., may have been removed or converted into other minerals. The second argument is even more remarkable, since the tourmalinization is nothing if not the "imprint" of the "mineral-charged fluids" on the "rocks through which (they) passed". The fact that the most prominent product of these fluids underlies the sulphide horizon, is surely significant. By comparison the albitization of the middle Aldridge above the sulphide horizon only requires that weak hydrothermal activity persisted while the overlying rocks were being laid down. What is more, the form of the alteration and the "style" of structures underlying the sulphides strongly resembles those in other deposits widely believed to be fumarolic, or exhalative, in origin.

(3) "Study of igneous activity", wrote Freeze, "reveals that the only known lavas and their associated pyroclastics were extruded only after many thousands of feet of sedi-

ments had accumulated above the ore zone bed," and "Even the oldest of the dated Moyie intrusions intrudes beds of Aldridge sediments that are younger than the ore zone beds". "The recent discovery", Freeze went on, "of a late Precambrian granite pluton... in the area establishes the fact that the magmas capable of differentiation were active in the area about the time the ore body was formed as postulated by Swanson and Gunning (1945)".

The first part of the argument seems inconclusive since, as noted before, the evidence that all the Moyie intrusions post-date the formation of the ore zone horizon seems less than convincing. The second part of the argument presupposes that the ore, like the granite pluton, is of late Precambrian age.

(4) Freeze also drew attention to the zoning of metals in the Sullivan deposit and argued that the pattern is more easily reconciled with an epigenetic–hydrothermal theory of origin than one involving sedimentation.

The metal zoning depicted in plan is, of course, equally suggestive of a fumarolic or "syngenetic–hydrothermal" origin. It is unfortunate that Freeze does not describe the distribution of metals perpendicular to the bedding. It would be interesting to see how the arrangement of arsenic, iron, lead and zinc compare with the pattern at, say, Rammelsberg (see below).

(5) In the text of the paper, as distinct from the conclusion from which the foregoing points were taken, Freeze made much of the fact that the mineralization in the "tin zone fracture" is discordant. Strangely, he failed to appreciate, or acknowledge, the significance of the observation that the tin zone fracture is developed in normal and tourmalinized *footwall* sediments only and is, "found to pass upward through the main ore zone with diminished intensity, becoming a relatively insignificant structure as it approaches the hangingwall".

It would be interesting to know more about the composition of the Aldridge sediments, since turbidites typically contain an appreciable percentage of volcanic rock fragments.

Deposits associated with volcanic and calcareous rock and greywacke

The deposits discussed in this section refer to the three-component system greywacke–limestone and shale–spilitic volcanics shown in Fig. 8a and b, but it should be remembered that examples are likely to reach in the third dimension, as it were, toward one of the other three systems, notably that made up of greywacke–chert–spilitic volcanics.

Examples of the greywacke–limestone–spilite system are relatively rare and perhaps that is to be expected, because they represent an offshoot toward another major class of deposits. In this respect, they are. in a sense aberrant. For those reasons, it might be advisable to begin with examples most like those last reviewed, namely, occurrences in greywacke–shale with only minor contributions of calcareous material.

Meggen

The Meggen deposit is situated in the Sauerland, southeast of the Ruhr industrial area in the German Federal Republic (=West Germany) (Fig. 5). The pyritic layer is the stratigraphic equivalent of the Massenkalk limestone of late Middle Devonian (Stringocephalus zone) age (Fig. 19).

Although the structural setting is complex, detailed structural studies indicate that the two surviving ore bodies were originally part of a single horizon with an average thickness of 5 or 6 m. The principal sulphides are pyrite (around 70% by weight), sphalerite (12%) accompanied by minor amounts of galena (0.6%), and barite (0.3%) with traces of chalcopyrite, and tetrahedrite. Barite also overlies the pyritic body, thickening around the margins where the sulphide layer thins. Further afield, the barite horizon passes laterally into bituminous limestone and shale.

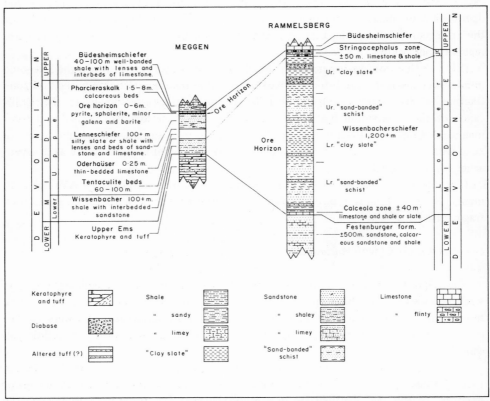

Fig. 19. Comparison of the stratigraphic column at Meggen and Rammelsberg (both Federal Republic of Germany. (After Ehrenberg, 1954; Kraume, 1955 and Ridge, 1962; with permission *Geol. Jahrb.*)

Since studies of Meggen have played such an important part in the evolution of genetic theories regarding the origin of massive sulphides, it is worth summarizing the current views. According to Ehrenberg and co-workers (1954), a shallow depression on the Meggen swell was separated from the upper Middle Devonian Massenkalk sea. The magma chambers which had fed the volcanism of the Lower Devonian upper Ems continued to release hydrothermal fluids, some of which reached the floor of the barred Meggen basin. Fumarolic activity was concentrated in two centres and an exceedingly shallow cone of sulphides was formed over each. The composition of the fumarolic solutions then changed and barite was deposited over and around the sulphides. Subsequent deformation and erosion put the finishing touches on the surviving relationships.

Rammelsberg

The stratigraphy at Rammelsberg near Goslar, at the northwestern end of the Harz Mountains (Fig. 5), is also summarized in Fig. 19. A number of differences between the country rocks enclosing this and the previous example are readily identifiable. The Rammelsberg host rocks occupy a stratigraphically lower position than those at Meggen. The Rammelsberg succession is thicker, consisting for the most part of fine-grained detrital rocks (and their low-grade metamorphic equivalents). Evidence of contemporaneous (or penecontemporaneous) volcanism is weaker at Rammelsberg than at Meggen and consists of thin layers of green, fine-grained material regarded as altered tuffs which are interbedded with the meta-sediments of the Wissenbacherschiefer. Unlike Meggen, which displays only one cycle in the compositional variation of the mineralized zone, Rammelsberg displays two. These are summarized in Table III which shows that both series exhibit the same sequence, from base-metal sulphides to barite, as Meggen. They also display the stratigraphic zoning of base-metals noted in connection with deposits found in volcanic rocks.

Some time ago E.R. Lea (oral personal communication, 1973) proposed to the writer that the massive sulphide deposits in the Balmat—Edwards district in northern New York state (Fig. 4) possess features which suggest that they are intermediate between massive sulphides in volcanic and sedimentary rocks with eugeosynclinal affiliations, on the one hand, and Mississippi Valley-type lead—zinc (copper) deposits found in miogeosynclinal or shelf limestones, on the other. In light of the nature of the present enquiry, it would seem desirable to inquire into this intriguing suggestion.

Balmat and Edwards district

According to the most recent geological account of the Balmat and Edwards deposits in upper New York (Lea and Dill, 1968), they occur in a series of metamorphosed dolomitic marbles in the Precambrian Grenville series. The stratigraphic succession is summarized in Fig. 20.

TABLE III

The stratigraphic zoning of sulphide and other minerals at Rammelsberg

UNIT or LITHOLOGY	PRINCIPAL METALLIC ELEMENTS	INTERPRETED SEQUENCE of EVENTS
Clay slate		Fumarolic activity ended and diagenesis and lithification, etc., began.
Liegendes Erzvorkommen	Ba	Final stage of fumarolic activity. Solutions poor in base metals.
Banderz	↑	Diminishing fumarolic activity gave rise to interbedded clay slate and base-metal sulphides.
Altes Lager	Ba Pb Zn Cu Fe	Strong flow of solutions which changed in composition progressively so as to give rise to the succession shown at the left.
Clay-slate	↑	Clay deposition occurred
Grauerkorper	Ba ↑	Barium-rich, base-metal poor waning activity caused deposition of barite.
Clay-slate		Brief interruption in hot-spring activity occurred
Neues Lager	Ba Pb Zn Cu Fe ↑	First strong inflow of hydrothermal solutions with appreciable metallic content. Composition changed gradually, forming the metal zoning shown. Composition of solutions changed appreciably.
Kneist	Si ↑	Early, silica-rich solutions released.
Clay-slate		Temporary cessation in supply of solutions.
Hangendes Erzvorkommen	As	Fumarolic activity began with relatively weak inflow of hydrothermal solutions into a number of shallow depressions in the sea floor of the time.

Based on Kraume, 1955; and Ridge, 1962.

The present writer has tried to extend Lea's suggestion that the deposits resemble in some respects massive sulphide mineralization in eugeosynclinal rocks by proposing a possible volcanic "parent" for a few of the metamorphic host rocks.

The quartz—mica—feldspar gneisses which lie above and below the dolomitic sequence are reminiscent, for example, of the feldspathic rocks found in numerous massive sulphide camps (see Table II, after Gilmour, 1971). It has been suggested that Unit 2 (which consists of chlorite, feldspar, mica and pyrite and thus resembles the beds of altered tuff at Rammelsberg) might also be tuffaceous, although there is no solid evidence for this interpretation. The proposal that Units 10 and 13, consisting for the most part of amphibolite and tremolite, respectively, might represent metamorphosed igneous material seems to be on slightly firmer ground. The term "lamprophyre" is something of a catch-all in geological literature and, since Lea and Dill noted that the lamprophyre bodies in Unit 14a are "sill-like" (i.e. presumably conformable), it is not inconceivabel that they might have originated as tuffs. Finally, one might speculate about a possible volcanic source for some of the contaminants in the impure dolomitic and calcite marbles. All of these proposals represent conjecture at best, but perhaps the implied questions will spur efforts to support or contradict them.

After reading a draft of the foregoing, Lea replied that he thought the suggestion that some of the units in the Balmat—Edwards district might be metamorphosed volcanic rocks seems reasonable (E.R. Lea, written personal communication, 1974). Lea went on

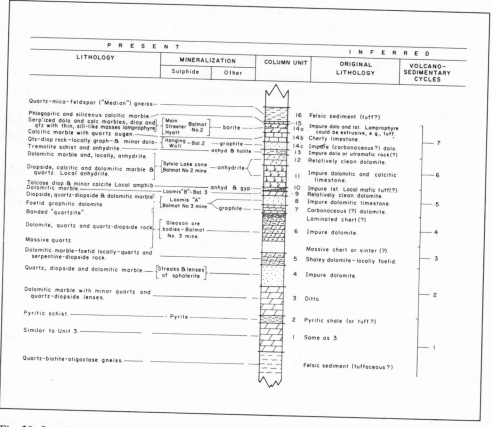

Fig. 20. Stratigraphic column in the Balmat and Edwards district, New York, showing the position of the principal ore deposits and inferred or interpreted original lithology. (After Lea and Dill, 1968; with permission Can. Inst. Min. Metall.)

to note the presence in the district of considerable thicknesses of amphibolite, which could be regarded either as metamorphosed volcanic or sedimentary rocks and concluded that he favoured the former view.

Mineralization, both sulphide and non-sulphide, has been included in Fig. 20. The resemblance, in a general way, to the ore horizon at Meggen or to the mineralization in one of the two cycles at Rammelsberg is very intriguing. Not only would it be interesting to know if there is an appreciable amount of arsenic in the pyritic schist which forms Unit 2, but it is also tempting to speculate on the possible equivalence of the siliceous zones in Unit 6 and the "kneist" at Rammelsberg. The most remarkable similarity in the comparison, however, is the tendency for barite and anhydrite to be associated with and to overly the massive sulphides in the Balmat area. The presence of graphite in a number of horizons, and the reference by Lea and Dill to halite and natural gas, suggest an intriguing connection with the reported bituminous facies of the Meggen horizon.

Lea and Dill referred to the alternations of dolomitic and silicated units. Thus Units 1, 3, 5, 7, 9 and 12 are identified as "dolomitic marbles with very minor quartz and silicates", whereas Units 2, 4, 6, 8, 10, 11, and 13 through 15 are described as "predominantly silicated members of various types". This suggestion too, was adopted as a point of departure and seven "volcano–sedimentary cycles" were tentatively proposed (right-hand column, Fig. 20). In general, the separation into paired dolomitic and "silicated" units, as proposed by Lea and Dill, were employed to define the limits of the cycles. Some departures were made, however, particularly toward the top of the succession where the units are relatively thin and the regularity in the alternation of dolomitic and silicated units tends to break down.

Other examples of massive sulphides in transitional host rocks, consisting of both metamorphosed volcanic deposits and calcareous sediments, may be presented by the pyritic lead–zinc deposits of the so-called Falun type in central Sweden (see Fig. 5 for approximate location).

Finally, one last example of the trend suggested by the Balmat and Edwards district might be noted.

The Tynagh deposit

Descriptions of the Tynagh deposit (notably Derry et al., 1965), located in County Galway in the Republic of Ireland (Fig. 5), indicate fairly clearly that the host rocks and the mineralization resemble both Mississippi Valley-type deposits and massive sulphides in eugeosynclinal rocks, although features of the former seem to predominate.

The ore body consists of a "boat-shaped mass" of weathered, or "secondary", ore and an underlying concentration of "primary" sulphides (Derry et al., op. cit.). Both occur in a prism of reef limestone which, in turn, is enclosed by argillaceous limestones of Lower Carboniferous (or Mississippian) age. Tuff and iron formation are also found in the argillaceous limestones at approximately the same horizon as the mineralized reef (Fig.

21). According to Derry et al., no intrusive igneous rocks are known to occur in the general area.

The portion of the reef which carries base metals, ranges up to 600 ft. in thickness. Concentrations are not confined to any particular horizon, or horizons, and minor amounts of sulphides have been observed in the "Lower Muddy Limestone" beneath the reef. Listed in order of decreasing abundance, metallic minerals present include pyrite, sphalerite, galena and tetrahedrite. Microscopic grains of chalcopyrite are disseminated throughout the reef and, toward the base of the reef and in the immediate underlying limestone in one or two places, chalcopyrite is present in megascopic particles. Besides

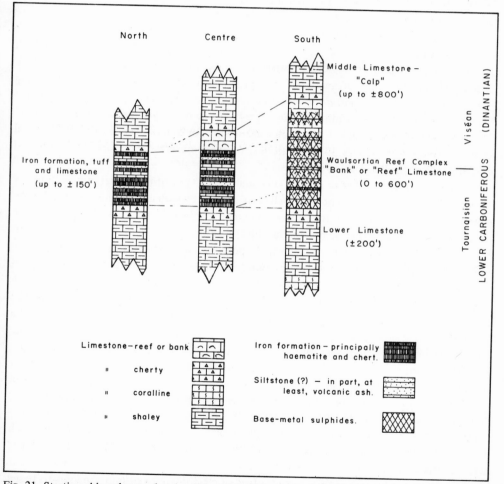

Fig. 21. Stratigraphic columns showing the relationship of the nonferrous base-metal sulphide deposit and banded iron formation at Tynagh, County Galway, Eire. (After Derry et al., 1965; with permission Econ. Geol.)

calcite, the gangue includes barite, which Derry et al. estimated locally exceeds 40% (presumably by volume) of the whole.

CONCLUSION

If the three-component diagrams of Fig. 9b and 9c, were placed side by side (sharing the "edge" greywacke—volcanic rocks), the field covered by both would be equivalent to the lower, centre-right portion of Fig. 1.

The examples of deposits and districts reviewed above might then be superimposed upon them, as shown in Fig. 22.

The principal point the writer has tried to make, of course, is simply that many, if not most, major groups of mineral deposits grade into other groups through occurrences with intermediate, or transitional, characteristics. As a consequence, dismay should not be evinced at the discovery of awkward examples which refuse to fit neatly into existing "pigeonholes". On the contrary, such occurrences merit special study in the hope that they may increase understanding of the deposits in the major groups they appear to join.

A complementary point is to urge once again the adoption of classifications of mineral deposits which, as nearly as possible, are based on the observable features of the mineral deposits rather than their inferred genesis.

It might be objected that to declare such-and-such a type of mineralization as typically associated with, say, eugeosynclinal rocks is a genetic classification or, at least, relies on genetic criteria because the term "eugeosynclinal" involves theories concerning the mode of origin of the host rocks, if not of the mineralization itself. There is some justification for this assertion, inasmuch as genetic hypothesis and observation are inextricably intertwined in geologic terminology. The objection misses a basic truth, however, which is that terms like "igneous" or "eugeosynclinal", although invested with genetic significance, are really "code words" for, respectively, "rocks of specific physical characteristics (i.e., composed of glass or crystals) *believed* to have consolidated from a melt", or, "a distinctive rock suite (characterized by the abundance of greywacke, soda-rich volcanic rocks, etc.) *thought* to have formed during the early stages of mountain-building". Genetic theories have been devised to account for the existence of these objects, or groups of objects, and then, in order to identify the objects conveniently, a term has been coined (e.g. "igneous") which, unfortunately, is derived from the putative origin, rather than the actual or observed characteristics of the objects. It seems to the writer particularly unfortunate that so many geological phenomena are commonly identified by such hypothetical genetic names instead of the physical characteristics from which the theories were derived.

On occasion, when advocating the adoption of classifications of mineral deposits based on the rock types in which they are found, the writer has been challenged to account for the associations of certain types of mineral deposits with particular host rocks, or - to

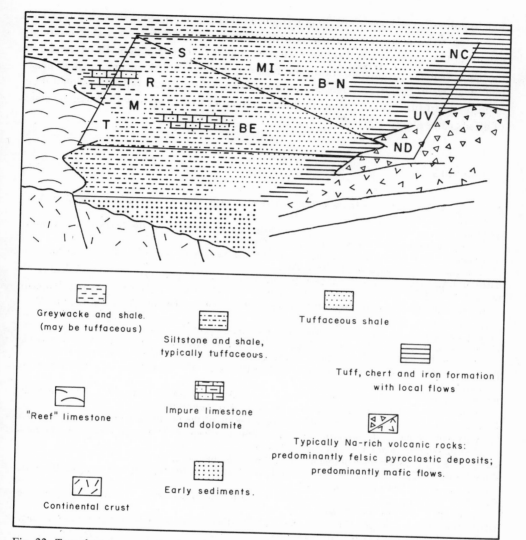

Fig. 22. Two three-component, triangular diagrams containing cited examples of massive sulphides superimposed on a portion of the diagrammatic cross-section, Fig. 1. Abbreviations: *BE* = Balmat and Edwards district; *B–N* = Bathurst–Newcastle district; *M* = Meggen deposit; *MI* = Mount Isa deposit; *NC* = North Coldstream; *ND* = Noranda district; *R* = Rammelsberg deposit; *S* = Sullivan deposit; *T* = Tynagh deposit; *UV* = United Verde deposit.

adopt the genetic terminology - particular stages in the evolution of a crustal segment, as if the inability to explain these associations amounted to a fatal defect in the basic approach. This objection merely illustrates again the tendency of many geologists to think in excessively genetic terms and the inability to distinguish between observation and hypothesis. Geologists have long recognized that certain igneous rocks are typically associated with particular sediments without *knowing* how these associations originated: all that the writer and others have advocated is that mineral deposits be treated in the same manner. If mineralization were regarded in an actualistic manner, many "puzzling" features of major occurrences would seem perfectly natural: only then would it be appropriate to turn to the task of formulating theories to explain how the observed state of affairs came about.

REFERENCES

Alcock, F.J., 1941. Jacquet River and Tetagouche River map-areas, New Brunswick, *Can. Dep. Min. Resour., Min. Geol. Branch, Geol. Surv. Mem.*, 227.

Anderson, C.A. and Creasey, S.C., 1958. Geology and ore deposits of the Jerome area, Yavapai County, Arizona. *U.S. Geol. Surv. Prof. Pap.*, 308.

Anderson, C.A. and Nash, J.T., 1972. Geology of the massive sulphide deposits at Jerome, Arizona – a reinterpretation. *Econ. Geol.*, 67: 845–863.

Bartrum, J.A., 1936. Spilitic rocks of New Zealand. *Geol. Mag.*, 73: 414–423.

Bennett, E.M., 1965. Lead–zinc–silver and copper deposits of Mount Isa. In: J. McAndrew (Editor), *Geology of Australian Ore Deposits*. Aus. Inst. Min. Metall., Melbourne, Vic,, 2nd ed., pp. 233–246.

Bilibin, Yu.A., 1955. *Metallogenic Provinces and Epochs*. Gosgeoltekhizdat, Moscow.

Bilibin, Yu.A., 1968. *Metallogenic Provinces and Metallogenic Epochs*. Geological Bulletin. Dep. Geol. Queens College, City Univ. New York, N.Y. (English translation by Eugene A. Alexandrov).

Boldy, J., 1968. Geological observations on the Delbridge massive sulphide deposit. ˌ*Can. Inst. Min. Metall. Bull.*, pp. 1–10.

Boyle, K.W. and Davies, J.L., 1964. Geology of the Austin Brook and Brunswick No. 6 sulphide deposits, Gloucester County, New Brunswick. *Geol. Surv. Can. Pap.*, 63-24.

Callahan, W.H., 1967. Some spatial and temporal aspects of the localization of Mississippi Valley–Appalachian-type ore deposits. In: J.S. Brown (Editor), *Genesis of Stratiform Lead–Zinc–Barite–Fluorite Deposits. Econ. Geol. Monogr.*, 3: 14–19.

Campbell, J.D., 1958. *En echelon* folding. *Econ. Geol.*, 53: 448–472.

Carter, S.R., 1953. Mount Isa mines. In: A.B. Edwards (Editor), *Geology of Australian Ore Deposits*. Aust. Inst. Min. Metall., Melbourne, Vic., pp. 361–377.

Collins, H.F., 1922. The igneous rocks of the Province of Huelva and the genesis of the pyritic orebodies. *Trans. Inst. Min. Metall.*, 31: 61–105.

Derry, D.R., Clark, C.R. and Gillat, N., 1965. The Northgate base-metal deposit at Tynagh, County Galway, Ireland. *Econ. Geol.*, 60: 1218–1237.

Descarreaux, J., 1973. A petrochemical study of the Abitibi volcanic belt and its bearing on the occurrences of massive sulphide ores. *Can. Inst. Min. Metall. Bull.*, pp. 61–69.

De Sitter, L.U., 1956. *Structural Geology*. McGraw-Hill, New York, N.Y., 552 pp.

Dewey, J.F. and Bird, J.M., 1970. Mountain belts and the new global tectonics. *J. Geophys. Res.*, 75: 2625–2647.

Dietz, R.S., 1964. Sudbury structure as an astrobleme. *J. Geol.*, 72: 412–414.

Diller, J.S., 1906. Description of the Redding Quadrangle, California. *U.S. Geol. Surv. Geol. Atlas*, Folio 138.

Dresser, J.A. and Denis, T.C., 1949. Geology of Quebec, volume III, economic geology. *Que. Dep. Min. Geol. Rep.*, 20.

Ehrenberg, H., Pilger, A. and Schröder, F. (with contributions by Goebel, E. and Wild, K., 1954. Das Schwefelkies–Zinkblende–Schwerspatlager von Meggen (Westfalen). *Geol. Jahrb.*, 12: 352 pp.

Freeze, A.C., 1966. On the origin of the Sullivan orebody, Kimberly, B.C. In: *Tectonic History and Mineral Deposits of the Western Cordillera*. Can. Inst. Min. Metall., Montreal, Que. pp. 263–294.

Giblin, P.E., 1964. Burchell Lake area. *Ont. Dep. Min. Geol. Rep.*, No. 19.

Gilluly, J., 1935. Keratophyres of eastern Oregon and the spilite problem. *Am. J. Sci.*, 29: 225–252.

Gilmour, P., 1962. Notes on a non-genetic classification of copper deposits. *Econ. Geol.*, 57: 450–455.

Gilmour, P., 1965. The origin of the massive sulphide mineralization in the Noranda district, Northwestern Quebec. *Proc. Geol. Assoc. Can.*, 16: 63–81.

Gilmour, P., 1966. A non-genetic classification of copper deposits with special reference to massive sulphides. Unpublished text of a paper presented to the Arizona Geological Society.

Gilmour, P., 1971. Strata-bound massive sulphide deposits – a review. *Econ. Geol.*, 66: 1239–1244.

Gilmour, P. and Still, A.R., 1968. The geology of the Iron King mine. In: J.D. Ridge (Editor), *Ore Deposits of the United States, 1933–1967*. Am. Inst. Min. Metall., New York, N.Y., pp. 1239–1257.

Graton, L.C., 1910. The occurrence of copper in Shasta County, California. *U.S. Geol. Surv. Bull.*, 430-B: 71–111.

Gross, W.H. and Sijpkens, J.P., 1965. The cosmic origin of mineral deposits. *Trans. Can. Inst. Min. Metall.*, 68: 25–29.

Heinrichs, E.W., 1966. *The Geology of Carbonatites*. Wiley, New York, N.Y., 555 pp., formerly published by Rand McNally, New York, N.Y.

Heyl, A.V., Lyons, E.J., Agnew, A.F. and Behre Jr., C.H., 1955. Lead–zinc–copper resources and general geology of the Upper Mississippi Valley districsts. *U.S. Geol. Surv. Bull.*, 1015-G: 227–245.

Holyk, W., 1956. Mineralization and structural relations in northern New Brunswick. *Ann. Conv., Prospectors Developers Assoc., Toronto, March*, 1956.

Horikoshi, E., 1969. Volcanic activity related to the formation of the kuroko-type deposits in the Kosaka district, Japan. *Miner. Deposita*, 4: 321–345.

Hutchinson, R.W., 1973. Volcanogenic sulphide deposits and their metallogenic significance. *Econ. Geol.*, 68: 1223–1246.

Hutchinson, R.W., Ridler, R.H. and Suffel, G.G., 1971. Metallogenic relationships in the Abitibi belt, Canada. A model for Archaean metallogeny. *Trans. Can. Inst. Min. Metall.*, 74: 106–115.

Irvine, T.N. and Baragar, W.R.A., 1971. A guide to the chemical classification of the common volcanic rocks. *Can. J. Earth Sci.*, 8: 523–548.

James, A.H., 1971. Hypothetical diagrams of several porphyry copper deposits. *Econ. Geol.*, 66: 43–47.

Kinkel, A.R., Hall, W.E. and Albers, J.P., 1956. Geology and base-metal deposits of West Shasta copper–zinc district, Shasta County, California. *U.S. Geol. Surv. Prof. Pap.*, 285.

Kinkel, A.R., 1962. Observations on the pyrite deposits of the Huelva district, Spain, and their relation to volcanism. *Econ. Geol.*, 57: 1071–1080.

Kraume, E., 1955. Die Erzlager des Rammelsberges bei Goslar. *Geol. Jahrb.*, 18: 394 pp.

Lea, E.R. and Dill Jr., D.B., 1968. Zinc deposits of the Balmat–Edwards district, New York. In: J.D. Ridge (Editor), *Ore Deposits of the United States, 1933–1967*. Am. Inst. Min. Eng., New York, N.Y., pp. 20–48.

Lea, E.R. and Rancourt, C., 1958. Geology of the Brunswick Mining and Smelting ore bodies, Gloucester County, N.B. *Trans. Can. Inst. Min. Metall.*, 61: 95–105.

Lowell, J.D. and Guilbert, J.M. 1970. Lateral and vertical alteration-mineralization zoning in porphyry copper deposits. *Econ. Geol.*, 65: 373–408.

McAllister, A.L., 1960. Massive sulphide deposits in New Brunswick. *Trans. Can. Inst. Min. Metall.*, 54: 88–97.

McCartney, W.D., 1965. Metallogeny of post-Precambrian geosynclines. In: E.R.W. Neale (Editor), *Some Guides to Mineral Exploration. Geol. Surv. Can. Pap.*, 65-62.

McCartney, W.D. and Potter, R.R., 1962. Mineralization as related to structural deformation, igneous activity and sedimentation in folded geosynclines. *Can. Min. J.*, 83: 83–87.

Mendelsohn, F., 1959. Structure of the Roan Antelope deposit. *Trans. Inst. Min. Metall.*, 68: 229–263.

Moss, F.A. and Gilmour, P., 1960. *Geological Map of the Bathurst District, New Brunswick*. In four sheets. Scale: 1 inch equals 1 mile. (Unpublished).

Price, P., 1948. Horne mine. In: *Structural Geology of Canadian Ore Deposits*. Can. Inst. Min. Metall., Montreal, Que., pp. 763–772.

Price, P., 1953. Noranda Mines Ltd. (Horne mine). In: *Geology and Mineral Deposits of Northwestern Quebec, Guidebook for Field Trip No. 10*. Geol. Soc. Am.–Geol. Assoc. Can., pp. 13–14.

Price, P. and Bancroft, W.L., 1948. Waite Amulet mine: Waite section. In: *Structural Geology of Canadian Ore Deposits*. Can. Inst. Min. Metall., Montreal, Que., pp. 748–756.

Pye, E.G., 1957. Geology in the Manitouwadge area. *Ont. Dep. Min. Ann. Rep.*, 66: part 8.

Read, H.H., 1943. Meditations on granite. *Proc. Geol. Assoc.*, 54: 64–85.

Ridge, J.D., 1962. *European Ore Deposits of Suggested Syngenetic Origin*. Am. Inst. Min., Metall., New York, N.Y., 37 pp.

Rosen-Spence, A. de, 1969. Genèse de roches a cordierite–anthophyllite des fisements-cuprozincifères de la région de Rouyn–Noranda, Quebec, Canada. *Can. J. Earth Sci.*, 6: 1340–1345.

Sillitoe, R.H., 1973. Environments of formation of volcanogenic massive sulphide deposits. *Econ. Geol.*, 68: 1321–1325.

Skinner, R., 1956. *Tetagouche Lakes, Restigouche, Gloucester and Northumberland Counties, New Brunswick*. Can. Dep. Min. Tech. Surv. Geol. Surv. Can. Preliminary Map 55–32, Map with marginal notes.

Spence, C.D., 1967. The Noranda area. In: *Guidebook of Canadian Institute of Mining and Metallurgy Centenial Field Excursion, Northwestern Quebec–Northern Ontario*. Can. Inst. Min. Metall., Montreal, Que, pp. 36–40.

Stanton, R.L., 1959. Mineralogical features and possible mode of emplacement of the Brunswick Mining and Smelting ore bodies, Gloucester County, New Brunswick. *Trans. Can. Inst. Min. Metall.*, 62: 337–349.

Stanton, R.L., 1962. Elemental constitution of the Black Star ore bodies, Mount Isa, Queensland, and its interpretation. *Trans. Inst. Min. Metall.*, 72: 69–124.

Suffel, G.G., 1948. Waite Amulet mine: Amulet section. In: *Structural Geology of Canadian Ore Deposits*. Can. Inst. Min. Metall., Montreal, Que, pp. 757–763.

Swanson, C.O. and Gunning, H.C., 1945. Geology of the Sullivan mine. *Trans. Can. Inst. Min. Metall.*, 48: 645–667.

Titley, S.R. and Hicks, C.L., 1966. *Geology of the Porphyry Copper Deposits, Southwestern North America*. Univ. Arizona, Tucson, Ariz., 287 pp.

Turner, F.J. and Verhoogen, J., 1960. *Igneous and Metamorphic Petrology*. McGraw-Hill, New York, N.Y., 694 pp.

Wargo, J.G., 1960. A proposed classification scheme for pyroclastic rocks. *Ariz. Geol. Soc. Dig.*, 3: 71–74.

Williams, D., 1934. The geology of the Rio Tinto mines, Spain. *Trans. Inst. Min. Metall.*, 43: 593–640.

Williams, D., 1962. Further reflections on the origin of the porphyries and ores of Rio Tinto, Spain. *Trans. Inst. Min. Metall.*, 71: 265–266.

Williams, H., 1966. Geology and mineral deposits of the Chisel Lake map-area. *Geol. Surv. Can. Mem.*, 342.

Wilson, I.F., 1955. Geology and mineral deposits of the Boleo copper district, Baja California, Mexico. *U.S. Geol. Surv. Prof. Pap.*, 273.

Wilson, M.E., 1941. Noranda district, Quebec. *Geol. Surv. Can. Mem.*, 229.

Wilson, M.E., 1948. Structural features of the Noranda–Rouyn area. In: *Structural Geology of Canadian Ore Deposits*. Can. Inst. Min. Metall., Montreal, Que., pp. 672–683.

Chapter 5

DEVELOPMENT OF SYNGENETIC IDEAS IN AUSTRALIA

HADDON F. KING

INTRODUCTION

Serious interest in an alternative to the long established and well entrenched theory of magmatic–hydrothermal replacement origin of stratiform sulphide orebodies dates in Australia from late 1951. It was the direct result of a re-study of the Broken Hill district in western New South Wales by The Zinc Corporation Limited with the objective of finding a new ore deposit to replace the one being mined out. The study and associated exploration commenced in 1946 and are still in progress 28 years later though to date the effort has been abortive. (Cf. Chapter 6, Vol. 4, by Both and Rutland.)

The geological outcome is, however, of worldwide interest and has been so recognised. This chapter endeavours to recount the events and stages of this development in geological thinking rather than to present arguments. Neutrality is, however, neither claimed nor sought. It would perhaps have been better written at a later date — since the concepts are still far from fully developed — and by someone other than one of the principal participants. It may, however, have some value in showing how the situation is seen by such a participant 20 odd years later.

BACKGROUND AND SYNOPSIS

As in most of the rest of the world, ideas on ore genesis in Australia prior to the 1930's relied largely on exotic deep-seated and therefore magmatic sources for the constituents of orebodies and also for the constituents of unusual layered rocks such as the iron formations which were both common and conspicuous in Western Australia.

By the middle thirties, virtually all metal concentrations were attributed to hydrothermal replacement (of pre-existing constituents by ore minerals) brought about by hot dilute aqueous solutions emanating from molten magmas, usually granitic.

One by one, as they were studied, a hydrothermal origin was proposed for all the major ore concentrations, even iron ore. In hindsight from one who accepted this interpretation for the first half of his professional life, the theory of hydrothermal replacement was impregnable partly because the postulate of particle-by-particle replacement and deposition only in favourable situations made the theory a fit for (almost) all the internal features of ore deposits and partly because it had become conventional not to examine — and not to be expected to examine — questions of the appropriateness or even the existence of the presumed granitic parent, of paths of ingress and egress, and of what happened to the material presumed to be displaced even where this was quantitatively very large. Under these conditions the theory could without noticeable discomfort be fitted to virtually any orebody *as seen from the standpoint of a mine geologist.*

And so it was that, when the theory came to be questioned, it was on broad grounds of exploratory significance, in relation to a particular deposit and in a region that was well enough mapped, to eliminate most of the usual unknowns. The deposit was the lead—silver—zinc concentration at Broken Hill, New South Wales, famous for decades for its size, its richness, its minerals and its complex geology as well as for its economic and technical importance. In addition, the 1950 publication by Gustafson et al. had made it a classic — perhaps the classic — example of selective hydrothermal replacement under stratigraphic and structural control.

The 1953 suggestion were that, in relation to the Broken Hill deposits, the magmatic—hydrothermal replacement theory involved tacit and critical assumptions of questionable validity and that an alternative of contemporaneous emplacement should be considered. Allied with this was the thought that such deposition being indigenous could be related to its environment. The reaction in the form of papers supporting a late high-temperature emplacement continued for more than a decade.

Whilst this one-sided debate was in progress there were other developments of interest. In 1954 in central New South Wales arose a suggestion that the association between base metal mineralisation there and its limestone/volcanic environment was so close as to appear genetic. Then there was the discovery, in the southeastern corner of the Northern Territory, in 1956 of a lead—zinc deposit devoid of any of the more dramatic geological features — igneous activity, strong folding, metamorphism — traditionally associated with ore occurrence. This was followed by another publication putting the hydrothermal theory in question on principally stratigraphic grounds.

Another result of the new interest in the stratiform ore deposits was the production in bacterial culture of some ore minerals previously attributed to high-temperature processes and the eventual formation of the Baas Becking Geobiological Laboratory to investigate the role of micro-organisms in the formation of stratiform base metal concentrations.[1]

In the 21 years since the Broken Hill publication, all the other stratiform deposits — known and since discovered — have come to be regarded as contemporaneous, i.e. as belonging where they are found.

In the technical and scientific sense the most solid of the contributions to this theme has been the publication of *Ore Petrology* by Stanton (1972b) following about 15 years of studies and laboratory experiments.

Finally this chapter attempts to look at the gaps in our knowledge which have become clearer as our understanding of other aspects has improved.

---

[1] Editors note: for bacterial processes, see Chapter 6, Vol. 2, by Trudinger.

CHRONOLOGY OF EVENTS

The principal events in this new approach to the genesis of stratiform ore deposits came along as follows:

1953

(a) Publication of *Geology of Australian Ore Deposits* with the paper (King and Thomson, 1953) containing a suggestion that, as applied to Broken Hill, the hydrothermal replacement theory involved the assumption that the texture of the ore minerals was original and depositional and that an alternative mode of formation "should be considered, even sought". Preference was expressed for a syngenetic emplacement of the ore constituents.

(b) Two papers on ore environments (King, 1953, and Knight, 1953) postulating a genetic association between stratiform ore deposits and their environments.

(c) In retrospect, "Geological Structure of Tasmania" is also of interest: "All the important orebodies of copper, lead–zinc and lead–silver. . . occur in the eugeosynclinal facies of the Dundas (Cambrian), and seem to prefer that particular horizon or lithofacies" (Carey, 1953, pp. 1127–1128).

1954

Ph.D. Thesis, University of Sydney, followed by two publications (Stanton, 1955a,b) reporting an association between base metal mineralisation and volcanic/limestone environments in New South Wales which was so close as to appear genetic and which was apparent only on the regional but not on the mine scale.

1955

In "Sedimentary environment as a control of uranium mineralization in the Katherine–Darwin region, Northern Territory" (Condon and Walpole, 1955) it was noted that there appeared to be a genetic association between uranium occurrences and algal bioherms.

1956

(a) Discovery of a lead–zinc deposit of major dimensions which was a perfect fit for the "earlier more simple form" visualised for Broken Hill 3 years earlier. The H.Y.C. deposit, in the southeast corner of the Northern Territory, was in a basin of Proterozoic dolomitic sediments undeformed, devoid of any plausible granitic parent and so little metamorphosed that the sulphide minerals are of exceptional fineness and not amenable to recovery by flotation (Cotton, 1965).

(b) "Uralite dolerite dykes in relation to the Broken Hill Lode". On the basis that these late and undisputedly postmetamorphic dykes had been marginally replaced by ore sulphides and manganese minerals, it was concluded that ". . .the ore cannot have been recrystallized by high-temperature metamorphism and its textures. . . must be primary features of deposition" (Stillwell and Edwards, 1956).

1957

"Ore genesis — the source bed concept" (Knight, 1957). Where the Broken Hill papers relied for their unorthodox conclusion on a long intimate knowledge of one mineralised region, this paper draws on the broad stratigraphic similarities of a number of deposits in four continents. It was proposed "that sulphide orebodies in the great majority of mining fields are, or were derived from, sulphide accumulations that were deposited contemporaneously with other sedimentary components. . .".

1958

"Notes on ore occurrences in highly metamorphosed Precambrian rocks" (King, 1958) compares Broken Hill, New South Wales, Calumet, Ontario, Montauban-les-Mines, Quebec, and Balmat, New York. "The conventional magmatic—hydrothermal concept is so poor a fit for the available evidence. . . that it is no longer acceptable as a basis for ore search in these rocks". Instead it was proposed that "the ore constituents were all deposited in the sediments at the time of formation. . . The present character of the orebodies is attributable to successive modification in the course of a long geological history. . .".

1959

(a) "Petrology of the Broken Hill lode and its bearing on ore genesis" (Stillwell, 1959) refers to the 1953 papers on Broken Hill as having created "a state of perplexed thought on the origin of the Broken Hill lode. . . which is as unfortunate as it is unsatisfactory. . ." after 74 years of observation "by many acutely observant engineers and geologists".

After 6 years of laboratory study of Broken Hill mineralised material — the 1956 paper (Stillwell and Edwards, 1956) was an interim result — and which was additional to intermittent studies reaching back to 1920, the conclusion was "A check on some of the speculations. . . removes all speculations which connect the lode formation with the metamorphism of the district" including "the view of King and Thomson (1953) that the ore occurred as a simple conformable lead—zinc deposit before the district was folded and metamorphosed" (pp. 4 and 5). "The time of ore deposition has been determined to be later than the intrusion of the dolerite dykes by the alteration of the dikes caused by the mineralising agencies" (p. 44). "The (petrological) evidence is compelling and cannot be set aside as proposed by King (1958)" (p. 5).

"It requires a return to the earlier conceptions of the Broken Hill lode as deposited by a high-temperature hydrothermal process subsequent to the folding and metamorphism".

(b) L.G.M. Baas Becking, working in the laboratories of the Bureau of Mineral Resources, Canberra, in a letter to this author in July 1959 remarked that in 1958 he had "promised to tackle the bacterial formation of sulphides. . . it seems quite promising. . . I obtained covellite, argentite and galena but not yet sphalerite or chalcopyrite. . .". "Mind you these simple experiments don't prove anything yet; cultures in bottles are circus lions. But they upset accepted theory considerably. . .".

The work was published in 1961 (Baas Becking and Moore, 1961). This work was followed up by others in 1961–1962 (Temple, 1964) and the eventual result of this was the formation in 1965, three years after Baas Becking's death, of the Baas Becking Geobiological Laboratory (supported by the Bureau of Mineral Resources, the Commonwealth Scientific and Industrial Research Organization and the Australian Mineral Industries Research Association) to study the possible role of micro-organisms in the formation of stratiform sulphide deposits. An outline is given in "The origins and aims of the Baas Becking Laboratory" (King, 1967). Since then the research has been joined by other departments of the C.S.I.R.O., now forming a broad front of research into the genesis of stratiform ore deposits. The Baas Becking Laboratory is now commencing a study of microfossils in Precambrian mineralised environments (see 1973, below).

1960

(a) A two year review of the Broken Hill geological data led to the conclusion (Thomas, 1960, 1961) that the Broken Hill deposit occurred in a distinctive lithological environment, two miles by twenty miles at outcrop, which also included all the other unusual layered rock-types of obscure origin. The inference was that the lead–zinc deposit belonged to this environment and did not just happen to occur within it.

(b) "Recrystallization of lead and zinc sulphides". A simple experiment by one of the companies supporting the Baas Becking research using precipitated lead and zinc sulphides showed that in three weeks appreciable crystallization developed at 200°C and strong crystallization at 400°C (Nixon, 1961).

(c) "Review of evidence of genesis of Mt. Isa orebodies" (Fisher, 1960, p. 110). A "point-by-point" assessment of the evidence for and against "syngenetic" and "hydrothermal" origin. ". . .comes out rather strongly in favour of the syngenetic cum re-crystallization theory for the lead–zinc orebodies". ". . .for the copper lodes. . . the assessment supports fairly strongly a hydrothermal origin. . .".

"The main objection to a syngenetic hypothesis is the lack of details of a satisfactory mechanism for depositing mixed sulphides in a marine environment".

1961

(a) "Formation of chalcopyrite. . . in cold solution" (Roberts, 1961). Pyrrhotite–chalcocite blocks with polished surfaces, left in distilled water for five days at room temperature, developed "a brass-yellow deposit on the surface of the chalcocite". "The newly formed mineral was identified by X-ray diffraction as chalcopyrite ($CuFeS_2$)" (p. 561).

(b) "Petrology of the zinc lode. . . Broken Hill, New South Wales" (Segnit, 1961, p. 87). "It was concluded that large-scale metasomatism was not widespread. . . and that rocks and lode were contemporaneously metamorphosed".

1963

"The low-temperature synthesis. . . of chalcopyrite and bornite" (Roberts, 1963). Chalcopyrite prepared by combining precipitates of CuS and FeS at room temperature.

1965

(a) "An environmental view of Broken Hill ore occurrence" (Carruthers, 1965) summarises the information available and together with the 1953 publications presents the modern view of the Broken Hill ore deposit.

(b) "Lead–zinc–silver and copper deposits of Mount Isa" (Bennet, 1965, p. 245) presents a comprehensive modern view of the ore deposit and concludes "the sulphides have formed. . . as the result of precipitation contemporaneous with the deposition of the sediments and represents an integral part of the rock forming minerals. Thus the orebodies have been subjected to the same. . . history as the enclosing rocks. . .".

(c) "Lead–zinc ore deposits of Australia" (King, 1965a) i.a. compares the treatment of these deposits in the first and second editions of *Geology of Australian Ore Deposits* (Edwards, 1953; McAndrew, 1965). In the first, they were all regarded as magmatic–hydrothermal. In the interval, the assumptions about the significance of mineral textures had been in question. In the second, "this article and some of the descriptions of individual deposits no longer assume mineral textures to be original and depositional. Freed of this restriction, it becomes possible to attach genetic significance to broad features and relationships such as age, attitude to enclosing rocks, lithological environment, . . .all – let it be noted – environmental in character" (p. 28). "Viewed in this light they (the deposits) appear as normal, if unusual, products of magmatism, sedimentation, biogenic agencies and metamorphism" (p. 29).

In the context of age "One other pertinent generalisation is of a worldwide character. The conformable bodies of iron, copper, lead–zinc and phosphate are not distributed at random. . . The writer seems to see a broad depositional pattern in which iron concentrations appear first, then copper, followed by lead–zinc and then phosphate". ". . .the maximum iron deposition is in the Archaean–Proterozoic (and) lead–zinc concentrations are most abundant in the Proterozoic–Palaeozoic".

"This suggests a geochemical pattern on the widest scale, perhaps related to increase of certain types of organic activity. Since many of the lead–zinc deposits are closely associated (spatially at least) with iron either as oxide or sulphide, there is already a suggestion here that they may share a common mode of origin with iron and other conformable concentrations" (p. 26).

(d) "The sedimentary concept in mineral exploration" (King, 1965b). ". . .this approach promises to lift ideas of ore occurrence out of the realm of miracles and put it, for the first time, where it belongs – in the scheme of natural events. In this view ore occurrence should become not only understandable but, within the limits imposed by its comparative rarity, perhaps even predictable" (p. 32).

1966

"The banded iron formations at Broken Hill, Australia, and their relationship to the lead–zinc bodies" (Richards, 1966) represents "an oblique – and for this reason probably unique – approach to a better understanding of the ore deposit" (King, 1973, p. 1371).

1968

(a) "Retextured Sediments" (Elliston, 1968) suggests a new mode of genesis for the porphyry-like rocks near Tennant Creek, central Australia, and that the genesis of many igneous-looking rocks... should be re-examined. So far only the simpler known mechanisms of crystal growth and cooling of melts have been considered. There is a need also to consider surface chemistry as a factor in retexturing of sedimentary rocks into crystalline rocks of non-igneous origin.

(b) "Geological research: a Broken Hill perspective" (King, 1968). Over its 80 year life, research has variously aided, impeded, or both, understanding of the ore occurrence. For this there are many reasons, including emphasis on small-scale laboratory work; nature of assumptions employed in interpretation of results; fragmentation of geological history; failure to realise that differing conclusions are due to interpretation, not differences of fact.

The situation is reminiscent of the six blind men who went to see the elephant and likened it to a wall (the flank), a spear (the tusk), a snake (the trunk), a tree (the knee), a fan (the ear), a rope (the tail) and thereafter "disputed loud and long", "about an elephant not one of them has seen" (p. 211).

1970

(a) Presidential address to geological section, Australian and New Zealand Association for the Advancement of Science (Fisher, 1970): ores are the result of geochemical cycling and recycling of igneous rocks to sediments to igneous rocks.

(b) "The Broken Hill ore deposit: an early note on its genesis". A German geologist, A.W. Stelzner of Freiberg, as early as 1894 (from examination of ore specimens) "was convinced that the ore occurrences of Broken Hill represented a bed which had been formed simultaneously with the surrounding country rock and hence was of sedimentary origin" (King, 1970, p. 191).

(c) "The iron formations of the Precambrian Hamersley Group, Western Australia" (Trendall and Blockley, 1970) describes a situation perhaps of fundamental general significance to the formation of stratiform metal concentrations. "Some parts of the Hamersley Group exhibit a degree of lateral continuity in their lithostratigraphic detail unparalleled in the sediments of any other recorded depositional basin, of any age. Within the group, there are at least 100,000 lithostratigraphic (mesoband) boundaries (on a scale of centimetres) available for correlation... over 20,000 square miles" (p. 65). Microband correlations (on a scale of millimetres) are illustrated over distances of 19, 92, 145 and 185 miles.

1972

(a) "A preliminary account of chemical relationships between sulphide lode and 'banded iron formation' at Broken Hill, New South Wales" (Stanton, 1972a). Evidence of chemical constitution indicates "quite unequivocally that the two materials are consanguineous" (p. 1128).

(b) *Ore Petrology* by R.L. Stanton (1972b). This book followed more than 15 years of sustained study, research and experiment concerned with ore textures, composition and environment. One reviewer (Professor David Williams) said in part. . . "This scholarly and stimulating book. . . (provides) the most exhilarating and informative conspectus on ore petrology yet available in the English language" (Williams, 1972).

Stanton's 25 most relevant papers appear in the list of references.

(c) "Biogenic sulfide ores: a feasibility study" by members of the Baas Becking Geobiological Laboratory (Trudinger et al., 1972).

1973

(a) "The nickel sulphide deposits of Kambalda, Western Australia" (Ross and Hopkins, 1973). Three quarters of the nickel occurs in concentrations at the contact of ultramafic and basaltic sequences. "A sub-aqueous extrusive origin is proposed for most if not all of the sequence including the contact sulphides" (p. 119). Sulphide-bearing shales and chert also occur within the ultramafic sequence.

(b) "Volcanogenic Cu–Pb–Zn mineralization in the Mons Cupri district, West Pilbara" (Miller, 1973, p. 95). Strata-bound Pb–Zn layer above disseminated Cu–Zn mineralisation within a chloritic pipe.

1974

"Microfossils from the Middle Precambrian McArthur Group, Northern Territory, Australia" (Croxford et al., 1973, and Muir, 1974) announces the discovery of microfossils in dolomite and ore at the H.Y.C. lead–zinc deposit.

In addition to these, there have been, over the years, a number of geological contributions touching more or less closely the question of contemporaneous deposition of sulphide ore minerals in layered sequences which for one reason or another did not initiate any new thinking or participate directly in the development of the new concept of stratiform orebodies.

Nevertheless they are of considerable historical and human as well as scientific interest. The ones which are closely identified with Australia are chronologically as follows:

1893

"The earliest suggestion that a major Australian sulphide ore deposit might be sedimentary was made by Dr. E.D. Peters in a report on the Mount Lyell copper deposit" (King, 1965b, after Gregory, 1905).

1922

"Geology of the Broken Hill District" (Andrews et al., 1922) shows that the team could not agree about interpretation of their observations and even included a chapter headed "Rocks concerning the origin of which there is no consensus of opinion". The report contains a regional cross-section presumably drawn by Andrews which correlates similar rock-types as stratigraphic layers over distances of many miles, an interpretation not apparent from the formal report. Whilst modern mapping does not support this interpretation in detail, the section suggests that Andrews' thinking about the nature of the layered rocks was far ahead of his time.

1927

"Notes on the geology of the Pinnacles Mine and district" (Turner, 1927) contains the results of some of the most meticulous and objective mapping known to the writer. It portrays the several Pinnacles lead–zinc lodes as stratigraphic layers, differing in metal content and ratios as the Broken Hill lode itself was portrayed (Gustafson et al., 1950) 12 years later.

1942

In a Ph.D. thesis at Harvard University, H.C. Burrell (1942) who was mine geologist at The Zinc Corporation, Broken Hill, 1934–1936 and later a member of Gustafson's team proposed that in the interests of "good fit" the manganese and lime of the Broken Hill ore layers should be assigned a sedimentary origin (p. 512). He notes that "in the light of modern understanding of sulphide ore deposition. . . there is little doubt that the commercial metals. . . were brought to their present position by solutions from an outside source" (p. 498). In retrospect it seems that if one constituent should be syngenetic the way is opened for all to be so.

1949

Paul Ramdohr spent some time with the Mineragraphic Section of the C.S.I.R.O., Melbourne, and whilst there studied Broken Hill ore specimens. Unknown to Broken Hill geologists he reached the conclusion that the mineral textures were the result of metamorphism at high temperature of originally low-temperature mineralisation (Cf. 1959, above) (Ramdohr, 1951).

1959

"The Oroya Shoot. . . at Kalgoorlie" (Tomich, 1959) broaches the question of possible sedimentary origin for stratigraphically disposed gold ore shoots on the eastern flank of the Kalgoorlie goldfield. The significance of these occurrences and the implications of their stratigraphic form have not yet been followed up. They even raise the possibility that if any of the orebodies of this classic hydrothermal deposit should have sedimentary affiliations all the constituents must be originally sedimentary.[1]

---

[1] A stratiform volcanogenic origin is now proposed (Tomich, 1974).

This chronological account must also take notice of some overseas developments since it will already be apparent that the Australian thinking did not develop in isolation.

Despite an unfavourable reception of the 1953 and 1957 ideas, thinking elsewhere commenced to move in the same direction. The more significant events known to me in the first decade after 1951 are as follows:

1954

"Reflections on prospecting and ore genesis in Northern Rhodesia" (Garlick, 1954). The essential conclusions that the presumed granitic parent was older than the sediments and that copper deposits were sedimentary became known to Knight and King in 1950 and 1951.[1]

1955–circa 1959

Exploratory work in the Noranda district (P.Q., Canada) by R.C.J. Edwards and C.D. Spence led to the development of a volcanic-sedimentary concept of ore occurrence (unpublished) in lieu of the prevailing magmatic–replacement theory.[2]

Exploratory geological work in the Bathurst district (N.B., Canada) involving stratigraphic and contemporaneous ideas was being done by Walter Holyk (Holyk, 1956) and by Frank Moss and Paul Gilmour (unpublished).[3]

1961

"Further reflections on the origin of the porphyries and ores of Rio Tinto, Spain" (Williams, 1961), in which David Williams came to "recant" earlier views in favour of a volcanic-sedimentary origin.

GROWTH OF IDEAS

It is interesting then to trace the growth of ideas over this 23-year period. It is perhaps best seen as occurring in sixteen steps, the first and last being principally questions and, in between, new interpretations, ideas, experiments and discoveries.

The first was the observation, in the context of Broken Hill, that "Previous attempts to reach an understanding of the Broken Hill deposit appear to rest on inter-related hypotheses which involve more or less tacit assumptions" including "The sulphides were transported by hydrothermal solutions and deposited for the first time in the location and in the form in which they now appear" (King and Thomson, 1953, p. 570). "There is therefore doubt of a hydrothermal mode of emplacement" (p. 570) and "an alternative mode of deposition. . . must be considered, even sought" (p. 569). These led on to "The present ore deposits are thought of as having evolved from an earlier more simple form

---

[1,2,3] Editor's notes: [1] see Chapter 6, Vol. 6, by Fleischer et al.; [2] Chapter 5 by Sangster and Scott, Vol. 6, summarizes the Canadian Precambrian strata-bound ores; [3] see also Chapter 4 by Gilmour, this Vol., and Chapter 3 by Ruitenberg, Vol. 5.

and as having acquired their present complex form during. . . folding and metamorphism" (p. 572). See also Ramdohr (1951). And to "Into these (sediments) lead and zinc sulphides were early introduced in some manner at present unknown but probably influenced, at least, by syngenetic factors" (p. 574).

It should perhaps be mentioned that at the time this paper went to press, early 1952, and for several years after (see Fisher, 1960) no mechanism for such an origin was known.

The second step is represented by the four papers (1953, 1954, 1955) on relationship of ore concentrations to their environment. It will be realised that the magmatic–hydrothermal theory did not contemplate an environmental relationship and, within the framework of this theory, most answers to genetic questions were based on mineral textures.

The third step, consonant with the new environmental thinking, was utilising the broad attitudinal relationships of a number of orebodies — in various parts of the world — as the basis of a new concept in ore genesis (Knight, 1957).

Fourth, and geologically closer to Broken Hill, a comparison of the features of five similar ore deposits and environments shows that (a) the genetic problem of Broken Hill is not unique and (b) the problem of origin of conformable igneous-looking rocks as being fundamentally similar to that of stratiform ore deposits (King, 1958).

Fifth, we have the production in bacterial cultures of some ore sulphides (in a form identifiable by X-ray diffraction), some of which were previously assigned origins at elevated temperatures (Baas Becking and Moore, 1961).

The sixth step was in two almost simultaneous parts. The demonstration that precipitated sulphides acquired crystal form *in three weeks* at a modest temperature and the more unexpected discovery that chalcopyrite and bornite could form in aqueous solutions *at room temperature* (Nixon, 1961, and Roberts, 1961 and 1963).

Seventh, we have the accumulating results of Stanton's studies and experiments indicating "that mineral textures in ores reflect the effect of crystallization, not the sequence of introduction; that these textures may readily be changed by low-temperature laboratory procedures; and that for some stratiform ores it is demonstrable that the ore minerals are present as additions to the silicate and other components and not as substitutes for some of them" (King, 1973).

Then eighth in our sequence of steps we have the conclusion *from petrological study* that metamorphism not replacement is the mechanism of formation of important Broken Hill minerals (Segnit, 1961). (See also in Vol. 4 Chapter 6 by Both and Rutland.)

Ninth we should put the environmental interpretation of the Broken Hill ore occurrence with the lead–zinc layers sharing a recognisable environmental unit with the other six or so unusual layered rock types of obscure and controversial origins (Thomas, 1961).

Then tenth we notice that most of the questions asked about Broken Hill also present themselves in much the same form in the much simpler environment of Mount Isa and that the answers lead to a similar conclusion: the sulphides are probably or conceivably syngenetic but no mechanism is known (Fisher, 1960).

Eleventh we should notice a forecast that the syngenetic concept offers a prospect of making ore occurrence "understandable. . . even predictable" (King, 1965b).

The twelfth step, contemporaneous with the one before, reveals world patterns of ore occurrences in relation to geological age suggesting a relationship to major geochemical and geobiological conditions (Pereira and Dixon, 1965, and King, 1965b).

The thirteenth, and the result of geobiological experiments commenced 14 years earlier, advances three conclusions (a) "there are few geochemical factors which, by themselves, would prevent the process (of formation of biogenic sulphide ores) from occurring in sedimentary environments", (b) "modern biochemical studies. . . suggest that sulphate-reducing bacteria may have developed in the Precambrian", and (c) quantitative considerations suggest that "while the rate of sulphate reduction and carbon fixation are sufficient to account for the Roan Antelope and Kupferschiefer deposits, certain specific conditions would be necessary for the biological production of other stratiform ores such as those at Mount Isa and McArthur River" (Trudinger et al., 1972, p. 1114).

Fourteenth is *Ore Petrology*, a text book based, as the first of its kind, on all the above concepts (Stanton, 1972b).

Fifteenth we have the recycling paper (Fisher, 1970) which takes ore occurrence out of the realm of local, especially local magmatic, events and makes it part of earth history.

And sixteenth, in the continuing series of steps, the discovery that the sulphide ore and some of the sediments at McArthur River contain microfossils (Croxford et al., 1973, and Muir, 1974).

Lastly, so far, we have a step that has been seen but not taken: the possibility that sediments can be "retextured" to resemble igneous rocks (Elliston, 1968). With this goes the much wider implication that most of our "igneous" rocks are presumed to have been molten "because of their crystalline texture". "If this presumption should be wrong — and it is almost certainly significantly in error — then we do not know how or where to draw the line between crystalline textures of molten and non-molten origin. Until we do, we do not know the nature of some of the rocks we are dealing with, i.e. of the ore environment" (King, 1973, p. 1373). The sedimentary/evolutionary concept therefore looks like disturbing another well established facet of geology: igneous petrology itself.

BROKEN HILL

Because it was the Broken Hill lead—zinc deposit which initiated the questioning that led to the development of a syngenetic concept — for itself, for many other individual deposits and for stratiform deposits as a special category of metal concentrations (as well as to this paper) — and because it was around the Broken Hill deposit that nearly all the effort to quash this unorthodox thinking was made, a brief summary of the story is appropriate here. It is, after all, part of the history of growth of geological ideas.

First we must look at the level of geological activity in Broken Hill up to 1951, when this story commences. Passing over early and very brief geological contributions, the first geological report was in 1894 (Jaquet, 1894). Even at that time the shape and south plunge of the southern part of the lode had been recognised. In the following 25 years Douglas Mawson did much pioneer work in the area (circa 1910, Mawson, 1912) and Broken Hill had a visit and a comment from Moore (1916) "selective replacement of particular beds by sulphide and gangue minerals transported by hydrothermal solutions" (King and Thomson, 1953). The next was a 3-year study by E.C. Andrews and associates (Andrews et al., 1922) which was so good that the work is still of interest and the information still of value.

After another 12 years during which only minor geological work was done, The Zinc Corporation, a southern mine, appointed a geologist and later sponsored a more ambitious effort — the Central Geological Survey, 1936–1939, under J.K. Gustafson.

When geological and exploratory activity was revived in 1946, it commenced, in the mine area, virtually (except for new exposures) where the Central Geological Survey had left off and, in the district, virtually with Andrews 1922 report and maps. (The 1927 Pinnacles mapping has already been specially mentioned.)

As at 1939 (Gustafson et al., 1950) the Broken Hill deposit was known as consisting of two 4-mile long blade-shaped folded stratigraphic layers of high-grade coarse grained to exceedingly coarse galena—sphalerite ore with two minor stratigraphically higher and more zincy bodies in its southern range. The two principal ore layers each had distinctive lead/silver/zinc ratios and differences in calcite—manganese—fluorite content. These ore layers were separated by bands of sillimanite gneiss. The deposit occurred in an environment of inter-layered sillimanite—garnet gneiss, amphibolite, granite gneiss, a garnet granulite and thin bands of (manganese-bearing) iron formation.

The concept of genesis then offered was that the lode minerals (gangue as well as sulphides) had been derived as hot solutions from an underlying (and hypothetical) granite batholith. Transported by these hot solutions the ore constituents were emplaced in favourable beds where the beds came within a belt of tight folding and attenuation. The ore emplacement was regarded as the latest (non-supergene) geological event of the region and as having taken place after all igneous activity, metamorphism, folding and faulting had subsided. The ore was therefore emplaced in its present environment, as it is now.

The size, richness, persistence and fantastic selectivity of the replacement (extending even to portions of the favourable beds isolated by tight folding) had made the deposit one of the greatest if not the greatest of its genetic type.

At this stage it should be mentioned that interest in the geology and genesis of the deposit was mainly in its possible value as a guide to ore. Since the favourable beds, by definition, were recognisable only where they were mineralised; since — for reasons which need not be discussed here — there was no sustainable genetic relationship to folding and faulting; since "wall rock alteration" was local and minor; and since the granitic parent was only hypothetical, there were no criteria of ore proximity other than ore itself. It seemed that within the scope of the magmatic/hydrothermal replacement theory there could be none.

Examining this theory as it applied to the Broken Hill deposit, there was no even plausible channel of entry or egress of the very large quantity of solution postulated as necessary; no sign of the very considerable quantity of rock-forming material displaced to make way for the sulphides and other lode minerals. The whole event had to be supposed to have taken place in hard rock leaving no trace.

Then there was the selectivity of the replacement. Lead, zinc and silver exhibited such persistent relationships that in any mine (and to a lesser extent in the deposit as a whole) the stratigraphic position of an ore layer could be determined by assay. The presence or absence of manganese silicates, calcite, fluorite could also be highly diagnostic. And this amongst six layers, with "total quantities and proportions available (appearing) to have been those which the various beds would together accept" (King and Thomson, 1953).

Next there were the "bedded iron formations", with a minor content of sulphides, manganese-bearing like the lode and unlike every other rock type, accepted as a sediment, occurring in the near vicinity of the lode both areally and stratigraphically.

Finally, the regional mapping showed that all the fifty or so occurrences of mineralisation of the Broken Hill type were stratigraphic in habit and that nearly all of them had a spatial association with one or more of the unusual rock types of the lode vicinity. Thus the conformable habit and the lithological associations were regional, not unique.

These formed the basis of the 1953 doubts, questions and suggestions.

It is not necessary to deal here with the many papers written between 1955 and 1962 by F.L. Stillwell and A.B. Edwards and their coworkers in direct or indirect support of a high-temperature late emplacement of the Broken Hill lode. Some comment may, however, be looked for on the two contributions — 1956 and 1959 — mentioned in the Chronology of events.

Replacement of portions of dolerite dykes had been known and accepted since 1910 (Geological Subcommittee, 1910). One of the 1953 Broken Hill papers mentions a 1910 observation that "ore was found replacing the dyke not only within the orebody but for 20 or 30 feet east into the footwall of the lode" (Black, 1953, p. 657). So the facts had been agreed for some to many years.

Therefore the conclusion given in these two papers is a matter of interpretation. "This interpretation denies the possibility that the relationship could be the result of later modification of the ore and dykes" (Carruthers, 1965). This possibility was implicit in the 1953 sequence of events "now visualised" (King and Thomson, 1953, p. 574):

(1) Sedimentation (including ore constituents), etc.

(2) Deformation, metamorphism, etc.

(3) Entry of the dolerite dykes.

An objective description of the dyke/ore relationship based on 100 exposures was published in 1968 (Watson, 1968).

There is perhaps no better indication of the strength of hydrothermal opinion at the time that this 1956 interpretation of the situation was still being pressed in 1962. Its implications were that a single relationship could be used — to the exclusion of all other geological information — to determine the time and mode of formation of the whole ore deposit, with wide implications as to the nature of the environment when ore emplacement (first) occurred.

The latest comment on the Broken Hill situation is in the 1973 Distinguished Lecture, Society of Economic Geologists (Derry, 1973), which also contains the suggestion that "most of the metamorphosed layers associated with the ore represent a dominantly volcanic succession". This has also been suggested at a conference in Leicester in 1967 (Williams, 1967).

In the context of this chapter, two other aspects of Broken Hill geology are of interest. First, why did not this new thinking arise earlier? Second, what lessons do we learn from this change of thinking?

To the first question, I feel part of the answer lies in lack of continuity. Essential features of the shape of the lode had emerged by the time of the first geological report (Jaquet, 1894); as has been seen, the importance of stratigraphy had been recognised by Andrews et al. (1922) and Turner (1927). Stratigraphic continuity and reliability was a principal result of the 1936—1939 study and the basis of the first comprehensive interpretation of the ore occurrence (Gustafson et al., 1950). Syngenetic factors were proposed by one of the members of this team (Burrell, 1942).

So that all the elements of the present interpretation had been present for many years and in the minds of several people. None of these people, however, remained with the problem long enough or was able to do sufficient mapping to appreciate how wide and general were these stratigraphic patterns.

The other part of the answer is that in the prevailing climate of geological opinion it was not easy to imagine (though Cay Burrell achieved it) that an ore with crystals up to several inches across and pegmatitic in part could have at one time been a nondescript sedimentary slime. And when that time came it was the worshippers of the magnificent ore textures who most stoutly defended the high temperature faith.

Second, of the many lessons learned from this geological experience, two deserve mention here. The first is that the ore deposit forms part of a distinctive observable environment containing a number of unusual rock types. Thus the genesis of the ore is only one of several problems of petrogenesis presented by this geological unit. The geological situation needs to be studied in its entirety. The "research" approach of selecting "key" elements or relationships for study and then drawing broad conclusions from the results is a waste of time and misleading. The second is that having developed an environmental view of the deposit one comes to see the tight folding, the metamorphism, the mineral textures and the petrofabrics as distractions rather than as essential features. It is necessary to see past or through these superimposed features to discern the simple underlying patterns. Nevertheless old ideas like old trees die slowly and some of these features continued for years (Lewis et al., 1965, Hobbs, 1966) to be interpreted as supporting a late emplacement of the ore (cf. Chapter 6, Vol. 4, by Both and Rutland).

As part of the approach to simplicity and as an attempt to present a stratiform orebody "in a meaningful form to scientists" generally, B.W. Hawkins produced a quantitative model of the Broken Hill lode in chemical rather than geological terms (Hawkins, 1968).

Two final notes on Broken Hill. It was interesting to discover in 1969, after the debate appeared to have died down, that in 1894 a German geologist, A.W. Stelzner, from study of ore specimens, had decided that the deposit – principally on the basis of its manganese content – was a recrystallized sediment (King, 1970).

Even earlier, and unknown until 1974, was a comment in "The Barrier Silver and Tin Fields in 1888" (by a special correspondent of the Adelaide newspapers): "The continuity of the orebody... 2 miles in extent... in no case has the ore been found to run across the strata" (Anonymous, 1888, p. 16).

OTHER DEPOSITS[1]

Of all the other stratiform base metal deposits in Australia only two – the King Island Scheelite and the Renison Bell "sill-like" lodes – are now (or still?) attributed to replacement. All typical lead–zinc–copper stratiform bodies are regarded as having been original members of the layered sequence.

Two other instances of stratiform deposition deserve special mention: Mount Isa and the West Australian nickel ores.

Prior to the Gustafson et al. publication on Broken Hill in 1950, Mount Isa was undoubtedly the best studied and best described example of hydrothermal replacement in Australia (Grondijs and Schouten, 1937, and Blanchard and Hall, 1942).

If any lead–zinc ore deposit in the world – H.Y.C. and perhaps also Sullivan, B.C., excepted – ought to have led the way to a contemporaneous concept, Mount Isa should have done. "The stratification of the minerals is so perfect, down to the minutest detail, that one's first conclusion is of a syngenetic sedimentary origin". "Examination of the microscopical structure, however, leaves no doubt as to an epigenetic origin" (Grondijs and Schouten, 1937, p. 409). These authors also mentioned that in the ore layers "the mineralization has resulted in *the total disappearance* (author's italics) of the slate" (p. 443) and that some of the pyrite might be syngenetic (as, they said, has been proposed for Meggen, Rammelsberg and Mansfeld) though they regarded this statement "more a reluctant admission that it has been impossible to prove the contrary by microscopic study than an expression of the authors' opinion" (p. 447). Thereafter orthodoxy prevailed until a sedimentary origin was suggested in 1960 (Fisher, 1960). Even so a sedi-

[1] Editor's note: a number of other chapters discuss either the development of hypotheses of ore genesis or/and contrast older with the more recently developed syngenetic theories, e.g. in Vol. 6 the chapters by Vokes, Fleischer et al., and Jung and Knitzschke.

mentary origin for the lead and zinc and a hydrothermal origin for the adjacent copper bodies (in the same stratigraphic horizon) was maintained in 1961. Four years later, as already mentioned, a contemporaneous origin was advanced for all the sulphides.

The West Australian nickel ores were immediately seen to be occurring mainly (in most instances) at the contact of the ultrabasic layers and underlying basalt. Orthodoxy required that the ultrabasics should be intrusive and the nickel sulphide concentrations attributable to crystal settling in the melt. Whilst this view is still held for some deposits and for a minor portion of the nickel in all deposits, in most areas the ultrabasics are now thought to be flows (allowing the black shales and cherts to be normal interface sediments). One of the deposits in a different environment is spatially associated with layers of banded iron formation thereby acquiring a flavour of closer identity with sedimentary conditions.

Of the remainder, it is also interesting to mention that Mount Lyell, judged in 1893 to be sedimentary, commenced in 1965 — with its neighbours Renison Bell and Rosebery[1] — to move towards syngeneric ideas (Solomon, 1965).

GAPS IN THE GEOLOGICAL STORY

Despite the progress toward understanding achieved by a concept which permits the utilisation of *all* available geological information, there are still big gaps in our comprehension even awareness of the events of the geological story. At this stage three are visualised in the immediate context of stratiform deposits: the extent of biological involvement in ore (and other rock) formation in these early ore-bearing rocks, Precambrian and Palaeozoic; the mode of concentration; and the nature of the source of the ore constituents. Looming beyond these is the interpretation of crystalline textures in silicate rocks. If these should be open to re-interpretation, layered sequences now regarded as igneous could possibly come within the range of this discussion.

The first is new in ore geology. The topic is important because it could conceivably provide stratigraphic and environmental guides to metal concentrations. Enough is already known to suggest that micro-organisms were far more prevalent in early times than was previously thought.

The second has been with us since the ore concentrations were first thought to be contemporaneous. The lateral as well as the vertical changes in concentration are sharp. Vertically they could be due to variations in relative rate of deposition of ore-forming and of ordinary detrital material. Laterally the situation is more complicated and here it may be necessary to consider another factor: preservation. We see only the part that has been preserved.

Source of the metals is a problem nearly everywhere. In a relatively few volcanic

---

[1] A volcanic–sedimentary origin is now proposed (Brathwaite, 1974).

environments the source is explicable or plausible. Everywhere else it is unknown, at best conjectural. The extraordinary lateral continuity of the layering in the Hamersley iron formations poses a problem of source and distribution to which (to me) only one answer makes sense: dust[1]. It has been interesting to discover that a dust source had also been considered on different grounds (Carey, 1973) and that a stratigraphic continuity of comparable scale had been observed in the Pritchard—Aldridge formation of British Columbia where also a suspicion of dust had been entertained (F.R. Edmunds, personal communication, 1973, and Huebschman, 1972).

If there should be such a source of layered constituents, we have a new range of possibilities to explore.

The last is at first sight outside the range of this discussion. But many of our Precambrian mineralised regions contain layered sequences in rocks of "igneous" texture. A wide range of rock types are "presumed to be igneous because of their crystalline texture. If this presumption should be wrong — and it is almost significantly in error — then we do not know where to draw the line between crystalline textures of molten or non-molten origin" (King, 1973). Drawn differently from present ideas, some sequences now regarded as igneous could possibly come within the range of this discussion.

SUMMARY AND CONCLUSION

Stimulated by the geology of some of the world's greatest ore deposits we have moved, in somewhat more than 20 years, from acceptance of certain major ore deposits as being independent and isolated phenomena produced by hypothetical and local deep-seated events to regarding a wide range of deposits as contemporaneous or indigenous and as related to their environments, even to the age of the rocks.

This achievement so far is, however, very sketchy and as indicated leaves us still with some big gaps. Those who are impatient either scientifically or professionally for more precise interpretation might remember Bligh's philosophy on his apparently endless open boat voyage: not to think too much about the remaining distance but rather about how far we have already come.

REFERENCES AND BIBLIOGRAPHY

Andrews, E.C. et al., 1922. Geology of the Broken Hill District. *Geol. Surv. N.S.W. Mem.*, 8.
Anonymous, 1888. *The Barrier Silver and Tin Fields in 1888,* by a special correspondent. W.K. Thomas & Co., Adelaide, reproduced by Libraries Board of South Australia, Adelaide, 1970.
Baas Becking, L.G.M. and Moore, D., 1961. Biogenic sulphides. *Econ. Geol.,* 56.

---

[1] Partly because no part of the 300 mile × 100 mile basin appears to be nearer the source of constituents than any other part.

Bennett, E.M., 1965. Lead–zinc–silver and copper deposits of Mount Isa. In: J. McAndrew (Editor), *Geology of Australian Ore Deposits – 8th Comm. Min. Met. Congr., Melb.*, 2nd ed.

Black, A., 1953. Broken Hill South Mine. In: A.B. Edwards (Editor), *Geology of Australian Ore Deposits – 5th Emp. Min. Met. Congr., Melb.*, 1st ed.

Blanchard, R. and Hall, G., 1942. Rock deformation and mineralization at Mount Isa. *Proc. Aust. Inst. Min. Met.*, 125.

Braithwaite, R.L., 1974. The geology and origin of the Rosebery ore deposit, Tasmania. *Econ. Geol.*, 69 (7).

Burrell, H.C., 1942. *Ore Types at Broken Hill, Australia.* Ph.D. Thesis, Harvard University, unpublished.

Carey, S.W., 1953. The geological structure of Tasmania in relation to mineralization. In: A.B. Edwards (Editor), *Geology of Australian Ore Deposits – 5th Emp. Min. Met. Congr., Melb.*, 1st ed.

Carey, S.W., 1973. Non-uniformitarianism. *4th Bertrand Russell Memorial Lecture, Flinders University, Adelaide.*

Carruthers, D.S., 1965. An environmental view of Broken Hill ore occurrence. In: J. McAndrew (Editor), *Geology of Australian Ore Deposits – 8th Comm. Min. Met. Congr., Melb.*, 2nd ed.

Condon, M.A. and Walpole, B.P., 1955. Sedimentary environment as a control of uranium mineralization in the Katherine–Darwin region, Northern Territory. *Bur. Miner. Resour. Aust., Rec.*, 1955–1956.

Cotton, R.E., 1965. H.Y.C. lead–zinc–silver ore deposit, Macarthur River. In: J. McAndrew (Editor), *Geology of Australian Ore Deposits – 8th Comm. Min. Met. Congr., Melb.*, 2nd ed.

Croxford, N.J.W., Vanecek, J., Muit, M.D. and Plumb, K.A., 1973. Micro-organisms of Carpentarian (Precambrian) age from the Amelia Dolomite, Northern Territory, Australia. *Nature*, 245 (5419).

Derry, D.R., 1973. Distinguished lectures in applied geology. *Econ. Geol.*, 68.

Edwards, A.B. (Editor), 1953. *Geology of Australian Ore Deposits – 5th Emp. Min. Met. Congr., Melb.*

Elliston, J., 1968. Retextured sediments. *Int. Geol. Congr., 23rd, Prague, Rep. Sess.*, 8.

Fisher, N.H., 1960. Review of evidence of genesis of Mt. Isa orebodies. *Int. Geol. Congr., 21st, Copenh., Sess. Norden*, 16.

Fisher, N.H., 1970. Presidential address to geological section, A.N.Z.A.A.S.

Garlick, W.G., 1954. Reflections on prospecting and ore genesis in Northern Rhodesia. *Trans. Inst. Min. Met., Lond.*, 63.

Geological Sub-committee of the Late Scientific Society of Broken Hill, 1910. Geology of the Broken Hill lode. *Trans. Aust. Inst. Min. Eng.*, 15.

Gilmour, P., 1962. Note on a non-genetic classification of copper deposits. *Econ. Geol.*, 57.

Gregory, J.W., 1905. The Mount Lyell Mining Field, Tasmania, with some account of the geology of other pyrite orebodies. *Trans. Aust. Inst. Min. Eng.*, 10.

Grondijs, H.F. and Schouten, C., 1937. A study of Mount Isa ores. *Econ. Geol.*, 32.

Gustafson, J.K., 1939. *Geological Investigation in Broken Hill.* Final Rep., Central Geological Survey, unpublished.

Gustafson, J.K., Burrell, H.C. and Garretty, M.D., 1950. Geology of the Broken Hill deposit, Broken Hill, New South Wales. *Bull. Geol. Soc. Am.*, 61.

Hawkins, B.W., 1968. A quantitative chemical model of the Broken Hill lead–zinc deposit. *Proc. Aust. Inst. Min. Met.*, 227.

Hobbs, B.E., 1966. The structural environment of the northern part of the Broken Hill orebody. *J. Geol. Soc. Aust.*, 13 (2).

Holyk, W., 1956. Relate structure and geology at Bathurst, New Brunswick. *Northern Miner*, March 1956.

Huebschman, R.P., 1972. Unpublished M.Sc. Thesis, University of Montana.

Jaquet, J.B., 1894. Geology of the Broken Hill lode and the Barrier Ranges Mineral Field. Department of Mines and Agriculture – *Mem. Geol. Surv., N.S.W.*, 5.

King, H.F., 1953. In: *Proc. 5th Emp. Min. Met. Congr., Melb.*, 12.

King, H.F. and Thomson, B.P., 1953. The geology of the Broken Hill District. In: A.B. Edwards

(Editor), *Geology of Australian Ore Deposits – 5th Emp. Min. Met. Congr., Melb.*, 1st ed.

King, H.F., 1958. Notes on ore occurrences in highly metamorphosed Precambrian rocks. In: *F.L. Stillwell Anniversary Volume – Aust. Inst. Min. Met., Melb.*

King, H.F., 1965a. Lead–zinc ore deposits of Australia. In: J. McAndrew (Editor), *Geology of Australian Ore Deposits – 8th Comm. Min. Met. Congr., Melb.*, 2nd ed.

King, H.F., 1965b. The sedimentary concept in mineral exploration. *Expl. Min. Geol., 2 – 8th Comm. Min. Met. Congr., Melb.*

King, H.F., 1967. The origins and aims of the Baas Becking Laboratory. *Miner. Deposita, 2.*

King, H.F., 1968. Geological Research: a Broken Hill perspective. In: *Broken Hill Mines – Aust. Inst. Min. Met., Melb.*

King, H.F., 1970. The Broken Hill deposit: an early note on its genesis. *Miner. Deposita, 5.*

King, H.F., 1973. Some antipodean thoughts about ore, 1971. *Econ. Geol., 68.*

King, H.F., 1974. Stratiform and stratabound ore deposits in Australia. *Circum-Pacific Resour. Energy Conf., Honolulu* (in preparation).

Knight, C.L., 1953. In: *Proc. 5th Emp. Min. Met. Congr., Melb.*, 12.

Knight, C.L., 1957. Ore genesis – the source bed concept. *Econ. Geol., 52.*

Lewis, B.R., Forward, P.S. and Roberts, J.B., 1965. Geology of the Broken Hill lode, reinterpreted. In: J. McAndrew (Editor), *Geology of Australian Ore Deposits – 8th Comm. Min. Met. Congr., Melb.*, 2nd ed.

McAndrew, J., 1965. *Geology of Australian Ore Deposits, 1965. 8th Comm. Min. Met. Congr., Melb.*

Mawson, Douglas, 1912. Geological Investigations in the Broken Hill Area. *Mem. R. Soc. S.A., 2.*

Miller, L.J., 1973. Volcanogenic Cu–Pb–Zn mineralization in the Mons Cupri district, West Pilbara. *Aust. Inst. Min. Met. Western Aust. Conf., 1973.* (To be published in full in *Economic Geology of Australia and Papua New Guinea*, 1974.)

Moore, E.S., 1916. Observations on the geology of the Broken Hill lode, N.S.W. *Econ. Geol., 2.*

Muir, M.D., 1974. Microfossils from the Middle Precambrian McArthur Group, Northern Territory, Australia. *Origins of Life, 5.*

Nixon, J.C., 1961. Recrystallization of lead and zinc sulphides. *Nature, 192 (4802).*

Pereira, J. and Dixon, C.J., 1965. Evolutionary trends in ore deposition. *Trans. Inst. Min. Met., Lond.*, 74.

Ramdohr, P., 1951. Die Lagerstätte von Broken Hill in New South Wales im Lichte der neuen geologischen Erkenntnisse und erzmikroskopischer Untersuchungen. *Heidelb. Beitr. Miner. Petrog., 2 (4).*

Richards, S.M., 1966. The banded iron formations at Broken Hill, Australia, and their relationship to the lead–zinc bodies. *Econ. Geol., 61.*

Roberts, W.M.B., 1961. Formation of chalcopyrite by reaction between chalcocite and pyrrhotite in cold solution. *Nature, 191 (4788).*

Roberts, W.M.B., 1963. The low temperature synthesis in aqueous solution of chalcopyrite and bornite. *Econ. Geol., 58.*

Ross, J.R. and Hopkins, G.M.F., 1973. The nickel sulphide deposits of Kambalda, Western Australia. *Aust. Inst. Min. Met. Western Aust. Conf., 1973.* (To be published in full in *Economic Geology of Australia and Papua New Guinea*, 1974.)

Segnit, E.R., 1961. Petrology of the zinc lode, New Broken Hill Consolidated, Broken Hill. New South Wales. *Proc. Aust. Inst. Min. Met.*, 199 (discussion, 201).

Solomon, M., 1965. Geology and mineralization of Tasmania. In: J. McAndrew (Editor), *Geology of Australian Ore Deposits – 8th Comm. Min. Met. Congr., Melb.*, 1, 2nd ed.

Stanton, R.L., 1954. *Lower Palaeozoic Mineralisation and its Environment near Bathurst, New South Wales.* Ph.D. Thesis, University of Sydney, unpublished.

Stanton, R.L., 1955a. Lower Palaeozoic mineralization near Bathurst, New South Wales. *Econ. Geol.*, 50.

Stanton, R.L., 1955b. The genetic relationship between limestone, volcanic rocks and certain ore deposits. *Aust. J. Sci.*, 17 (5).

Stanton, R.L., 1958. Abundances of copper, zinc and lead in some sulphide deposits. *J. Geol., 66.*

Stanton, R.L., 1959. Mineralogical features and possible mode of emplacement of the Brunswick mining and smelting orebodies, Gloucester County, New Brunswick. Symposium of the genesis of massive sulphide deposits. *Trans. Can. Inst. Mining, 62.*

Stanton, R.L., 1960a. General features of the conformable "pyritic" orebodies. I. Field Association. *Trans. Can. Inst. Mining, 63.*

Stanton, R.L., 1960b. General features of the conformable "pyritic" orebodies. II. Mineralogical features. *Trans. Can. Inst. Min., 63.*

Stanton, R.L., 1962. Elemental constitution of the Black Star orebodies, Mount Isa, Queensland. *Trans. Inst. Min. Met., Lond., 72.*

Stanton, R.L., 1963a. Constitutional features of the Mount Isa sulphide ores and its interpretation. *Proc. Aust. Inst. Min. Met., 205.*

Stanton, R.L., 1963b. Elemental constitution of the Black Star orebodies, Mount Isa, Queensland, and its interpretation. Discussion, *Trans. Inst. Min. Met., Lond., 73.*

Stanton, R.L., 1964a. Textures of stratiform ores. *Nature, 202 (4928).*

Stanton, R.L., 1964b. Compositions and textures of conformable ores as guides to their formation. *8th Comm. Min. Met. Congr., Melb.*

Stanton, R.L., 1964c. Mineral interfaces in stratiform ores. *Trans. Inst. Min. Met., Lond., 74.*

Stanton, R.L., 1966. The composition of stratiform ores as guides to depositional processes. *Trans. Inst. Min. Met., Lond., 75.*

Stanton, R.L., 1968. Annealing of single-phase polycrystalline sulphide ores. *Nature, 217* (with Helen Gorman).

Stanton, R.L., 1970a. Sulphides in sediments. Contribution of three thousand words to *Encyclopaedia of Earth Sciences,* F.W. Fairbridge (Editor), Columbia University.

Stanton, R.L., 1970b. Experimental modification of naturally deformed galena crystals and their grain boundaries. Contribution to a mineralogical symposium on *Crystal Growth in Solid Media* at the University of Durham, January, 1970. *Min. Mag.*

Stanton, R.L., 1972a. A preliminary account of chemical relationships between sulphide lode and "banded iron formation" at Broken Hill, New South Wales. *Econ. Geol., 67.*

Stanton, R.L., 1972b. *Ore Petrology.* McGraw-Hill New York, N.Y.

Stanton, R.L., 1973. Evolution of thought on the evolution of mineralization in the Tasman geosyncline. Contribution to Geological Society of Australia Volume in honour of Professor Dorothy Hill.

Stanton, R.L. and Rafter, T.A., 1966. The isotopic constitution of sulphur in some stratiform lead—zinc sulphide ores. *Miner. Deposita, 1.*

Stanton, R.L. and Rafter, T.A., 1967. Sulfur isotope ratios in co-existing galena and sphalerite from Broken Hill, New South Wales. *Econ. Geol., 62.*

Stanton, R.L. and Richards, S.M., 1961. The abundance of lead, zinc, copper and silver at Broken Hill. *Proc. Aust. Inst. Min. Met., 198.*

Stanton, R.L. and Richards, S.M., 1963. The abundance of lead, zinc, copper and silver at Broken Hill. Discussion, *Proc. Aust. Inst. Min. Met., 206.*

Stanton, R.L. and Russell, R.D., 1959. Anomalous leads and the emplacement of lead sulphide ores. *Econ. Geol., 54.*

Stanton, R.L. and Willey, H.G., 1970. Natural work-hardening in galena, and its experimental reduction. *Econ. Geol., 65.*

Stanton, R.L. and Willey, H.G., 1971. Recrystallization softening and hardening in sphalerite and galena. *Econ. Geol., 66.*

Stillwell, F.L., 1959. Petrology of the Broken Hill lode and its bearing on ore genesis. *Proc. Aust. Inst. Min. Met., 190.*

Stillwell, F.L. and Edwards, A.B., 1956. Uralite dolerite dykes in relation to the Broken Hill lode. *Proc. Aust. Inst. Min. Met., 178.*

Temple, K.L., 1964. Syngenesis of sulfide ores: An evaluation of biochemical aspects. *Econ. Geol., 59.*

Thomas, W.N., 1960. Broken Hill ore occurrence reinterpreted. *Consolidated Zinc Pty. Ltd., Rep.* (unpublished).

Thomas, W.N., 1961. Environment aspects of ore occurrence at Broken Hill — a contribution to a symposium on syngenesis, A.N.Z.A.A.S. Conf., Brisbane, unpublished.

Tomich, S.A., 1959. The Oroya Shoot and its relantionship to other flatly plunging ore pipes at Kalgoorlie. *Proc. Aust. Inst. Min. Met.,* 190.

Tomich, S.A., 1974. A new look at Kalgoorlie Golden Mile geology. *Proc. Aust. Inst. Min. Metall.,* 251.

Trendall, A.F. and Blockley, J.G., 1970. The iron formations of the Precambrian Hamersley Group, Western Australia. *Bull., Geol. Survey Western Aust.,* 119.

Trudinger, P.A., Lambert, I.B. and Skyring, G.W., 1972. Biogenic sulfide ores: a feasibility study. *Econ. Geol.,* 67.

Turner, W.J., 1927. Notes on the geology of the Pinnacles Mine and District. *Proc. Aust. Inst. Min. Met.,* 68.

Watson, D.P., 1968. Structures and field relations of epidiorite dykes in the Broken Hill orebody. *Proc. Aust. Inst. Min. Met.,* 227.

Williams, D., 1961. Further reflections on the origin of the porphyries and ores of Rio Tinto, Spain. *Trans. Inst. Min. Met., Lond.,* 71.

Williams, D., 1967. Discussion, Geological significance of stratiform ore deposits. *Proc. Inter-Univ. Geol. Congr., 15th, Univ. Leicester, 1967.*

Williams, D., 1972. Review of "Ore Petrology" by R.L. Stanton. *Trans. Inst. Min. Met., Lond., Bull.,* 789.

Chapter 6

ORIGIN, DEVELOPMENT, AND CHANGES IN CONCEPTS OF SYNGENETIC ORE DEPOSITS AS SEEN BY NORTH AMERICAN GEOLOGISTS[1]

JOHN DREW RIDGE

INTRODUCTION

The term "syngenesis" was first introduced just over 50 years ago by Fersman (1922) and was applied by him to the formation, or stage of accumulation, of unconsolidated sediments in place, including those changes that affect detrital particles still in movement in the waters of a depositional basin (Gary et al., 1972). In 1933, in the fourth and last edition of his "Mineral Deposits", Waldemar Lindgren defined syngenetic deposits as including magmatic segregations or accumulations of useful minerals formed by processes of differentiation in magmas, generally at considerable depth below the surface. Examples of this class given by Lindgren were masses of chromite in peridotite or of titanic iron ore in anorthosite; with the latter of his two examples, some modern workers would not be in agreement. Lindgren further said that syngenetic deposits included sedimentary beds because they have the same tabular, sheet-like, or flat lenticular form of magmatic cumulates; they are horizontal in attitude if they have not been affected by later folding or faulting. He pointed out that, parallel to their bedding, their extent could be measured in miles, and he gave as examples the Clinton hematite ores of the Appalachians or the Jurassic iron ores of France. He added that each bed usually thins out in wedge-shaped form and might be replaced by others at slightly different horizons. He also includes in this category beds of coal, rocksalt, anhydrite, and gypsum. Immediately following this discussion of syngenetic deposits, Lindgren briefly touches on epigenetic deposits with which he lists the sphalerite deposits of the Joplin district and the galena deposits of southeastern Missouri. Not all geologists today would agree with so categorizing these two deposits.

In discussing lead and zinc deposits in sedimentary rocks, Lindgren (1933) described what are now designated as Mississippi Valley-type deposits under this heading because of the uncertainty in his mind as to their origin. He did say that the earliest interpretation of

---

[1] A topical index is given on pp; 294–297.

them as marine deposits had been generally abandoned since it was recognized that, even if the metals had been derived from primary ocean sediments, the finely divided sulfides must have been concentrated and redeposited. He considered the epigenetic nature of these deposits to be clear. Although the term telethermal had not yet been put forward as a name for these deposits, Lindgren does quote Beyschlag, Krusch, and Vogt as defining them as telemagmatic, a term much more definitive as to the origin of the ore fluids than the more indefinite one — telethermal.

Lindgren (1933) pointed out that galena and sphalerite were of widespread occurrence in many limestones and dolomites far from regions of deep fissuring and igneous action, and that the source of the sulfides in these deposits well may have been sedimentary beds, usually Paleozoic in age. The waters that collected these sulfides probably were chlorine-rich and of meteoric origin, with it having been more likely that they worked their way downward into the formations surrounding the area of ultimate deposition and then worked their way upward (he made no mention of their having been heated in this process), and deposited their loads in the areas where the sulfides are now found. He quoted with approval Bastin's suggestion that the necessary sulfur to precipitate the metals from the chlorine-rich waters in which they were carried came from the reduction of sulfate sulfur by anerobic bacteria. As for the theory that the ores of this type were deposited by magmatic solutions, he argued that one of the strongest arguments against it was that ores of this type never have been traced in depth, that no avenues of solution entry had ever been found. Yet two pages later he remarked that he could not deny that, in many countries, transition types appear to connect these distinct non-igneous types with deposits of igneous affiliations.

In discussing sedimentary sulfide deposits, Lindgren (1933) said that "the evidence is scant as to the sedimentary deposition on a large scale of sulphides other than pyrite or marcasite". He also said that he found "little evidence for extensive sedimentary beds of pyrite".

In his section on the relation of volcanoes and lava-flows to mineral deposits, he gave no indication that he considered as possible the development of ore deposits through the medium of hydrothermal fluids or volcanic exhalations reaching the sea floor and there depositing their loads to produce massive or heavily disseminated sulfide ore bodies. He mentioned the Kuroko deposits of Japan, but only as "black ores" and said nothing as to how they originated. Although he discussed the deposits at Rammelsberg at some length and said that they were thought, by some geologists, to be of sedimentary origin, he finally concluded that the deposit may have been formed by ascending solutions derived from the Brocken granite only 3 miles away. He reached no conclusions about Meggen and pointed out that it was thought to be either sedimentary or a replacement of lime-stone. In his summary on the Mansfeld deposits, he remarked that they had been considered as having been formed by a sedimentary process and that the arguments for a syngenetic origin are very strong. He believed that Mansfeld was not an ordinary deposit precipitated from seawater, and thought that it was laid down in a shallow sea that was full of decaying vegetable and animal remains, and into which probably were discharged cupriferous waters from the surrounding littoral, most likely sulfate solutions derived from the eruptives and the ore deposits of Early Permian Epochs. He quoted Schneiderhöhn as saying that the copper sulfides were formed syngenetically as mixed iron—copper-sulfide gels under the influence of bacteria in the sulfur cycle.

Lindgren (1933) did not discuss the effects of diagenesis on syngenetic deposits; in fact, I cannot find that he used the word at all. Only a few years before, in 1920, Schuchert had complained that the word — diagenesis — was only rarely employed in the United States. Thus, the possibility of sedimentary pyrite being replaced by copper or lead and zinc sulfides during the diagenetic stage to produce workable ore deposits is not even mentioned by Lindgren. Since the term — diagenesis — goes back at least to 1893 and Walther's "Einleitung in die Geologie als historische Wissenschaft", it is rather surprising that it was not in more common use in North America by 1933. This is especially true because, in Lindgren's time and before, it was much more common for North American geologists to read German easily than it is today. Certainly by 1939, Twenhofel was defining diagenesis as "all modifications that sediments undergo between deposition and lithification under conditions of pressure and temperature that are normal to the surface or outer parts of the crust". He also included under diagenesis those changes, not katamorphic in character, that take place after lithification under near-surface temperatures and pressures to cause delithification. He also added that many of the modifications before lithification are due to organic activity. With this much general use of the concept of diagenesis, it is rather surprising that Lindgren (1933) did not consider it in relation to the formation of syngenetic ores.

Lindgren (1933) devoted only six pages to "regionally metamorphosed sulphide deposits" and mentioned a few mines in which he though the textures of the ore minerals had been appreciable changed by metamorphism, but only one of these does he designate as certainly syngenetic in origin — the Åmmeberg mine in Sweden. He did not, however, say why he considered Åmmeberg to be syngenetic.

Although this summary of necessity only touches briefly on this distinguished author's views, this much is necessary because essentially every American geologist trained in the 1920's, 1930's and early 1940's depended on Lindgren's textbook for his introduction to economic geology. It is not surprising, therefore, that most of them were less than enthusiastic for the idea that important ore deposits, other than bedded iron ores, had been produced by syngenetic processes. And it would be even more unusual for one of them to have thought much about volcanic exhalations, even if he could have defined the term, as providers of the metallic content in the large number of ore bodies that are considered today to be of such origin. Thus, the next step in this paper will be the consideration, year by year, of the changes in North American ideas on the formation of ore deposits, with emphasis on the growing recognition of the possibility of syngenetic (and diagenetic) processes having been responsible for many of the valuable ore deposits of the world. Although the papers cited were published mainly in North America, they may deal with deposits in any part of the world and may have been written by geologists from anywhere in the world and may concern a wide variety of deposits in many countries.

CHRONOLOGICAL OUTLINE OF THE DEVELOPMENTS IN CONCEPTS OF SYNGENETIC ORES

The Lindgren Volume

In *1933, the A.I.M.E. published the Lindgren Volume,* a tribute to the work and influence of Waldemar Lindgren on the study of ore geology on the North American continent. Only a very small fraction of that volume is devoted to deposits that were, or might have been, formed by sedimentary processes. F.L. Hess discussed sedimentary deposits of uranium and vanadium, pointing out that the deposits, all in sedimentary rocks, contained minerals that were produced in them by secondary processes. He said that "what these minerals (the source minerals of the secondary uranium and radium minerals) were and what their original source may have been can only be surmized, for no original mineral and no vein carrying a group of minerals, the weathering of which would produce any considerable quantity of carnotite, or closely related tyuyamunite, has been found". From this it would seem that Hess did not consider that the source material of the secondary radioactive deposits was sedimentary in origin.

In the same volume, Finch wrote on the sedimentary copper deposits of the Western States and, in this study, considered only copper deposits in clastic sediments that are dominately red in color. Nevertheless, although occurring in such beds, the individual strata that contain the copper seldom are red. He pointed out that these deposits had no physical connection with Mesozoic or Tertiary igneous centers and that they had no relationship with hypogene copper deposits of any age. He appeared to believe that the copper was transported in solution or in fragments, deposited in part as native copper through the aid of decaying organic matter, in part as copper sulfate that was later, during deep burial, converted to chalcocite by hydrogen sulfide produced by the coalification of plant debris incorporated in the clastic sediments. He suggested, in one sentence, that the ore of the Boleo mine might have been formed in this manner. Later weathering was thought to have converted much of the chalcocite to carbonate, which remained in the red beds, or back to copper sulfate, that was carried away in solution.

Except for such deposits as those of manganese, phosphates, saline minerals, and gold placer deposits, the Lindgren Volume contains no further contributions on the formation of ores by sedimentary processes.

1931–1937. Early work

Bell (1931) produced the first paper that discussed the Pine Point deposits in the Northwest Territories in any detail. These deposits are now considered, by a majority of geologists, to have been formed by near-surface processes, but Bell concluded (1931, p. 624) that "the hydrothermal theory of origin" (of course he meant formation) "of the primary mineralization at Pine Point harmonizes more readily with the information so far available than does either of the other hypotheses: (1) descending or laterally moving ground water; and (2) ascending ground water". Apparently a sedimentary formation of the ores, plus their later remobilization, had not occurred to Bell as a possible explanation for their geologic features.

Bastin (1933) took up the problem of the origin of native copper and chalcocite types of ore deposits. He recognized five types: (1) chalcocite zones of downward enrichment; (2) Red beds copper deposits; (3) Mansfeld or Kupferschiefer ores; (4) Lake Superior and Corocoro varieties; and *(5) the Kennecott deposit. He considered that type 4 resulted from the removal of iron and sulfur from ore fluids of magmatic origin by reaction with wall rocks; he suggested that type 5 might have formed in a similar way. The downward enrichment deposits, type 1 he thought were formed in the standard manner so well explained by Emmons. The copper of the red beds deposits he considered to have been obtained from the weathering of standard (hydrothermal) copper deposits and to have been transported as copper sulfide. He had recourse to decaying organic material to reduce the sulfur in sulfate ions to sulfide ion. He would have had the Mansfeld type 3 deposits formed from copper sulfate-bearing ground waters entering into the bottom muds of shallow but extensive lagoons where the necessary hydrogen sulfide was provided by the bacterial reduction of the sulfates. Thus, only for type 3 did he consider probable anything that might be called a sedimentary mode of formation.

Lindgren (1935) considered the problem of magmatic versus meteoric waters and reached the conclusion that the tendency was to minimize too much the quantity and distribution of meteoric waters in the earth's crust. He suggested that meteoric waters may reach to depths of 8000 to possibly over 10,000 ft in sedimentary rocks but that in igneous and metamorphic rocks depths of 3000 ft probably were a maximum for surface-water penetration. In only a few places are ground waters to be considered enterily of connate origin. Salinity in ground waters increases with depth, but the distribution and composition of such waters may now be far different from what it was when ore deposits in these rocks were emplaced. Despite this belief in the deep penetration of meteoric waters, Lindgren thought that magmatic waters were the principal agency in the formation of ore deposits. He concluded that what he called the telemagmatic lead–zinc deposits (now normally referred to as Mississippi-Valley type) were formed by a mixture of magmatic and meteoric waters, a conclusion with which I for one would not quarrel. He, however, related them to epithermal deposits, which I would not do. All this has no direct relationship to sedimentary deposits, except to show that Lindgren did not consider Mississippi Valley-type deposits as remobilized or regenerated syngenetic deposits as is now done for so many of this type.

Fischer (1937) discussed the problem of copper, vanadium—uranium, and silver deposits in the southwestern United States. He reported the presence of chalcocite pseudomorphous after plant fossils; this suggested to him that the copper sulfide was introduced prior to deep burial. The faults and folds in the area Fischer believed to be later than the mineralization and to show no genetic relationship to the ores. He thought that the metallic elements now in the ores were deposited from dilute solutions by reactions involving organisms, but that the present minerals of the deposits were epigenetic. He considered that the ultimate source of the metals was igneous rocks or hydrothermal veins older than the rocks in which the ore minerals are now enclosed. Because no primary deposits of these metals are within a 100 miles of the southwest Colorado carnotite region, he assumes that the source rocks need not have been rich primary deposits but to have had ore minerals minutely, but widely, disseminated through them. He does not think that the metals were transported mechanically but were concentrated from dilute solutions by organisms or organic material that reduced the sulfur and the copper of the solutions. As to the details of the process, Fischer offers no definite suggestions. In discussing this paper, *Koeberlin (1938)* suggests that the ore materials may have entered the basins of deposition as volcanic debris given off by batholithic masses that approached within moderate distances of the then existing surface. From this debris, the various metals were later derived by surface waters and were concentrated in the sedimentary basins in question.

1937–1950. An hiatus

Essentially nothing was published *in North America between 1937 and 1942* to suggest that syngenesis was an important manner of the formation of ore deposits. Although for more than 200 years the Mansfeld Kupferschiefer had been thought by many German geologists to have had a syngenetic manner of formation, an American geologist, C.H. White (1942), visited them and expressed the opinion that they were epigenetic. White points out that the majority opinion, at the time he visited the mines (1932), was that the copper (and other ore elements) were deposited from sea water and were, therefore, truly syngenetic. Other German geologists, however, considered the ores to have been deposited epigenetically by ground waters that gathered their metal content from adjacent Permian eruptive rocks, moved up the faults in the Mansfeld region, and spread into the shale where deposition took place through the reducing effect of bituminous matter. White (1942) put forward the following items as favoring an epigenetic manner of formation and a magmatic source for the ore fluids:

(1) The ores are concentrated near the base of the shale and near the faults, indicating that the ore came from beneath the shale and entered it along the faults.

(2) The solutions that removed the iron from the underlying sandstone (the Weissliegende) seems to have been guided by the lower surface of the shale as it moved up dip.

(3) The Weissliegende originally must have been an upward continuation of the now underlying Rotliegende, since the bleaching to produce Weissliegende rock cuts indiscriminately through cross-bedding, bedding planes, and other structures and is not, therefore, confined to particular beds.

(4) The red volumes in the calcareous beds overlying the ore-containing shale suggest that the copper-bearing solutions were unable to penetrate into the carbonate rocks.

(5) Ore is present locally in the sandstone beneath the shale and in fractures in veins cutting shale and underlying sandstone (Rücken) and cannot have been syngenetically formed.

All of these features indicate that much, if not all, of the ore was epigenetic in its present positions but does not prove that the ore fluids came from magmatic sources. It may show no more than that at least a considerable fraction of an originally syngenetic ore was remobilized after the rocks of the sequence had been lithified. Nevertheless, White (1942) concludes that the ore fluids brought in all the ore, and that they came from great depths through east—west fractures. He suggests that the east—west veins in the Harz Mountains probably were formed contemporaneously with the Mansfeld ores and were cogenetic with them. Because the Harz Mountain veins are almost certainly Hercynian, they hardly can be contemporaneous with ores in Permian rocks. Thus, White (1942) seems to have done little to fix the ultimate source of the ore fluids that deposited all, or at least the definitely epigenetic part, of the Mansfeld ores.

In North America, the war years and the first 5 years of the post-war period produced little work to forward the concept that syngenesis was an important manner of ore formation. *In 1950, Fischer* again discussed the uranium-bearing sandstone deposits of the Colorado Plateau in which the ores are restricted to a few stratigraphic zones along which the ores are widely, but spottily, distributed, even though the ores do not follow the beds in detail. Fischer rejects a truly syngenetic manner of formation for the ores, as he does a magmatic one. He believes that the ores (mainly those of the carnotite type) were precipitated from ground-water solutions after the enclosing sandstone formations had been lithified but before the region was structurally deformed. Sedimentary structures, however, appear to have controlled the movement of the solutions, and much of the ore has replaced fossil (not decaying) wood. Bleached rock adjoining the ores indicates that the ore fluids could readily reduce iron. Fischer (1950) does not, however, consider where the ground-water solutions acquired their metal content.

In 1950, Collins summarized a translation of a detailed paper by Kinoshita, published in 1943 and written entirely in Japanese, in which the author expressed his belief that Kuroko-type ores and the volcanic rocks that enclose an appreciable fraction of them came from the same magma reservoir. Kinoshita (1943) was not certain if the ore fluids were gases or liquids when they passed through the fractured eruptive rocks. He did, however, refer to these transporting materials as low-temperature, alkaline hydrothermal-solutions from which the ores were directly deposited by replacement of the host rocks. His evidence for this statement consists of:

(1) Crystallized ore minerals replaced country rock.

(2) Crystals of ore minerals cut across sedimentary laminae.

(3) Barite is found in igneous rocks.

(4) The ore contains remnants of rock structures such as stratification, as well as phenocrysts, breccia, and fossils.

These are not overwhelmingly compelling evidences that the ores could not have been deposited from hydrothermal fluids that reached the sea floor before depositing their loads of metallic and gangue minerals. The present favor that the volcanic-exhalative method of ore formation finds among Japanese geologists (as well as many from other lands) indicates that these reasons were not convincing to them either.

1951–1956. Syngenesis versus epigenesis

Although *Garlick and Brummer* were working in the Rhodesian (now Zambian) Copper Belt[1] when they *published their 1951 paper*, it appeared in *Economic Geology* and had an impressive effect on the thinking of North American geologists. In the early work on the Copper Belt, it was assumed that the ores were deposited from hydrothermal solutions that came from the same magma-chamber as the rocks of the so-called Younger granite that, as its name would indicate, was considered to be younger than the sediments that contain the ores. The work of these authors showed with considerable certainty that the Younger granite, in its various aspects, consisted of variations of the Older granite, with the ores, where they are in rocks that are in direct contact with granite, being in rocks that lie on erosion surfaces of granite. This relationship requires either that the ores were introduced with the sediments or deposited by ore fluids that came from sources younger than the sediments. Such sources, of course, may have included ground water as well as hydrothermal fluids of magmatic origin. Garlick and Brummer later were led to the opinion that this relationship of ores to Older granite requires that the ores were deposited with the sediments in which they are now located, although some migration of the syngenetic ores may later have been achieved through remobilization (a term not yet of sufficient status to be included in the American Geological Institute's 1972 Glossary of Geology or Amstutz's Glossary of Mining Geology).

Davis (1954) obtained his doctorate in South Africa and is now a member of the Mining Geology faculty of the Imperial College in London. However, he published his work on the formation of the Roan Antelope copper deposit in Economic Geology where it influenced to an appreciable extent the thinking of North American geologists on stratiform deposits. Davis points out that the source of the ores is not certain for either the epigenetic or syngenetic hypotheses. For the former, resort must be had to an unknown source at depth; for the latter, appeal must be made to the generality of the rocks surrounding the basin of Roan sedimentation, and no more than tiny amounts of copper

---

[1] Editor's note: see Chapter 6, Vol. 6, by Fleischer et al.

are known in these rocks. Davis finds that the arrangement of the sulfides does not agree with his ideas of how hydrothermal fluids might produce zoning. Firstly, the sulfides are, in his opinion, unaccompanied by gangue minerals. Secondly, deposition over 5 miles of strike was highly uniform and lacked irregular low-grade or high-grade areas that were developed in relation to structural features. He is not convinced that the prevailing temperatures were sufficiently high to account for the sulfide textures that he considers were produced by unmixing of solid solutions, nor does he believe that such a mechanism is compatible with what he knows of ore deposition. The zoning of the deposit might, conceivably, have been produced by successive mineralizing surges along different levels in the ore body, but he does not believe that a single ore fluid could possibly produce the arrangement now seen.

In favor of a syngenetic formation, Davis sees indications that the various shales were deposited in a body of water not very deep nor greatly varied in depth. Nevertheless, a theory of detrital deposition of the ores he considers to be untenable because the zoning cannot be explained by such a process, even aided by diagenesis. On the contrary, a bacterial origin finds much favor in his mind. Although the Roan shales are nonfossiliferous, graphitic material has been recognized in them. The abundant preservation of detrital feldspar in the rocks of the basin suggests that the environment was arid and that the waters in the shallow basin of sedimentation were concentrated by evaporation. He suggests that the precipitation of metallic compounds by obscure chemical powers of primeval forms of life might result in colloidal (?) globules of copper and iron sulfides in various proportions that could be controlled by the environment to produce a zonal pattern. (To this reader, this explanation seems as difficult to reconcile with reality as that of surges of ore fluids. No question exists but that the zoning relationship has been used to find new ore, but this does not prove that the zoning was developed in a particular manner.)

Davis sees the stratigraphic limitation of the ore as a definite plus for the syngenetic hypothesis. He thinks that the sulfides in the Copper Belt cut off so sharply within a greater thickness of homogeneous rock that no epigenetic hypothesis can explain this relationship. A killing off of the bacteria during the sedimentary process, however, could easily do this; such destruction of the bacteria could be achieved without any change in the physical characteristics of the rock concerned. He appeals to reconstitution during the burial metamorphism that has affected the Roan beds to explain the ore textures once thought to have been caused by unmixing. He also thinks that the metamorphic mechanism would account for what he says is the essential contemporaneity of the three main copper sulfides (chalcopyrite, bornite, and chalcocite). This metamorphism, he believes, would have raised the rock temperatures high enough to have produced the intergrowths of these minerals. The veins of sulfides that cut the rocks (but do not extend far above or below the ore zone) could also be explained as metamorphic rather than hydrothermal effects.

W.C. White and Wright (1954) discuss the major chalcocite-minor native copper deposits of White Pine, in the upper peninsula of Michigan. They point out that the copper-bearing beds at White Pine are in the lower 20–25 ft of the Nonesuch shale of Late Keweenawan Age. They divide the cupriferous zone into four stratigraphic units that are, from bottom to top, a sandstone, a shale, a second sandstone, and a second shale. They consider that the sequence and distribution of these sedimentary facies was caused by two submergences, separated by the emergence of a deltaic area. Practically all of the copper minerals are in the two shales, although in a small area near the major White Pine fault, they are abundant in the two sandstones. The thicker the bed, the greater the amount of copper it contains, and the thickest areas are those in hollows away from the main channels of the delta. As the sand content goes up in the shales, the copper content goes down. These copper-bearing portions of the shales are from 1 to 2 ft thick and cover areas of several square miles. They accept the concept that the copper minerals in the sandstones near the fault can be explained by hydrothermal transportation, but they consider that the distribution of copper minerals in the two shales is completely inde-pendent of local structure, faults, and rock permeability. Since the control on the deposi-tion of copper minerals in the shales is lithologic and stratigraphic, they argue that the copper minerals were dropped into, or precipitated within, the original mud. They say, however, that the copper minerals, however, may have replaced, with exquisite detail, something else that was so deposited. This last they conceive as having been accomplished by downward diffusion into the uppermost few inches of mud of copper-bearing solu-tions that attacked pyrite or calcium carbonate. The authors are convinced that the syngenetic method of formation is much the more probable than an epigenetic one and that the first variation of the syngenetic hypothesis is the more likely.

1957. Source beds

With the publication (1957) of Knight's paper on source beds (for ore materials), geologists in North America seem for the first time to have begun to consider that large numbers and varieties of ore deposits might have been actually or ultimately syngenetic in origin. Knight says: "the source bed concept postulated that all sulfide ore bodies of the majority of fields are derived from sulfides that were deposited syngenetically at one particular horizon of the sedimentary basin constituting the field, and that the sulfides subsequently migrated in varying (sic) degree under the influence of rise in temperature in the rock environment". He considers that the two most important causes of rise in temperature would be deep burial (resulting in strong metamorphism or granitization) and granite intrusion. If neither of these factors affected a field, the original stratigraphic control of the source bed would be clearly evident. Knight found that, in part 6 of the Report of the 18th International Geological Congress, the country rocks of 108 sulfide deposits were described and that 56 were said to be in limestone or dolomite and, for 7 more, carbonate rocks were reported in the immediate vicinity. This seemed to Knight to be far out of proportion to the percentage of carbonate rock in the earth's crust, and that lead-zinc deposits, at least, therefore were originally deposited in, or had never been disturbed from, such a sedimentary environment.

Knight's paper produced a flood of replies in the pages of *Economic Geology* over 3 years. In these discussions, Knight's original idea seems to have been rather completely buried under a mass of material extraneous to it but not extraneous to the concept of the syngenetic formation of ore deposits of many types. In those remarks particularly critical of Knight's concept, numerous deposits are cited as being of definitely epigenetic origin despite Knight's suggestions that they might have been derived from source beds. Perhaps surprisingly, perhaps not, many of these undoubtedly epigenetic deposits (in the opinion of some of the discussants of Knight's paper) are now considered by some to most geologists actually to have been formed syngenetically.

(I do not know of any deposit that Knight lists as having been derived from a source bed that today is considered to have been so formed. Many of them, however, are thought to have been deposited originally as part of the sedimentary column in which they are found. Thus, the following are listed by the discussants as undoubted examples of epigenetic ore deposition — Noranda, all deposits in the Bathurst—Newcastle area, and Mount Isa. Each one of these deposits today would be considered by a major fraction of geologists to have been formed syngenetically, though not with the sulfides having been derived from a source bed. Thus, Knight did not put forward a concept that was accepted by anyone, but it did cause many geologists to have second thoughts about the origin of stratiform ore bodies, and a considerable fraction of these to conclude that the ores were syngenetic, in the broad sense at least, and were not epigenetic at all. In short, Knight was a catalyst in the reaction that caused the development of a wide range of ideas concerning syngenetic ore deposition in North America, if not in the rest of the world.)

Mid 1950's. Uranium in ancient conglomerates

In the middle 1950's, interest in Canada in the possible sedimentary formation of ore deposits was brought to a high level by the discovery of the uranium—thorium deposits in the conglomerates of the Quirke Lake—Elliot Lake (Blind River) area, and by that of the massive sulfide deposits of the Bathurst—Newcastle area of New Brunswick. First, to consider the uranium deposits, *Joubin (1954)* published a paper on their geology but said little as to the manner in which they were emplaced. In *1956, Roscoe's paper* suggested two possible origins for the deposits. The first of these was a modified placer hypothesis, similar in a general way to that commonly held for the gold—uranium ores of the Rand. For the Blind River area, Roscoe conceived that the conglomerate originally contained, in addition to the quartz pebbles, detrial hematite, ilmenite, magnetite, rutile, titanite, epidote, monazite, zircon, and many other heavy minerals, which may have included brannerite and uraninite and pyrite. The present textures of the uranium—thorium minerals show that, if the materials of which they are composed were detrital, they must have been remobilized and recrystallized during diagenesis and/or metamorphism but with little change in bulk composition except for the sulfur introduced to convert the iron oxides to pyrite. Uranium—thorium minerals can have been brought into the conglomerates only if the distance of transport was short and the atmospheric conditions were such that the radioactive elements could not have been oxidized. Further, monazite, which is much more resistant to transport and oxidation than any possible detrital uranium minerals, has a urianum-thorium ratio of no more than 1/3. Only if, in the source areas of the detrital minerals, uranium minerals were more abundant and/or more resistant to transportation than the thorium minerals, placer deposits, with uranium more abundant than thorium, could have been formed. Roscoe finds this idea unreasonable and can only suggest that uranium must have been introduced by hydrothermal or ground-water solutions. He objects to a straight hydrothermal hypothesis because the preference of hydrothermal fluids for the conglomerates could be explained only by a greater permeability of the conglomerates over the remainder of the host formation, and he does not think it likely that the gravels were much more permeable than the adjacent sands, (but he does not, of course, know this to be true).

In *Roscoe and Steacy's 1958 paper*, Roscoe had changed his opinion and believed that "the hypothesis that both the thorium and uranium in the Blind River ores ... is of syngenetic origin (*sic*), appears to be more in accord with the geologic facts then the rival hypothesis that the uranium is of epigenetic origin". Their arguments for this changed viewpoint are:

(1) Sharp and large contrasts in uranium content cannot be correlated with large differences in permeability, mainly because rocks with finer-grained material may contain more uranium than those with coarser.

(2) Most of the pyrite was originally detrital iron oxides that have been altered by introduced sulfur. Many of the pyrite grains have rounded shapes and well many have replaced detrital grains. (Why this argues against a hydrothermal formation of the pyrite is not clear — for ground water seems far less likely to bring in sulfur in large amounts than could hydrothermal fluids.) These authors add half a dozen more arguments in favor of the pyrite having replaced iron oxides, (but these do nothing more to demonstrate anything about the uranium for there is no doubt but that the pyrite is secondary).

(3) The uranium is not in uraninite to more than a small fraction of the total; therefore the brannerite must be a secondary mineral. Of course, its parent may have been detrital uraninite, but it seems unlikely that all the detrital uraninite would have disappeared in the remobilization process. In short, the problem was not solved by Roscoe and Steacy's work.

Further work on the Blind River deposits will be reported later.

Late 1950's. Massive sulfide deposits

1958. Volcanic-exhalative ores

Although *Oftedahl's paper (1958)* was not published on this continent, it has had so much influence on thought about ore deposit formation in North America that it should be summarized briefly here. The author modifies the old concept of the volcanic-exhalative origin of ore deposits to relate it to the formation of welded tuffs and similar silicic pyroclastics. He believes that many ore deposits are precipitated on contact with sea water by volcanic gases derived from the magma chambers of crystallizing granites. He quotes as present day examples of this process the filling, during the 1817 activity of Vesuvius, of a 3-ft wide fissure with hematite resulting from the reaction between gaseous ferric chloride and water vapor, and Helmqvist's observation (1955) in the Eolian Islands of compact blocks of obsidian, lying beneath the water table, having been nearly completely replaced by pyrite and of vesicles and pores in this volcanic rock having been filled with pyrite and marcasite. He points to pyrometasomatic (contact-pneumatolytic) ore deposits as examples of the reactions of similar gases on the rocks adjacent to plutonic igneous bodies, saying that "ore geologists nearly unanimously agree that these deposits are formed by interaction of the bedrock and high-temperature magmatic gases driven off during the last phases of crystallization within the plutonic body". (He does not mention, however, that such gases are under such high confining pressures that the water molecules are so closely packed together that they act as a solvent in the same way as liquid water and are, therefore, quite different in character and composition from the tenuous gases given off as near-surface volcanic exhalations. In this connection, the replacement of obsidian, just mentioned, is carried out in a water-rich solution beneath the water-table and not by gases or vapors as such.) Oftedahl thinks that such deposits as those of pyrite and magnetite in the Caledonides of Norway and Sweden, the Rio Tinto type of deposits, the Precambrian Fosen-type magnetite ores in Norway and the magnetite ore of Central Sweden, and the Lahn-Dill ores of Germany all are the products of precipitation of volcanic exhalations on contact with sea water. He also suggests, that the lead and zinc in Mississippi Valley-type deposits and the copper in the stratabound copper deposits of the Mansfeld type was introduced as volcanic exhalations during the formation of the sediments that now contain the ores. The epigenetic characteristics they now show would be due, by his concept, to later meteoric waters or metamorphism.

Until 1959, no massive sulfide deposits in North America had been described as having been formed by syngenetic processes. In that year, Friedman reported on the pyrrhotite-pyrite deposit at Samried Lake, some 28 miles north-northeast of the town of Blind River in Ontario. This deposit differs markedly from those of the Sudbury district in that it is essentially barren of base metal mineralization, although chalcopyrite and cubanite are present in minor amounts. The author considers that the original material was magnetite, plus iron-hydroxide gel, that accumulated in a marine clastic-sediment environment, probably on the slope zone of a shallow sea in an area where volcanic activity was prominent. The major materials discharged from the volcanoes were tuffs and lavas. The iron hydroxide gels and perhaps some of the magnetite was converted to pyrite by hydrogen sulfide that permeated the clastic sediments. After the consolidation of the sediments and the volcanic material diabase was injected. A post-diabase metamorphic cycle developed temperatures of 420–600°C and converted a considerable fraction of the pyrite to pyrrhotite. When the sulfides were brought to or near the surface, some pyrrhotite was reconverted to pyrite and altered to marcasite. Thus, here is the first example of a deposit formed by syngenetic processes and later changed by metamorphism to be reported in North America. Although this paper was published after that of Oftedahl (1958), Friedman (1959) does not refer to it in his list of references, nor did it command the almost universal acceptance it does today; see Ridge (1974) for references to papers arguing against its validity.

1959 also saw a *symposium held in Canada (Stanton, 1959)* of the genesis of massive sulfide deposits in which the formation of massive, base-metal sulfide deposits was held to have been primarily a result of syngenesis. Had such deposits never undergone anything more than the metamorphism due to burial, without any effects of differential movement, such deposits would easily be recognized. Whatever the complexities introduced by such metamorphism, however, if no differential movement occurred, such deposits still would easily be identified. The complexities caused by metamorphism and tectonics, however, would have given such deposits an over-print that would, at least superficially, show evidence of epigenesis. This is particularly true if the sulfides were remobilized or redistributed more readily than the non-sulfide minerals with which they originally were associated. The result of such remobilization is not only to change the arrangements of one syngenetic mineral to another but also, in some instances at least, to produce new minerals that are of higher-temperatures of formation any normal syngenetic process could develop. Thus, the major factor that favors the formation of massive sulfide bodies by syngenesis is the stratabound character that many of them have. To many geologists, stratiform bodies are less likely to have been formed by ore fluids introduced into the rock after lithification than to have been initially incorporated in the primary sediments at the same time as these were laid down. Stanton considers the New Brunswick ores to be sulfide metamorphic rocks that show, instead of a sulfide paragenetic sequence, a sulfide crystalloblastic series. The sulfur in the ores is thought to have been produced from SO_4^{2-} ions brought in with the volcanic emanations that also carried the metals. The presence of volcanic rocks in the stratigraphic column in the area is thought to add to the argument in favor of a volcanic—exhalative method of formation. Incidentally, the term volcanic—exhalative was not used in Stanton's paper although the concept of such a method of formation is clearly expressed.

1960. Syngenesis versus epigenesis

In 1960, Bain discussed in detail the distribution patterns of ore deposits in layered rocks and considered that they most probably were syngenetic in formation if they correspond in area, symmetry, and metal variation with rock facies patterns. On the contrary, epigenetic deposits are controlled by antecedent and contemporary tectonic structures or deviate systematically under their influence. He agrees with the concept that many deposits have metallization, initially syngenetic, redistributed slightly by ground water of varied origin; such deposits acquire a tectonic symmetry while retaining their syngenetic facies limitation. The controversial classes, that Bain discussed, include the shale—siltstone class, the alum shales of Sweden, the Kupferschiefer, and the Zairean—Zambian deposits and the conglomerate class, the South African gold fields, the Blind River uranium deposits, and the uranium of the Colorado Plateau. Bain considers that the bulk of the metals in the alum shales was deposited syngenetically but has been redistributed slightly. In the Kupferschiefer he thinks that the metals were leached from the red beds by ground water and were redeposited in the shale. He thinks that the Zambian deposits are epigenetic, without specifying the source of the solutions, because of the closer relationship of the ores to a tectonic facies pattern rather than to a sedimentary one. He also considers the Zairean (Katanga) ores to be epigenetic, obviously he says, and to have formed in the middle intensity range of hydrothermal activity. Bain's work on the Witwatersrand ores was done before the importance of carbon in localizing gold was fully realized, so his comments are confined to the ores in conglomerates. He appears to believe that the ores were formed syngenetically, because additional ores can best be found by determining and following the alluvial pattern of the conglomerates. His conclusions on the Orange Free State (Welkom) field are much the same as those for the Rand proper. At Blind River, he believes that the radioactive material was syngenetic to the conglomerate but was redistributed by solutions moving through those rocks. In the Colorado Plateau uranium deposits, Bain finds that permeability is a barrier to ore deposition because the solutions move too freely through the rocks. Ore is present mainly in rocks in which the natural permeability has largely been closed by hydrocarbons of various types and different ages.

1960. Uranium in ancient conglomerates

Derry (1960) further considered the problems of the formation of the Blind River ores. Although Davidson's 1957 paper put forward the concept that the uranium and thorium in the Blind River deposits had been introduced hydrothermally, Friedman in discussion in 1958, said that locally the uranium—thorium ratio of 1/13 in the Blind River ores fell within Davidson's value for this ratio in alluvial deposits, and provided some evidence for their alluvial formation. Derry's paper, however, makes the first major case for the placer formation of the uranium—thorium deposists of Blind River. His points are:

(1) The source of the clastic material of the deposits was from Archean basement to the northwest and the detrital minerals certainly include zircon and "normal" monazite.

(2) The conglomerates and associated quartzite were deposited as gravel and sand in broad river valleys or deltas near a shoreline that was advancing to the north, so younger beds overlapped older ones.

(3) Mineable ore is confined to certain quartz pebble conglomerate beds, none of which is more than 150 ft above the base of the formation. The best grades and widths of ore are where the conglomerate is thickest and pebble packing is at a maximum.

(4) Hydrothermal alteration has affected the Mississagi sediments and, in places the ore horizons, but Derry sees no evidence that these solutions added or removed any measurable amount from any uranium-bearing horizon.

To explain these conditions, Derry suggests that very little in the way of uranium-mineral grains was deposited with the detrital quartz but that appreciable uranium and thorium were brought into the area by ground water after the conglomerates had been lightly buried under sand or clay. This radioactive-element-bearing ground water travelled through underground channels in which he assumes that conditions were favorable to reducing the uranium and thorium from the plus-six to the plus-four state and to depositing them in rather unstable minerals as the forerunners of brannerite. Derry designates these as gummite, but, since this is a mixture of secondary uranium oxides, it seems unlikely that gummite was the parent of the brannerite now in the deposits. The conditions that reduced the uranium and thorium Derry considers probably provide the S_2^{2-} to convert the iron oxides to pyrite. After deep burial, whatever older uranium minerals were present would have been converted to uraninite, to uranothorite that replaced parts of the detrital monazite, or to brannerite, this last requiring the participation of titaniferous minerals (as Ramdohr suggested in what he called the "Pronto" reaction). Age determinations on the brannerite and uraninite give 1700–2000 m.y., whereas the monazite has an age of about 2500 m.y.

Derry believes that these changes could not have been caused by hydrothermal solutions, because the ages of the brannerite do not correspond to Huronian igneous activity. Further, the brannerite is as common on the north side of the Elliot syncline and farthest from the known Huronian igneous activity as it is at Pronto, which is near to the area of Huronian igneous rocks. Derry appears to believe that solutions that albitized quartzite and conglomerate at Pronto were not the same ones that formed the brannerite but did redissolve and redeposit some uranium. From this, he concludes that most of the uranium minerals and what he calls the granular pyrite were in their present form before the solutions arrived that caused albitization, chloritization, and the deposition of what he designates as later pyrite. Thus, Derry's concept seems to be that the uranium was almost contemporaneous, in its first form, with the formation of the conglomerates and that the changes that later affected it were primarily the result of burial and had been completed before hydrothermal solutions entered the uranium-bearing rock volumes. (It does not seem difficult to vary this scenario so that no uranium was present before hydrothermal fluids brought it in. In the conglomerates, this introduced uranium could have reacted with ilmenite to produce brannerite, and S_2^{2-}, also from the ore fluids, could have produced much pyrite, also from the iron oxides. Nevertheless, few geologists, familiar with the deposits, would consider the uranium to have been introduced hydrothermally.)

1960. Sulfur-isotope ratios

In *1960, Dechow* determined the sulfur-isotope ratios for over 150 sulfide and sulfate specimens from five ore bodies and other rocks in the Heath Steele area of New Brunswick. His results show that the ratios are close together, ranging from 21.82 to 22.02, as compared with the $^{32}S/^{34}S$ ratio of 22.22 in the Canyon Diablo meteorite. This indicates to Dechow that there has been no detectable fractionation either during hypogene mineralization or supergene enrichment (if any). He also believes that the enrichment in ^{34}S relative to the standard suggests that the sulfur in the metallic sulfides came from sulfates buried in an original source bed. He notes the presence of graphite in the ores and thinks that this adds evidence in favor of a sulfate-reduction formation for the sulfur. He thinks that the sulfur reduction took place in the presence of organic carbon at temperatures in excess of 600°C. He calculated that an isotopic exchange reaction between sulfide and sulfate under equilibrium conditions and the spread of ratios suggests that the source temperature was 700–800°C. Thus, he believes that the present deposits were formed hydrothermally but that the ore fluids were produced by the granitization of a source bed, reduction of sulfates, and mobilization of the resulting sulfides to form ore deposits at favorable loci. (This hypothesis is, of course, a modification of Knight's (1957) source bed concept that seems even farther removed from reality than that idea.)

The next paper, that of *Gavelin et al. (1960)* discusses the fractionation of sulfur isotopes in sulfide mineralization and reaches a conclusion that in general agrees with Dechow's that no detectable fractionation occurs during hypogene mineralization. Although the authors are not Americans, the paper appeared in Economic Geology and had a considerable influence on American thought. They said that various sulfides produced by a continuous mineralization process show no significant isotopic ratio differences when early and late minerals are compared. Neither did they find any correlation between isotopic composition and zonal arrangement of minerals and metals around centers from which the ore fluids moved outward. They believe, however, that a certain correlation between kind of wall-rock alteration and isotopic composition may be indicated. If a variation is found in a district, the range of variation appears to be wider in sulfides deposited at high temperatures than those emplaced at low. They found that supergene oxidation of sulfides to sulfates takes place without any change in isotopic composition but that, where supergene sulfides are formed from such sulfates, the isotopic ratios are displaced toward lighter sulfur. As could hardly be otherwise, co-formed hypogene sulfides and sulfates have significantly different isotopic ratios, with the sulfates containing an appreciable higher proportion of ^{34}S. The difference in ratio also is greater between sulfides and sulfates deposited at low temperatures than between those formed at higher temperatures. They also found that sulfur-isotope fractionation, particularly that happening at low temperatures, may be very local and may differ considerably from one site to another quite close by. Despite the positive statements proceeding the last one, the authors consider that, in an ore district where appreciable differences in isotopic proportions are established, the fractionation of sulfur appears to be the cause of these differences. (This hardly seems to conform to the authors' statement that sulfur isotope fractionation in hypogene deposition is unimportant where early and late minerals are compared.) They believe that a few determinations of isotope ratios of sulfur in sulfide minerals cannot give any definite information on the source of the sulfur. They add that, if it is possible for isotope fractionation to take place at higher temperatures of hypogene mineralization, variations of isotope composition in a given ore district provide no definite evidence that the sulfur in the deposit resulted from the mixture of sulfur from various sources.

In 1960, Kendall stated that the sphalerite in the Jefferson City deposits was penecontemporaneous with the deposition of the sediments, even though the sulfides were deposited in their present loci as replacements and open-space fillings in coarse dolomites, and as fracture fillings in fine dolomite. Kendall is particularly certain that a laminated clastic dolomite that was deposited in vugs originally was flat-lying; the dips that these clastic pockets now show are considered to have resulted from post-ore folding. Kendall believes that these small clastic bodies, the marked restriction of the ore to certain definite stratigraphic horizons, and its intimate association with sedimentary structures indicates that the ore was deposited essentially at the same time as the sediments, and that its present position in the rocks was the result of mobilization after the fractures and brecciated rock volumes had been developed. This was the first American attempt in any detail to demonstrate a sedimentary—remobilization origin for an ore deposit of the Mississippi Valley-type.

Ridge (1960) suggested that the appreciable variation of $^{32}S/^{34}S$ ratios in the minerals of sulfide ore deposits of magmatic hydrothermal origin may be due to processes operating within the ore fluids. Such process, he thought, include:

(1) Polymerization of metal-sulfur complexes, with the screening sulfur ions that are removed from such complexes being of higher $^{32}S/^{34}S$ ratios that the ratio in the complex prior to the polymerization stage in question.

(2) Oxidation of sulfur concomitant with the reduction of certain metal ions (such as Cu^{2+}, As^{5+}, Sb^{5+}, and Sn^{4+}) to a lower valance state, with the sulfur ions oxidized being of a lower $^{32}S/^{43}S$ ratio than the ratio of the complex prior to the oxidation in question.

(3) Probable preferential precipitation at higher temperatures of sulfide with higher $^{32}S/^{34}S$ ratios than the original averages for the ore solutions involved and at lower temperatures of sulfides with lower $^{32}S/^{34}S$ ratios than the original averages of the ore solutions concerned (this is analogous to the hydrothermal Leadville dolomite in which the $^{18}O/^{16}O$ ratios are lower, the higher the temperature of carbonate emplacement).

(4) The taking over of sulfur deposited at higher temperatures by sulfides deposited by replacement at lower temperatures (or occasionally the reverse) so that the $^{32}S/^{34}S$ ratios resulting in the guest minerals will be anomalously high (or low).

Ridge gave examples in his oral presentation of this abstract to show how possible combinations of these factors can explain many of the wide variations from the norm of 22.13 known in natural ores, particularly in low-temperature deposits. He also discussed the correspondence in isotopic behavior of oxygen in silicon—oxygen complexes in silicate minerals and of metal—sulfur complexes in hydrothermal solutions. This paper was not well received and never was published, but now, nearly 15 years later, some workers are reconsidering the concepts put forward in 1960.

1960. Uranium in ancient conglomerates

In 1960, Robertson and Steenland argued that the uranium minerals of Blind River, brannerite, uraninite, and a "monazite" complex are in detrital assemblages and are themselves detrital; this is contrary to previous opinions on the formation of the Blind River ores. They do, however, think that the radioactive minerals may have been modified or that some such material was introduced at about 1300 m.y. and 600 m.y. They consider the conglomerates themselves to be older than 1200 m.y. and probably older than 1700 m.y. They believe that there is no connection between the mafic dikes, non-detrital sulfides, and quartz veins, all of which are younger than the uranium-bearing minerals. With the exception of one minor mineralization type, these materials never are radioactive. They quote, with approval, Stieff and Stern's statement that the old radiogenic lead in the deposit may be from detrital minerals or pre-existing ore deposits in the basement complex. They believe the uranium minerals to be detrital because:

(1) They normally have a detrital aspect.

(2) They occur with minerals that usually are detrital and occur in the same way as these probably detrital minerals.

(3) The uraninite, with its coarsely crystalline character and high thorium content, seems to belong to pegmatitic and granitic uranium mineralization rather than to have come from hydrothermal veins where, they appear to assume, such uraninite cannot have come. They also think that the "monazite" complex (by others described as formed by replacement) and the brannerite (described as having come from reactions between uraninite and ilmenite or ilmenite and uranium-bearing solutions) probably were formed in a granite rather than by hydrothermal fluids.

(4) They report that no hydrothermal alteration effects, such as might be produced concomitantly with the hydrothermal introduction of uranium minerals, are present.

(5) No connection exists between uranium minerals and the sulfides, quartz veins, and mafic dikes.

(6) The ratios between ThO_2 and U_3O_8 differ in a way explicable by syngenetic formation but not by a hydrothermal one.

(7) They think it an unacceptable coincidence that uranium-bearing solutions should have sought out the ore-bearing conglomerate and to have avoided the numerous other conglomerates in the district. (All of these reasons are opinions and not sound evidence.)

1960. Syngenesis versus epigenesis

In 1960, Sales criticized papers by *Knight (1957) and Gray (1959),* the latter not published in North America, and pointed out that a hydrothermal origin is indicated for the Northern Rhodesian (now Zambian) Copperbelt deposits by the presence of post-Roan Age igneous rocks (the Roan rocks being the host rocks for the ores), and post-gabbro mineralization, the wide distribution of the ore in the various lower Roan lithologic types, and the zoning of copper-iron sulfide minerals and of various minor elements. He objects to Gray's concept of metamorphic waters having dissolved, transported, and re-deposited in the ores found in sedimentary rocks because the United States, Switzerland, and other areas have range after range of mountains that are made up of great thicknesses of folded, faulted, and squeezed sedimentary rocks of many types and ages yet fail to contain ore. He concludes that the absence of ores in such environments, under Gray's theory, must be because:

(1) The rocks involved in the mountain building did not, as original sediments, contain enough of the ore metals.

(2) The circulating metamorphic waters were present in insufficient quantity to affect ore-metal transportation.

(3) The water present was of the wrong chemical character to dissolve the ore metals.

He believes that high-temperature solutions of magmatic origin are far more reasona-ble as dissolvers, transporters, and depositors of ores than metamorphic waters that probably were neither as hot nor had as ready access to ore-forming metals as did the hydrothermal fluids in the residually molten portions of a magma at considerable to great depth. He emphasizes the often made point that the lack of igneous rocks, particularly in association with low-temperature deposits is no reason to deny (nor, of course, does it prove) the existence of magmatic sources of ore fluids beneath the present zone of observation.

Garlick's (1964) discussion of Sales' paper states that the principal evidence for the syngenetic theory is:

(1) The very even fine-grained dissemination of copper sulfides and pyrite is restricted to particular sedimentary horizons for distances measured in miles.

(2) The primary mineralization is pre-folding as the grade of the primary ore is inde-pendent of the drag folds.

(3) The distribution of copper–iron sulfides shows a direct relationship to cross-bedding, ripple marks, and pre-consolidation slump structures.

(4) The zonal arrangement of copper and iron sulfides, from chalcocite through born-ite and chalcopyrite to pyrite is invariably in sequence from shallow-water to deeper-water sediments and the zones of sulfides parallel the shore lines of the host formations.

(5) Evidence (unspecified) that the sulfide-bearing veins are due to lateral secretion from originally disseminated sulfides.

He further says that the syngenetic concept would have been impelled by this evidence even if there had been granites in the district younger than the Roan sediments. Garlick (1964) believes that many of what he refers to as Sales' so-called facts actually are misrepresentations of the data. He thinks that Sales (1960) is incorrect in believing that the syngenetic sulfide material was deposited in the Roan shale bed only. Garlick (1964) is of the opinion that the copper sulfides in graywacke, quartzite, and arkose are so con-sistently concentrated on foreset laminae, on bottomsets of crossbedding, and in hollows of ripples that they must have been introduced there syngenetically, (apparently as detrital particles or they would not have been concentrated in these situations). Garlick (1964) also claims that, in places where the folding in the ore shale was intense, the beds are arranged in alternating normal and overturned folds. Where such overturning occurs, he finds that the zonal arrangement of sulfides also is overturned so that the pyrite zone always remains adjacent to the (stratigraphic) Footwall conglomerate. From these facts, Garlick (1964) deduces that the stratigraphic zonal sequence was firmly established before the folding of the Roan ore shale. (The emplacement of most of the copper sulfides by replacement or open-space filling dictates that they were, at the earliest, diagenetic. Their presence, therefore, in the sedimentary structures just mentioned would be due to factors other than their physical response to running water. In short, Sales' data may not all be correct, but the opposing viewpoint does not meet all the problems either. Much more work is needed even in addition to that reported in Mendelsohn's volume.)

1960. Massive sulfide deposits

Stanton (1960) considers that conformable pyrite ore bodies should be thought of as an integral part of the geologic environment in which they occur. They are a normal, though minor, type of sediment (and sedimentary rock) and are produced in near-shore and shelf facies of intravolcanic units formed along the margins of eugeosynclines. He does not believe that this type of ore body ever is produced by replacement as an overprint on rocks to which they are genetically unrelated. Where such massive ores are contained in folded and fractured rocks, such epigenetic features as they contain are attributed to mechanical migration during orogenesis, the epigenetic structures being directly related to the response of sulfides to earth forces. Granted the validity of these premises, Stanton believes that prospecting such be based on paleogeography, with the delineation of sedimentary (and volcanic?) facies to be followed by the determination of geologic structures.

Stanton recognizes two main types of "pyritic" ore: (1) banded pyrite—sphalerite—galena—chalcopyrite ore; and (2) non-banded ore made up mainly of pyrrhotite and chalcopyrite, but containing appreciable pyrite and more magnetite than type 1. He believes that these two types are the results of the effects of differently oxygenated environments that prevailed during sedimentation, rather than to differences in source material and in pressure—temperature conditions of deposition. (How such differences account for the absence of sphalerite and galena in the ores of type 2 is not explained.) The textures in the banded ores he thinks were caused by slow, low-temperature segregation in situ during compaction, folding, and regional metamorphism rather than to replacement or high-temperature exsolution. Stanton does not think that deformation textures always can be identified in these ores. He believes the deposits lacking deformation textures were, at the time of deformation, cryptocrystalline mixtures in varied concentrations in fine sediment. Thus, the minerals in such ores were formed, not deformed, by tectonic activity. (As Huckleberry Finn remarked about Pilgrims Progress, these statements are interesting but steep.)

1961–1963. Sulfate-reducing bacteria[1]

Baas Becking and Moore (1961) produced a number of simple sulfides by biological methods. They used, in so far as bacteria were used, the bacterium *Desulphovibrio desulfuricans* (or *Clostridium desulfuricans*), which is anaerobic and sulfate reducing. It cannot survive at pH's below 4.2 but has been found at pH's well over 10. The experimental work was carried out under unnatural conditions in that lactate or acetate were the necessary hydrogen donors. Under conditions in which copper is present in any abundance as copper sulfate, these bacterial actions do not proceed, since copper sulfate is highly toxic to life. Using organic media and combinations of cuprous oxide to which they added either lepidocrocite or hematite, they never were able to form chalcopyrite or bornite and only obtained covellite (perhaps because of Eh–pH constraints in their experiments). These experiments, repeated under a variety of conditions, always had the same negative results so far as the iron–copper sulfides were concerned. They report that bacteria do not form chalcocite, although they do form digenite when cuprous oxide is subjected to sulfate reduction. Rather strangely, when a silver sulfide was produced at temperatures below 50°C, it was as argentite and not as acanthite. They point out that any geologist advocating the syngenetic production of metal sulfides does not have, as yet, enough experimental chemical evidence to justify his claims and that, with the heavy metals being as dissipated in the oceans as they are, concentration of them in sufficient quantity to produce an ore deposit is difficult to explain. This would appear to apply to the direct precipitation of these sulfides or to their diagenetically replacing syngenetic pyrite. They appear to be on strong ground, however, in regard to syngenetic pyrite, since they report that enormous quantities of ferrous sulfides occur in the black muds of estuaries and rather similar stagnant bodies of water. Almost certainly the S_2^{2-} needed for this iron sulfide was provided by the reduction of the S^{6+} of sulfate apparently originally to a form of troilite so that the original reduction was to S^{2-}. (The production of the S_2^{2-} of pyrite requires that the original S^{2-} must have been oxidized to S_2^{2-}. Such an oxidation would be impossible for the bacteria involved to carry out – certainly the iron sulfide in consolidated sediments is pyrite or marcasite and not troilite – so some other mechanism than that provided by the bacteria must be appealed to.)

Love and Zimmerman (1961) believe that the fine-grained pyrite phase (pyrite-I), that makes up about two-thirds of the sulfides in the Mount Isa area, is syngenetic–diagenetic in origin and that, therefore, over 60% of all sulfides in the district have a syngenetic–diagenetic origin. The outcrop length of the Mount Isa shale is about 30 miles, and pyrite is known to extend over 20 of those 30 miles and is in roughly the same stratigraphic horizon. This pyrite-I phase, they believe, was formed by the infilling of an hitherto unknown type of micro-organism, the relic form of which is directly related to the zonal

---

[1] Editor's note: see in Vol. 2 the chapters by Trudinger and by Saxby, on bacteria and organic matter, respectively.

structure found in the pyrite grains. They envision the sulfide needed to precipitate the pyrite as being obtained from H_2S produced by the organisms and the iron (and the ore metals) from solutions introduced into the sea by streams or by hydrothermal solutions or volcanic-exhalations. They also appear to believe (quoting Baas Becking and Moore (1961)) that the same mechanism (during sedimentation or no later than diagenesis) produced the other sulfides not much after the pyrite-I. The present paragenetic relationships among the various sulfides are thought to have resulted from metamorphism after burial, the mineralized zone then being classed as a "closed" hydrothermal cell. The mechanisms are thought to be those proposed by Gill (21st Int. Geol. Congr., pt. 16, p. 209–217) who showed that diffusion occurs in copper sulfides at temperatures above 400°C at rates sufficiently fast to be of interest as a possible mechanism for the modification of the primary minerals and mineral relationships in copper ore deposits. Gill says that the diffusion rates among FeS, PbS, ZnS, and possibly other sulfides, are less striking but that, nevertheless, sulfide diffusion well may be an important factor in the formation of ores without hydrothermal solutions; (just where the raw materials to be diffused come from is not stated). Gill's mechanism does not seem to be exactly a "closed" hydrothermal system since he talks of doing without hydrothermal solutions altogether, but Love and Zimmerman apparently think that diffusion would be important in their hydrothermal cell and that the textures, structures, and minerals (some of which would not have been primary-syngenetic) produced by these reactions would resemble closely those formed in "normal" hydrothermal deposits. Thus, a deposit of hydrothermal appearance could be produced without a deep-seated magma or channelways between the magma and the ore body. The authors suggest that, in addition to the ways already mentioned, the metals in the syngenetic deposit might have come from the direct weathering of an ore body exposed at the surface. (This process might be satisfactory for a considerable distance back into geological time, but eventually there would have to be the Ur-deposit — in the German sense — that had not been formed in this manner and must have been formed by "normal" hydrothermal means. If the Ur-deposit could be formed in this manner, so could many more in a long time sequence.)

Cheney and Jensen (1962) argue in refutation of Davidson's 1962 paper (that follows), that the low solubility of copper in sea water guarantees that the copper ions (mainly cupric) in that medium never would be in large enough concentration to kill off such sulfate-reducing bacteria as might be present. Instead, if the concentration of cupric ions rose above 3 ppm, the excess would be precipitated as copper sulfide (CuS) and the cupric ion-concentration would not increase. Instead, the copper dissolved from the rocks surrounding the copper-bearing terrain would be entrapped in clays and organic compounds. Particles of these materials would be transported seaward and deposited in the finest fraction sea-basin sediment, at maximum distances from the shore line. With this material would be included "amorphous, slimy, or colloidal sulfides". The presence of these particles in the sediment would provide a source of continuous replacement for the copper ion-concentration (3 ppm) that was deposited syngenetically or diagenetically with the iron-sulfide-containing slimes. Once these iron sulfides had been converted (diagenetically?) to pyrite, they would be replaced by copper ions to form a series chalcocite–bornite–chalcopyrite–pyrite from the shore side of the basin toward its center. These authors justify this statement on the assumption that less and less copper would be available as the area of deposition was farther and farther from the shore line. (The flaw in this reasoning is that the copper in chalcocite is cuprous copper, whereas that transported in the sediment particles is cupric. This would require a reduction of the cupric copper to cuprous for the deposition of chalcocite, then its partial reoxidation to cupric to permit the formation of bornite, and then a final and complete oxidation to cupric ion so chalcopyrite could be formed. It seems to me more probable that the greatest reduction of copper should be in the areas farthest from shore, so that the process recommended by Cheney and Jensen should result in a reverse of the zoning found in the Zambian deposits, for example.)

Davidson (1962) points out that paragenesis of the iron minerals in silts is the result of diagenetic processes in them; content of hydrogen sulfide in the bottom waters was not the determining factor. Further, the higher the ferric iron content of the sediments, normally the lower will be the pyrite content; this condition simply reflects the need for ferrous iron in pyrite — pyrite cannot form unless the ferric iron in the sediments is reduced to the ferrous state. (This argument, however, does not recognize the need to oxidize the sulfide ion of hydrous troilite to the S_2^{2-} of pyrite.) He deduces that, in modern bottom deposits, the material initially present are colloidal iron compounds and organic carbonaceous–sulfurous matter, and the iron and sulfur in these materials are diagenetically converted to pyrite by bacterial or other agencies. Further, since it appears that ground water is saturated with a very small amount of copper, less than 0.01 ppm, the hypothesis, that direct drainage from a cupriferous province would greatly augment the soluble copper content of the adjacent ocean, almost certainly is false. Thus, he finds it difficult to explain how copper can be enriched in, and dispersed throughout, sulfuretted bottom water in sufficient amounts to produce: (1) ore-shale deposits of wide lateral extent when the solubility of copper is known to be extremely small; (2) abnormal concentrations of copper notoriously are lethal to bacteria; and (3) it does not seem possible to obtain, by deposition from sea water, the marked lateral and vertical zoning of the various ore metals in such deposits. (It can be argued that the important factor is the amount of water available; if much water with little but constantly replenished copper is at hand, H_2S in sufficient amounts can precipitate a lot of copper sulfides.) Davidson considers that the replacement of the cartilaginous parts of fishes, of cell structures in fossil wood (which normally would decay before lithification), and of pyritized microorganisms (as at Mount Isa), by sulfide ore minerals probably was a two-stage process — first the organic materials were replaced diagenetically by pyrite, and much later the pyrite was replaced by the ore sulfides, derived from solutions not related to sedimentation or diagenesis. The author points out that, in hundreds of chemical analyses of modern deposits in the Black Sea, of present-day marine sediments, of black shales throughout the stratigraphic succession, and of a great range of marine sedimentary rocks nickel occurs consistently in excess of cobalt, often by a large amount. In contrast, the Zambian and Katangan Copper belt is the largest cobalt field in the world and is essentially devoid of nickel. He quotes Mendelsohn as saying that the cobalt/nickel ratio of the Copper belt pyrite falls within the range commonly accepted for hydrothermal pyrite and outside that accepted for the syngenetic variety. Similarly, the high uranium/thorium ratio is a barrier to believing that the radioactive minerals in the Witwatersrand were deposited in placers. Davidson, however, goes on to say that the source of the ore metals in these stratabound deposits in shales (or other rocks for that matter) is the result of the introduction of ore elements on a broad scale from igneous sources and their later concentration into ore deposits was caused by a variety of lateral secretion. This lateral secretion is caused by a regional redistribution of the "relatively" soluble ore elements in response to orogenic events, such as regional uplift and folding, accumulating the ore elements either in: (1) more permeable horizons such as the Witwaterstrand conglomerate; or (2) chemically receptive pyritic or carbonaceous strata.

In *1962, Love* considered the problem of biogenic primary sulfides in Permian Kupferschiefer and Marl Slate. He says that two theories have been put forward by earlier writers to explain where the metals come from in these rocks, granted they are syngenetically deposited. The first of these is, that the metals were leached from pre-existing humus soils or crusts of dried-up playa deposits. The second is the direct weathering of deposits of ore or of eruptive rocks; the metal-rich solutions so produced may have accumulated as deep ground waters, before rising again, in a period of greater rainfall, to enrich the sea of Kupferschiefer time. (Neither of these suggestion has much merit, principally because it is difficult to imagine how these sources would supply metal to the Kupferschiefer seas only during the short period of time over which the thin, metal-bearing beds would have accumulated. Further, these concepts must be augmented by explaining how the syngenetic pyrite is formed — it does not appear to have been precipitated directly as such — and how it was converted to the copper, lead, and zinc minerals present in the deposits.) Love quotes data to show that the concentrations of free ions of iron, copper, and other metals in true solution are very small. Most of the elements brought into the sea are in non-ionic association with organic and inorganic colloidal material, the latter usually being clay or iron oxide. He also says that, the presence of ferrous sulfide in normal quantities reduces copper-iron concentration to indetectability. He does believe, however, that copper ions could be transferred from the non-ionic condition to replace iron in ferrous sulfide. (Again, the problem of turning this mechanism on and off seems to be insoluble.) He believes that the presence of copper minerals in certain parts of fishes in the Kupferschiefer-type rocks indicates that the copper must have been added early-on in the sedimentation—diagenesis phase. (Davidson, 1962, argues against this necessity by pointing out that the copper mineral may have replaced syngenetic pyrite at an early stage.) At the very end of the paper, Love indicates the possibility of the ore fluids having been added as hydrothermal exhalations, à la Oftedahl, to the sea.

This concept seems far more reasonable as a syngenetic explanation of the ores of the Kupferschiefer-type than do the complex processes suggested to provide the ore materials from the surrounding land surface.

Sales (1962) summarizes nine assumptions, two of which are combined here, that are held by certain geologists who consider the Copperbelt ores to be syngenetic in origin:

(1) There is no post-Roan granite (or other obvious igneous source) from which hydrothermal solutions could have come — post-Roan gabbros are not thought to be a possible source since they show no consistent relationship to the copper deposits.

(2) Formerly, though not now, syngeneticists contended that the ores were confined to a single stratigraphic horizon in the Lower Roan beds.

(3) The impermeability of the beds containing the Copperbelt ores and the absence of through-going fissures into the basement rocks drastically limit at best the possibility of the ores having been brought in by ore fluids of magmatic or, for that matter, any origin.

(4) The ore bodies and the zoning in them of the copper-iron and copper sulfides parallel an ancient shore line.

(5) The occurrence of disseminated copper sulfides in the Roan beds indicates a genetic connection between sedimentation and ore-mineral deposition.

(6) The uniformity of copper assays in samples from the crests, limbs, and troughs of drag folds at Nkana is proof that the sulfides were present before folding took place.

(7) The sulfides are associated with, and their location controlled by, sedimentary structures.

(8) The ore minerals were deposited before the rocks were deformed or metamorphosed.

As to (1), granite may exist at depths, as the kersantite dikes at Nkana suggest, or the gabbro may have been the source of the ore fluids since, south from the Mufulira ore body as far as the Bwana Mkubwa mine, there is no ore in the favorable beds and gabbro is absent as well. As to (2), it is now recognized that Copperbelt ores are in a variety of host rocks, only one-sixth of the total copper being in the formation (Lower Banded shale) most favorable for the deposition of syngenetic copper. As to (3), no evidence is known to Sales to show that the beds are impermeable to ore solutions moving laterally through them. Sales thinks that the necessary channelways for fluid movement existed at the time of ore formation.

As to (4), Sales says that Garlick (1964) is the only author in "Geology of the Northern Rhodesian Copperbelt" to make this claim, and Sales says that Garlick's pattern that relates ore zoning to shore lines is not borne out by the facts. As to (5), that copper sulfides occur in sedimentary beds means nothing as to their origin — they could have been deposited in other ways than syngenetically and still occupy the positions they do today. As to (6), Garlick's position eliminates the possibility of sulfide mobilization during the orogenic period yet says that bornite—quartz stringers cutting through 50-ft. gabbro dikes were produced by remobilization of the original disseminated sulfides. As to (7), similarly to (5), a control of ore location by sedimentary structures means nothing as to formation; depostion may be so controlled during the sedimentary process or certain of the structures produced during the sedimentary cycle may have been more conducive to ore deposition than others. As to (8), the amount of metamorphism of the Roan beds is minor, and such evidence as veinlets cutting gabbro dikes may mean that pre-metamorphism ores were remobilized, but at least equally well it may mean that the ores are all post-metamorphism.

Gillson (1963) points out that the problem in the Copperbelt deposits is not the availability of copper in the rocks that surrounded the basin during its formation, since each cubic mile of eroded rock would have yielded about 1.4 million tons. The problem not solved by the syngenetists is how 500 million tons of copper (the amount in the deposits before erosion as estimated by Davidson (1965)) from such a diffuse source were concentrated in a quite thin layer of sediments and how, at the same time, the immeasurably larger amount of rock debris produced by the erosion that freed the copper was kept from almost infinitely diluting the copper when it was deposited in the sea. Gillson says that, if the ore was prefolding, there should be much evidence of this in: (1) ore beds displaced by post-mineral faulting; (2) deformable minerals such as chalcocite highly stretched; and (3) brittle minerals such as pyrite, carrollite, and linnaeite crushed and broken. He quotes Jordaan as saying that, although the beds are displaced by a zone of faulting, the assay hanging wall and footwall of the ore body are undisturbed. Gillson considers this relationship as positive evidence that the ore was introduced after faulting. The author believes that the thickening of the ore in the keels of the folds is at least as well explained by its having replaced favorable beds that were thickened and thinned in the folding as by their being primary or diagenetic deposits. This he considers further confirmed by Mendelsohn's statement that the ore shoots pass impartially, apart from attenuation and thickening, over folds and to considerable depths and that, although locally ore shoots or mineral zones parallel folds, this is exceptional. Gillson also points to the large-scale feldspathization and sericitization described by Darnley as indicating the large-scale introduction of material and that the material introduced included the ore elements as well as the wall-rock alteration additions. Gillson cites numerous examples to show that belief in syngenesis is not universal among Copperbelt geologists.

Hirst and Dunham (1963), from their studies of the Marl slate, concluded that the pyrite was formed early in the diagenetic stage, that it is mainly spheroidal, and that it is enclosed by bitumen. They consider that the average levels of Mo, Co, and Ni suggest slow sedimentation, whereas the antipathetic variation of these elements with the rate of deposition indicates that the ions of these elements are adsorbed on organic carbon and/or clay. They consider that the normal erosion of the adjacent land surface is sufficient to account for quantities of these elements in the slates, as well as for those of Zr, Rb, Sr, and Mn. The Sr and Mn are in the carbonates; the Rb and Zr are detrital. They consider it possible that a gradual change of source might account for the relatively higher Sr and Rb contents in the lower part of the Marl slate and for the increase of Mo, Co, and Ni toward the top. A fluctuating source, not the adjacent land surface, is required to provide the Pb, Zn, and Ba. They suggest three ways that might provide this type of source:

(1) Temporary exposure of highly mineralized outcrops. This might be applicable in Durham if that source were outcrops of the northern Pennines Pb—Zn—F—Ba deposits. They believe, however, that no possibility exists that the ore bodies exposed during the geologically brief period of weathering during Marl slate times could have provided the required amounts of the metals in question.

(2) The metals might have been introduced long after consolidation by hydrothermal solutions. The textural evidence, they think, is not against this, but the extremely persistent nature of the anomalously high Pb, Zn, and Cu in the slate is.

(3) Submarine springs introduced the metals intermittantly into the lagoon where the parent muds and silts of the Marl slate were being deposited. The well-mineralized veins in the Carboniferous rocks beneath the Permian of Durham do not extend upward into the Marl slate or into the Magnesian limestone that overlies it, but the fluids that deposited the Carboniferous ores may have continued upward into the Marl-slate lagoons. There the metals would have been fixed as sulfides where and when bottom conditions were stagnant. When the sedimentation changed to purer limestone of dolomite, the metals would have been widely dispersed.

They believe that the third hypothesis offers the least difficulty in its acceptance.

1963. Diplogenesis

Lovering (1963) attempts to remove some of the barriers (semantic in this case) to communication between syngenetists and epigenetists. He suggests that the term "diplogenetic" be used for ore deposits that are in part syngenetic and in part epigenetic. Such deposits would be, he thinks, ones in which certain anions (or cations) were introduced from outside the rock volume that now contains the ore and certain cations (or anions) were present in that rock volume since its original deposition, (deposition apparently including the diagenetic stage). He suggests that, in certain black cupriferous shales of the Mansfeld-type, the sulfur was deposited syngenetically (in pyrite) and the copper (and other ore metal cations) introduced later and epigenetically; thus, such deposits are diplogenetic in Lovering's use of the term. He also introduces the term lithogene as of the same magnitude as magmatic and syngenetic and as meaning the process of mobilizing ore elements from the solid rock and transporting and redepositing them elsewhere (i.e., lateral secretion). Such lithogene deposits could, he believes, be derived from syngenetic, diplogenetic, or epigenetic deposits through the action of metamorphic, hydrothermal, supergene, or other solutions. To express the distance over which the materials of a lithogene deposit had been moved, it could be described as a locally derived lithogene deposit or as a regionally derived lithogene deposit. He summarizes the characteristics of diplogenetic deposits as having:

(1) The space relationships characteristic of epigenetic deposits, but a composition

that indicates that an appreciable portion of the ore components were derived syngenetically, even though they have been incorporated into new mineral species.

(2) Certain constituents present in both the ore and the wall rock in similar concentrations (though how they become ore in this manner is not readily apparent).

(3) A wide range of isotopic ratios in associated sulfides. The characteristics of lithogene ore deposits Lovering describes as being: (1) composed of minerals later than the enclosing rock; (2) facies controlled; and (3) abnormally high in the chemical constituents characteristic of the associated facies. In lithogene deposits, it should be possible to demonstrate that the amount of ore and gangue in the mineral deposit is comparable to the amounts of these elements lost from the wall rock. As examples of diplogenetic deposits, Lovering appears to include Mount Isa (much of the sulfur in the ore minerals seemingly was originally in syngenetic pyrite) and the Kupferschiefer (on similar grounds). (It is surprising, if diplogenetic deposits are of such importance, that the author does not list them clearly and prominently, but he does not.)

1963. Water of compaction

Noble (1963) suggests that the water of compaction, when expelled from sediments by that process might contain metals in solution, derived from: (1) ions present prior to burial; and (2) ions acquired during diagenesis of the enclosing sediments. The amount so expelled could have been in huge volumes and, even if highly dilute in ore elements, might have contained sufficient metals to have formed such widely distributed deposits as those characteristic of the Colorado Plateau and the Mississippi Valley. This expulsion probably would take place largely through transmissive zones that become loci of major deposition. Numerous smaller deposits may have been formed by smaller-scale expulsions through zones that transmitted lesser volumes of ore fluid. The author believes that this method of ore-fluid production and movement is more reasonable than the penetration of consolidated sediments by exogenous fluids. (Essentially nothing is said as to how the ore elements are transported, or how they are deposited. How such deposits could form by this hypothesis in rocks that never were deeply buried and were completely lithified before the ore fluids entered is not discussed.)

1964. Uranium in ancient conglomerates[1]

Davidson (1964) bases his paper on a review of a Russian monograph on "Uranium in Ancient Conglomerates" that, in turn, is based entirely on western literature and was published by Gosatomizdat in Moscow. Davidson says that the two most valid arguments against the syngenetic origin of the Witwatersrand gold—uranium deposits, in his opinion, are:

(1) That the auriferous reefs are devoid of any significant concentration of the common refractory placer minerals such as ilmenite, rutile, zircon, and monazite that characterize alluvial gold deposits.

_____

[1] Editor's note: see in Vol. 7 the chapters by Pretorius on the Witwatersrand.

(2) That the principal uranium mineral, uraninite, does not survive as detrital grains in modern placers.

It has been argued by others, however, that these refractory heavy minerals actually do accompany the uraninite in genetically significant amounts.

Davidson lists the chemical analyses of three bulk samples of Rand ore, two of which are uranium-rich. If these heavy minerals just mentioned are abundant in association with uranium, then the two uranium-rich samples should show markedly enhanced tenors of iron, titania, zirconia, and the rare earths. These analyses are compared with the arithmetic mean for the analyses of about 100 Permian and Triassic sandstones. The Rand samples are appreciably lower in titania, somewhat lower in zirconia, and somewhat higher in iron (2.10% as compared with 1.30%). Rare earths were not determined in the sandstones but are recorded only as a trace in the two uranium-rich samples. Uranium was not determined in the sandstone but averaged about 0.081% in the uranium-rich samples. This evidence convinces Davidson that the Rand deposits cannot be placers. In ragard to point 2, the Russian monograph says that "uraninite is one of the rarest minerals of placer deposits". The rounded or ovoid grains of uraninite in the Rand deposits have been cited as proof of their detrital origin; Davidson says that grains of exactly similar habit have been found in epigenetic uranium ore bodies. Davidson, using Lovering's term, says that the brannerite in the Blind River deposits well may be diplogenetic in that the titanium now combined with it in brannerite was introduced directly from placer rutile and the uranium epigenetically. He also suggests that the common association of thorium-rich epigenetic uraninite with placer monazite suggests that some thorium was transferred from the monazite to the uraninite during some stage of the ore-forming process, including later metamorphism.

1964. Connate brines

Dunham (1964) was directly involved in the drilling of a hole that reached 2650 ft below the surface, the drilling having started close to the minimum of the larger of two gravity anomalies in the Alston ore field of northern England. This drilling encountered granite at 1281 ft., continued in that rock to a depth of 2650 ft., and, therefore, reached 1369 ft. into the granite. Fragments of granite were found in the overlying cyclothermic sediments, and the granite surface was weathered. Radioactive age-dating, moreover, indicates the granite to be Devonian in age. Obviously, therefore, it could not have been the source of the ores in the granite itself and in the overlying and unmetamorphosed Carboniferous sediments above it. Nevertheless, the vein-type mineralization, which extends through at least 1000 ft. of granite and through a cyclothermic Carboniferous sequence not less than 3500 ft. thick, must have been epigenetic in origin. He finds this distribution and arrangement of ore veins makes it impossible to believe in a syngenetic or diagenetic ore formation, even though some of the features of the ore are similar to other ore deposits that have been termed, by some, as syngenetic. Dunham believes that it is impossible not to postulate sources at depth for the ore fluids, but he adds that there is no good evidence as to their nature. He does, however, apparently dismiss the idea that, at even greater depths, there may be other, and younger, granites from which hydrothermal fluids might have come. Instead, he suggests that, in Early Permian time, juvenile waters, perhaps mixed with connate brines, were rising from extreme depth into fracture systems formed by Hercynian earth movements. Much of the load in these solutions was deposited in, or adjacent to, these fractures, but Dunham believes that an appreciable amount of them may have leaked into the Zechstein lagoon, and that their metal content may have been concentrated in the foul bottom that transiently existed when the Marl Slate—Kupferschiefer—Kupfermergel sediments were being deposited. (It seems, then, reasonable to assume that Dunham considers the metal sulfides of the Kupferschiefer were emplaced either syngenetically or diagenetically but by ore solutions directly related to those ore fluids that formed the deposits in the Northern Pennines and North Derbyshire. Why juvenile waters and their mineral load — acquired by some unstated process from some unspecified source — provide a more reasonable explanation for the fluorite—barite—galena—sphalerite ores of the north of England than would water-rich, magmatically derived fluids from similar depths is not explained.)

1964. Organic reefs

Garlick (1964) points out that organic reefs, specifically those composed of collenia-type stromatolites, act as traps for petroleum and should be especially favorable for the emplacement of epigenetic mineralization. In instances, however, where such bioherms are barren but adjacent beds, apparently less favorable for epigenetic mineralization, contain extensive stratiform mineralization, then the ores more probably are syngenetic than epigenetic, even though they may have been remobilized to give, at least locally, the appearance of epigenetic formation. The reason that such syngenetic deposits should be in such a relation to the reef material is explained by Garlick (based on work, which he cites, by Condon and Walpole) as being caused by sulfide precipitation in the zone of mixing of land-derived water and oceanic water at the edge of the continental shelf, where such a zone is narrowed and accentuated by establishment of an organic reef. Garlick gives examples of such occurrences from the Zambian Copper Belt, Zaire (Katanga), Katherine—Darwin, Australia, and Mount Isa, Australia. At Mount Isa, he interprets the silica—dolomite bodies, in which the copper ares are contained, as organic reefs. The much greater competence of these reefs, as opposed to that of the adjacent shales, caused the silica—dolomite rock to break, whereas the shales folded where earth forces affected the ore-bearing area. He quotes Stanton as saying that statistical analysis shows that these two ore host-rocks are constitutionally related and that the silica—dolomite material is the shoreward member of a continuous and contiguous zone of carbonate—shale sedimentation. He also considers that this sequence of copper, lead, zinc, and iron sulfides is comparable to those in the Kupferschiefer or the Zambian Copper Belt. He believes that the sharp separation (and it is sharp) of copper mineralization, on the one hand, and lead—zinc mineralization, on the other, is incompatible with either hydrothermal emplacement of both types of ore or of the hydrothermal introduction of copper in carbonate bodies immediately adjacent to syngenetic deposits of lead—zinc ores. He believes that the ore fluids were derived from the adjoining land surface and percolated through the reef, where they underwent a sudden change in salinity, pH, and temperature; they then passed seaward into the immediately adjacent euxinic basin. This marked change in water conditions, Garlick believes, can account for the clean separation between copper and lead—zinc minerals.

1964. Sulfate-reducing bacteria

Temple and LeRoux (1964a) tested the theory that the toxic effect of copper on sulfate-reducing bacteria (*Desulfovibrio*) can prevent them acting to precipitate copper sulfides. The culture that they used contained the bacteria, nutrient material, and varied amounts of aqueous solutions of copper sulfate. They indeed found that copper ions in concentrations of 3.0 mg/l sterilized the culture and stopped the reduction of the sulfate. If the cupric-ion concentration was held to about one tenth of that amount, full bacterial action went on. If a toxic level of copper-ion concentration was to be obtained, it was necessary to add enough copper sulfate to precipitate all of the sulfide that had been formed and to provide the necessary excess copper ions. In Temple and LeRoux's experiments, this ranged between 0.25 and 0.29% of copper sulfate. In the process of sulfide precipitation, the bacteria surrounded themselves with zones of copper sulfide that they had made themselves; these zones of sulfide protected them from all but massive doses of copper. (It would seem to follow, however, that, once the bacteria had effectively isolated themselves from the copper contained in the solutions, they would have a difficult time in obtaining the oxygen their metabolism required since they would, at the same time, apparently have isolated themselves from the sulfate ions as well.)

The authors consider that anaerobic zones in sedimentary basins are large enough to be useful in carrying out the biogenic precipitation of sulfide ores. Such basins, they think, also would acommodate large influxes of dissolved copper without cessation or diminution of bacterial sulfate reduction. Thus, they conclude that copper toxicity is no bar to the syngenetic deposition of copper sulfides in quantities large enough to form ore deposits.

Davidson (1965), in his discussion of this paper (cited under Temple and LeRoux), points out that the laboratory environment in which the experiments just described were carried out lack muds with relatively abundant soluble iron that would be in competition with a trace amount, less than 3 ppm, of ionic copper for the attention of the *Desulfovibrio*. He claims that the loss of gaseous H_2S from this system, capable of precipitating copper from the waters above it, was but an insignificant portion of the total sulfur. He holds that the content of ionic copper in the pore waters of the muds could not have exceeded 3 ppm without killing of the bacteria (or, what is the same thing, screening them with copper sulfide so they are no longer in contact with the pore fluid). Because of the much higher iron content than of copper in the pore fluids, the proportion of sulfur to become involved in iron sulfide (of some sort) would have been so high as to prevent the sulfide produced from becoming copper ore by any further diagenetic action. Davidson admits, although he considers it most unlikely, that the copper might have reached the sea floor in particle form; should that occur such copper might be ionically dissolved and so quickly precipitated by the bacteria that all sulfate might be used up without poisoning the bacteria. He also concedes that, in such a system, replacement by copper of bacteriogenic iron sulfide might have taken place. The main problem in such a system is not the bacteria poisoning themselves but screening themselves from further participation in reactions to produce copper sulfides or replacements of iron sulfides by copper. (Further, and more complex experiments would be needed to settle this problem.)

Temple and LeRoux (1946) used a column of 3% agar, with a sulsate-reducing culture at one end and a shale with (upper or lower) absorbed Fe^{3+}, Cu^{2+}, Pb^{2+}, or Zn^{2+} at the other. This system allowed the diffusion of soluble constituents. The absorbed metal ions were displaced, they presume by Na^+, in amounts that were characteristic of the metal ion and the absorbent. Desorbed metals were precipitated as sulfides in the upper part of the agar column. The percentage of the absorbed metal recovered as sulfide was greatest for Cu^{2+} and Zn^{2+} and least for Pb^{2+}. These authors consider that their system served as a model for a fresh-water system containing absorbed metals in juxtaposition with a saline system containing active sulfate-reducing bacteria. Natural materials might have a better absorbing ability than the experimental materials, thus making it possible for a greater concentration of metal to be absorbed than those obtained experimentally. The authors do not claim that these experiments evaluate the possibility of these conditions having taken place in the geologic past or present.

Temple (1964) points out, after considering in detail mechanisms by which syngenetic sulfides can be produced, that the chief objection to the process is that it explains too much. Because metal sulfides can be precipitated in natural-reducing systems, as his experiments indicate, it might be expected that ore bodies would be forming in many places and in large size in many of the potentially favorable sites. Sulfate-reducing bacteria are so widespread and so active that Temple believes that they should be presumed to be likely to form metal sulfides in quantities comparable to their formation of hydrogen sulfide, provided that metals are being added to the waters in question in the necessary amounts. It would seem to follow from this that metals are not being added to sulfide-producing waters in appreciable amounts in the geologic present. (It seems that this brings out the need for a system for introducing metals in large quantities into portions of the sea over short spaces of time, and the best concept that I know of to do this would be by the escape of hydrothermal fluids from the earth's crust to the floor. These fluids probably, however, would bring their own sulfur with them, because it is difficult to imagine that such fluids always would reach the sea in areas where sulfate-reducing bacteria were providing hydrogen sulfide in quantities sufficient for the metals, brought in by the fluids, to be converted to sulfide minerals in ore-producing quantities. Thus, the difficulty with the sulfate-reducing-bacteria source of hydrogen sulfide as the cause of stratiform ore deposition is not only the surprise that it does not happen more often, as Temple suggests, but also that surface sources ever provide enough metal for major ore deposits to form. Even if the metals are introduced by escaping hydrothermal fluids, it remains difficult to see how they could have, in many instances, reached areas where hydrogen sulfide was available. From this it would follow that, if appeal is made to hydrothermal fluids rather than to those derived from the surrounding land surface for the needed metals, sulfate-reducing bacteria would seem to have been of essentially no importance in the production of stratiform sulfide deposits. This is true, even if the escaping hydrothermal fluids form what are truly syngenetic deposits.)

1965. Massive sulfide deposits

Boyle (1965) believes that his work in the Bathurst—Newcastle area contradicts the conclusions reached by such authors as Kalliokoski (1965) and Stanton (1959, 1960) as to their origin. The Brunswick—Newcastle ores are located in an intricately and highly folded and sheared complex of sedimentary and volcanic rocks; the belt contains two types of deposits:

(1) Massive sulfide deposits of large dimensions and tonnages in Ordovician rocks in the southern part of the belt.

(2) Massive sulfide veins in somewhat less metamorphosed Silurian beds and in intrusive plugs that Boyle considers to be volcanic necks, mainly in the northern part of the belt.

He shows that the mineralogy and chemistry of the ores of the two types are identical, even though the metal ratios may be somewhat different, the major minerals being pyrite, pyrrhotite, chalcopyrite, arsenopyrite, galena, sphalerite, and tetrahedrite. The minerals in the vein deposits are coarser than in the massive bodies, and banding is common in the bedded deposits but rare in the vein types. Boyle considers that these deposits were formed at the same time (although, if the massive deposits are syngenetic, they are Ordovician and the vein deposits probably are Devonian). Since Boyle does not believe the massive deposits to be syngenetic, he considers both types to be of Devonian Age. He thinks that the massive deposits were formed during the late-stage shearing associated with the Devonian folding and metamorphism. He points out, the vein fractures in the north are the result of tectonic forces so oriented that they would produce shear zones and dragfolded zones along the grain of the country, that is, more or less parallel to the strike and dip of the rocks in the southern part. These structures are, therefore, conformable to the rocks in which they are found, and Boyle thinks that the massive deposits are considered to be conformable only because they were localized by such conformable structures. He has found that all the massive deposits he has studied (and these are most of the major bodies in the area) are localized in some structural site. He says that all the classic features of replacement and filling of dilatent zones are characteristic of the sulfide bodies and that a marked wall-rock alteration of chlorite, pyrite, and sericite exists. (Of course, those advocating the syngenetic formation of the deposits would say that the folding and shearing were most extreme in the sulfides and that the replacement and fracture filling phenomena and the wall-rock alteration resulted from remobilization during tectonism and metamorphism.) Boyle, however, has shown that some of the deposits, such as Larder U, are discordant on so impressive a scale as to be unlikely, if not impossible, to have been so placed by folding and shearing of a syngenetic deposit. Further, the marked envelope of wall-rock alteration that surrounds the average deposit is so well developed as to require the aid of fluid solutions and could not have been produced by tectonism alone. He is convinced that the banded, folded, and schistose appearance of some of the ores is due to replacement of structures developed in the folding of the sediments, and that the ore minerals entered the area both during and after the structural features had been developed. Because of this, some elongation and shearing of the early deposited sulfides would be expected. Nor does Boyle find that the ores follow the structural and recrystallization trends of their wall rocks, as they should if they were syngenetic. He also points out that the ore minerals may be quite fine-grained in areas where the footwall schists are coarse-grained. (It does not seem to me that all the points raised by Boyle can be explained away as the result of the tectonic remobilization of massive, syngenetic sulfide bodies, although most of them may be.)

1965. Sulfur-isotope ratios

J.S. Brown (1965) has examined data provided by Chow and Patterson and has come to the conclusion that an extraordinary resemblance exists between oceanic leads and the youngest normal (not "J" type) ore leads. This suggests to him, that the formation of lead-bearing ores is a process concomitant to marine sedimentation and that the principal ore fluid is essentially connate water concentrated by diagenetic processes. He believes that ore deposition from this fluid usually is stimulated by heat from igneous sources, but he considers that notable exceptions exist. He thinks that this deposition occurs long after original sedimentation but before metamorphic recrystallization. He considers that "J"-type leads results generally from selective leaching and comingling of radiogenic leads from much older rocks, such as Precambrian basement beneath Paleozoic sediments. Sawkins, in his discussion of Brown's statement points out that pelagic sediments contain about 40 ppm as opposed to 15 ppm in sedimentary rocks, so that it is difficult to compare meaningfully the chemistry of pelagic sediments with that of the bulk of the sedimentary rocks now exposed on the continents. He believes that the bulk of the ore elements in such pelagic sediments is provided by submarine volcanic activity (hydrothermal fluids) that have reached the sea floor. Therefore, the lead in manganese nodules and associated pelagic sediments probably did not come from the continents. Thus, the data quoted by Brown well may mean no more than that normal leads are related to volcanism or, more generally, to magmatic activity. Brown also says that the similarity of the leads in many German ore deposits indicates that the lead-depositing solutions were developed in the adjacent sediments. Sawkins argues that this similarity may have been due to the leads having been derived from some single deep-seated source and risen quickly in an ore fluid that followed various channels through the crust and deposited in the areas of known German ore bodies. Sawkins thinks that connate waters may supply reduced sulfur but that evidence is lacking that they are capable of collecting and transporting significant quantities if base metals.

1965. Strata-bound ores and evaporites

Davidson (1965) argues that strata-bound copper ores are regionally and stratigraphically associated with evaporites and that interstratal brines, derived by diagenesis and leaching of the salt deposits, removed copper from primary disseminated sulfides of magmatic origin in regions of higher geothermal gradient at depths of 2–3 miles. Where most copper-bearing, saline solutions reached shallower and cooler environments, they deposited the metal where chemical and structural conditions were favorable. (This is a modification of the lateral-secretion theory in which the source of the high-chloride and sodium-ion contents of fluid inclusions is explained not as initial components of connate waters but as materials picked up in transit.) Davidson believes that there is a growing body of evidence that the ore fluids are not derived from the same source as the metals they eventually deposit. He considers it quite possible that metallogenetic provinces and epochs are at least as much the results of the salinity and temperature of the deeper ground waters as of the geochemical idiosyncrasies of regional magmatic activity. Davidson does not go into detail on the chemistry of the collecting mechanism, nor does he postulate the deposition of ore minerals from it.

1965. Sulfur-isotope ratios

Dechow and Jensen (1965) determined the sulfur-isotope ratios in 550 sulfides taken from deposits in Rhodesia, Zambia, and Zaire (Katanga). These came from four different regions: (1) the Katanga (Lufilian) arc that contains the Copper Belt; (2) the region west of the Copper Belt; (3) a belt 150 miles by 450, the long dimension of which trends northwest–southeast, that is southwest of the Copper Belt; and (4) in Rhodesia. They conclude that a simple biogenic origin for the sulfur in most of the deposits studied is not supported by the $\delta^{34}S$ determinations made because:

(1) The spread in $\delta^{34}S$ values is too narrow to be diagnostic of biogenic sulfur, unless a redistribution and some homogenization of the sulfur isotopes took place during metamorphism.

(2) The enrichment of ^{32}S in the samples analyzed from the Zambian Copper Belt

they consider to be typical of biogenic fractionation if the sulfate source (assuming of course that there was one) had the isotopic composition of present-day sea water. They believe, however, that the spread in $\delta^{34}S$ values is too narrow to be diagnostic of biogenic H_2S. They suggest, however, that metamorphic homogenization may have narrowed this spread.

(3) A considerable difference exists between the $\delta^{34}S$ values of the sulfides in the copper deposits and those of the associated sediments.

Nor do these authors believe that the sulfur in the sulfides was of simple magmatic derivation:

(1) The spread in $\delta^{34}S$ values in the individual deposits is greater than that of un-doubtedly (their word) magmatic hydrothermal deposits.

(2) The average $\delta^{34}S$ value for many of the deposits is considerably removed from the 0‰ that is characteristic of undoubted (again their word) magmatic hydrothermal de-posits.

On the basis of these conclusions, they consider that the $\delta^{34}S$ results are neither clearly diagnostic of either hydrothermal or biogenic source for the sulfur. They suggest that regional metamorphism assisted in forming the deposits and in readjusting the $\delta^{34}S$ values. (Their arguments for this position are given in detail but are not completely convincing to me.)

1965. Volcanic-exhalative ores

Derry et al. (1965) studied the newly discovered Tynagh deposit in west-central Ireland[1] and concluded that two types of ore are contained in it. The first is a boat-shaped mass of residual ore, and the second is the primary sulfides that underlie the secondary ore. The ore area lies just north of a fault that dips toward the ore at 50°. The host rock is a Lower Carboniferous limestone reef structure; this rock grades laterally into a muddy limestone not far from the ore locality. In the muddy limestone is a volcanic-ash bed at essentially the same stratigraphic horizon as the reef. A bed of banded iron formation is interbedded in the muddy limestone; no intrusive rocks are known in the area. The primary ore minerals have colloform, banded, or concentric structures intermixed with some masses or veins of coarser sulfides. The authors believe that the coincidence of ore with reef material, ash, and iron formation is significant in the ore formation. They conceive of the ore as having been deposited on the sea floor from volcanic exhalations (solfateric solutions is their phrase); the ores were later somewhat redistributed after lithification.

The residual ore is a black mud in which most of the ore metals are present as finely divided sulfides, but slightly less than 30% of the total tonnage of the residual material is made up of oxides. The authors believe that the mud was formed by the removal in solution of calcium carbonate, leaving behind sulfides and the clay constituents of the

---

[1] Editor's note: see also in Vol. 5 the chapter by Evans on Irish base metals.

limestone. The residual material to some extent moved down into a deeping, fault-controlled gully. (An inconsistency exists in this description of the formation of the mud, because the authors describe the reef material, in which the ore is included, as having little bedding or argillaceous material.) Their statement that the ores were formed at low temperatures seems undeniable, but their suggestion that they may have originated contemporaneously with reef complex seems less well documented. They go on to say that the fault (that forms the south limit of the ore) may have had some influence in the first precipitation of the ore. (It is difficult to see how a fault, that could not develop until after the rocks had lithified, could have influenced the location of ores syngenetic with the host reef limestone.)

1965. Massive sulfide deposits

Hutchinson (1965) considers that the copper-bearing deposits in Cyprus have many geologic similarities to the massive sulfide deposits of Canada. (Of course, the Cyprus deposits are far younger and, in comparison with the Canadian ones, far less — almost un-metamorphosed.) The Cyprus deposits, near and at the top of a Cretaceous volcanic pile-igneous complex, have not been deeply buried, appreciably deformed, or much eroded. He believes that the ores were brought in by volcanic exhalations and deposited contemporaneously with the volcanics with which they are included. He thinks that the Canadian massive sulfides were initially formed in the same way but have been largely changed and remobilized by deformation, regional metamorphism, and later granitic intrusions. Such epigenetic features as they now possess he finds to have been caused by these processes. The differences in mineralogy and texture between the copper deposits of the two areas are the result of high pressures and temperatures (particularly the former) that produced the Canadian remobilization. Included in the Canadian remobilization effects are the wall-rock alterations that once were considered to have been caused by hydrothermal fluids. He feels that the proximity of these alterations to the sulfides are all-too-often only vaguely connected in space with the ores to have been produced by the same process that emplaced the ores. The structural control of Canadian deposits, once thought to have localized their initial deposition, Hutchinson believes merely to have provided volumes of low pressure to, and toward which, remobilized sulfides could flow.

In discussing this paper, Kavanagh says that, except for the Skouriotissa deposit, the unmetamorphosed Cyprus ores are elongated in a vertical direction and do not have the thin vertical dimension that volcanic-exhalative deposits should have. He says that all the deposits he knows on the island are located in the immediate vicinity of faults, and their forms appear to be controlled by these structural features — situations impossible for undeformed volcanic-exhalative deposits. (Thus, the mechanism of the formation of both Canadian and Cypriot deposits remains moot.)

Kalliokoski (1965) points out that many massive sulfide deposits in North America show what he calls "textures and structures that are deformational in origin". He contends that these features have been superimposed on the original mineralogical ones to such an extent that, for many deposits, it is impossible to determine the essential characteristics needed to elucidate the conditions of primary genesis. He also thinks that deformation has destroyed much of the evidence of (primary) wall-rock alteration, if it ever existed. He says that the characteristics of a metamorphosed ore are: (1) a granoblastic texture; (2) complex and paragenetically inconsistent grain-boundary relations; (3) the porphyroblastic growth of some minerals; (4) the reequilibration of others under metamorphic conditions; and (5) the migration of minerals such as chalcopyrite in a stress field. Kalliokoski presents the following evidence in support of his position:

(1) The ores were fine-grained originally (Boyle 1965, says they still are).

(2) He finds it impossible to deduce a sequential sequence of mineral deposition from the textures they display.

(3) He says that the $^{18}O/^{16}O$ equilibrium temperature in magnetite in an adjacent banded iron formation is 515°C (this temperature is not, of course, directly applicable to a sulfide deposit a mile from the magnetite occurrence).

(4) The ores may be of Ordovician age (this is possible only if they were initially genetic; if they were, they must be Ordovician; this is circular reasoning).

(5) The ores may have been deposited (or dumped) under slight lithostatic or hydrostatic loads, forming replacements in consolidated or unconsolidated sediments.

The author also discusses two Ontario mines, some Appalachian deposits, of which he devotes most space to Ducktown, and several other deposits such as Thompson Lake and Iron King. See Boyle (1965) for a contrary opinion.

1965. Sulfur-isotope ratios

Nielsen's (1965) paper has been quoted frequently enough in the North American literature that it is worthwhile to summarize it here. He emphasizes that, if the S^{6+} in marine water is reduced to S^{2-} (or S_2^{2-}, the sedimentary sulfides developed will contain a higher concentration of ^{32}S than does the original sulfate sulfur unless all the sulfate is reduced. (If complete reduction of the sulfate occurs, the $^{32}S/^{34}S$ ratio of the sulfide sulfur would be the same as that of the sulfate sulfur from which it came. The $^{32}S/^{34}S$ ratio in the sulfides precipitated from sea water is more likely to be on the lower side of the meteoritic standard ratio of 22.22 (i.e., $\delta^{34}S$ is positive), but it can be on the higher side (i.e., $\delta^{34}S$ is negative) if the $^{32}S/^{34}S$ ratio in the source water was lower than usual, or if only a small portion of the sulfur in the sulfate was reduced. Nielsen reports that the sulfides in marine sediments as a whole contain an excess of ^{32}S (i.e., $^{34}S \approx -5‰$). Under no circumstances can this process explain a lower $^{32}S/^{34}S$ ratio than that in the sulfur in the water from which the sulfide sulfur comes. All this follows because ^{32}S is more readily reduced than ^{34}S, the reverse of the condition that obtains if sulfur oxidized.)

Nielsen says that $^{32}S/^{34}S$ ratios in marine evaporites from the Cambrian through the

Devonian were lower than the present day ratio, that is, the $\delta^{34}S$ was even greater than the $\delta^{34}S$ today or $\approx +20\%_o$. From the middle to the end of the Paleozoic, the $\delta^{34}S$ of marine evaporites decreased to reach about $+11\%_o$ in the Upper Permian. Since then the $\delta^{34}S$ has steadily become more positive. Nielsen thinks that the abnormally high accumulation of the large quantities of biogenic sulfides in Early–Mid-Paleozoic sediments (slates or shales) accounts for the high positive $\delta^{34}S$ values in marine evaporites in that time span. When these Early–Mid-Paleozoic rocks were exposed by Caledonian and Variscan orogenies, much of the light sulfur in them was returned to the oceans as they were eroded. The large development of shales in Mesozoic times removed so much ^{32}S from sea water in sulfide form that the $\delta^{34}S$ values in the oceans have become continually more positive until they have reached the $\approx +20\%_o$ of today.

1965. Sulfate-reducing bacteria

Solomon (1965) in his studies of the sulfide mineralization at Mount Isa concluded that:

(1) The sulfide assemblages have an inherent bedded nature that seems characteristic of their formation as an integral part of the sedimentary accumulation. (This certainly cannot apply to the copper sulfides in the largely brecciated silica–dolomite rock.)

(2) Pyrite in its various forms reflects the depositional and post-depositional history of the deposits. (Of all the sulfides at Mount Isa, pyrite is the most likely to have been deposited syngenetically and to have undergone all the vicissitudes that the rocks have been exposed to since they were deposited.)

(3) Within the natural system, the process of nucleation and accretion of the constituent ions into dispersed sulfide precipitates is restricted to reducing environments significantly enriched in these ions. (Of course what Solomon means by "reducing environments" is ones in which sulfur is being reduced. No reduction can go on without a concomitant oxidation, so "reducing environment" should be qualified by saying reducing in respect to what element.)

(4) Coalescence of the dispersed phases into massive sulfide segregations continued until either grain size attained a dimension compatible with thermodynamic stability in the aqueous system or else rapid burial and diagenesis impeded free migration to sites of crystal growth. (This does not seem to be a very profound statement.)

(5) Further detailed empirical observations and comprehensive experimental studies of the system are necessary to confirm or modify the above mechanism.

(6) All sulfides, especially galena and pyrrhotite, may exhibit textural patterns produced during plastic deformation and flow in response to folding. (Or such minerals may have replaced ones that had been so affected.)

(7) The presence of recrystallized and partly recrystallized sulfide phases indicates the extreme importance of pressure in activating solid-state mechanisms of reconstitution and crystal growth at low temperatures. He says that their (the sulfides') subsequent response

to deformation has been preserved in the fine textures and structures that have implied mobility somewhat greater than that currently proposed for carbonates and far beyond that conceived for silicates. (Granted that this is true, if the sulfides were emplaced in the late stages of metamorphism, they could have developed essentially the same textures as sulfides that had been present in a sedimentary sulfide system from the time of deposition on. Saying, as Solomon does, that there is little evidence for a contemporaneous or subsequent increase in temperature may be true but it is not proof of anything.) He points out that the general lack of knowledge on the various factors that complicate the study of sulfide mobilization limits the present usefulness of theoretical studies.

(8) The broad sulfur-isotype distribution is indicative of a biogenic origin for the sulfur of the sulfide assemblages. (This last statement would seem to indicate that the sulfur came from one source and the metals from another, although he may mean that the metals were brought into, in solution or in particles of metal compounds, an environment where sulfur was being bacterially produced on a large scale. This probably would be an isolated sea-water basin, which would be unlikely to have the long, narrow configuration that a basin of deposition for the Mount Isa sulfides must have had. If the ores were formed syngenetically, it appears much more probable that the metals were introduced along a fracture or series of fractures, the long dimension of which was essentially parallel to the trend of the strata in the Mount Isa area. Since the shape of this basin must have been long and narrow, the chances of the needed sulfide ion in quantity having been produced by bacteria seems improbable. Rather it seems probable that, if the ores were formed by volcanic exhalations, the sulfur came in with the metals. That would require that the broad distribution of sulfur isotopes would have resulted from processes inherent in the hydrothermal fluids (or volcanic exhalations, the same thing in this case). If the metals are indeed carried in complexes with sulfur, the various oxidation and reduction reactions to which that sulfur is subject during the vicissitudes of its upward journey, readily could account for wide-spread and localized variations in sulfur isotope ratios. Certainly, it has not been proved that sulfur isotope ratios can be changed only by near surface oxidation and reduction of that element.)

Suffel (1965) compares the massive cupriferous deposits of northwestern Quebec, Cyprus, and Turkey and emphasizes the need to consider the total geologic environment before arriving at an explanation of their genesis. He points out that ores are not going to be found at volcanic interfaces unless ore-forming solutions reach them. He concludes that, in Cyprus and Turkey, the ophiolitic magma and the chamber from which it came formed essential parts of the ore-forming system. Not all deposits developed by ore fluids from such magma chambers will be on the sea floor; different types of ore can be formed at different levels below the submarine surface at any given time. Many of these deposits, he thinks, would be hydrothermal replacements that grade into fracture fillings at shallow horizons. He sees no necessity why a surficial or near-surface deposit should have exhalative—sedimentary beds (ore deposits?) above it. Such beds might never have formed or might have been removed by turbidity currents or submarine sliding (McArthur River?).

He suggests that it is probable that every period of tectonic activity and metamorphism has an effect on already existing ore deposits and has caused new (ore?) materials (fluids?) to rise from unknown depths. The depth at which an ore fluid originated and the level to which it would rise before ore materials began to deposit from it would depend on many factors. One factor that seems certain to him is that the fluid would follow some extensive zone of weakness. He thinks that the persistence of movement on such structures is well established, some probably having provided ore-fluid channelways from the earliest Precambrian to the present. He holds that some of the complications in ore genesis may be due to the repetition (in a given region) of ore formation, perhaps producing similar ores each time but more probably different ones. He believes it at least distinctly possible that some of the complications in Precambrian ores are due to the superimposition of new ore (or ores) on older ones, further confused by more or less remobilization of the older ore or ores. Suffel implies, therefore (though he does not specifically state) that volcanic-exhalative deposits on the sea floor are only one event in the cycle of ores of magmatic origin and that such sea floor deposits may never form or may be removed before they are lithified.

1966. Massive sulfide deposits

Anger et al. (1966) interpret their studies of the sulfur isotopes in the Rammelsberg ore-deposits of Germany as indicating a syngenetic hydrothermal origin for that ore. They conclude that all of the copper, lead, and zinc in the sulfides was derived from submarine springs but that only about half the pyrite came from that source. The sulfur in the remaining pyrite, especially that in concretions, they consider to have been formed from the reduction of marine sulfate by bacteria. They believe that the barium of the barite was introduced from the submarine springs but that the sulfate was derived from that in the sea water, although they think that about one-tenth of this sulfate may have come from hydrothermal sources. The $\delta^{34}S$ values of sphalerite, galena, and chalcopyrite show a distinct increase during the formation (syngenetically) of the main ores lenses. In the lower parts of the lenses, the $\delta^{34}S$ values range from about +7‰ and in the top part they range around +20‰. This means that the earlier sulfides contain a higher proportion of ^{32}S than do the later ones. This Anger and his colleagues explain as being due to the slightly higher mobility of ^{32}S in comparison with ^{34}S, so that ^{32}S ions are more readily received into sulfides being formed than is ^{34}S. (Of course, this assumes that the sulfides are being formed by the neutralization of an anion by a cation; this might be true in the bacterial formation of the sulfide ions and their combination with metals in solution in sea water. In submarine spring-water, however, the metals must have been carried as complexes, either with sulfur or with chlorine. If the metals are in sulfur complexes, the precipitation is affected by the removal of excess sulfurs from the complex. If, as appears normally to be the case, the sulfide ions removed are removed by oxidation to sulfate, the ions removed first would have higher proportions of ^{34}S than ^{32}S, thus explaining why the earlier sulfides at Rammelsberg would have a lower positive $\delta^{34}S$ than would those formed later when the available ^{32}S had been appreciably reduced.) They suggest that the total mass of barite requires a 2 km depth of sea water above the sea floor on which the barite was being deposited. (This depth of about 6550 ft seems excessive for the types of rocks in the deposit.) They conclude that the pyrite was produced by a diagenetic change of FeS to FeS_2. The $\delta^{34}S$ of pyrite is consistently lower than that of the corresponding $\delta^{34}S$ of the associated non-ferrous sulfides. (This does seem to confirm the authors' opinion that the bulk of the pyrite is of biogenic origin. Since, in the S_2^{2-} ion, half of the sulfur, at any given instant, has no charge or a valance of 0, it should have a higher $\delta^{34}S$ than a non-ferrous sulfide being formed at the same time and place. If, however, the pyrite was formed initially as biogenic FeS and was later converted diagenetically to FeS_2, the FeS first precipitated would have had a lower $\delta^{34}S$ than would the sulfate from which it was formed, since ^{32}S is more readily reduced than ^{34}S. The later addition of somewhat oxidized sulfur ion (S^0) to the FeS would form pyrite. This reaction requires something to oxidize the sulfur from H_2S (or HS^- or S^{2-}) to produce S^0; this most probably is Fe^{3+}, which is concomitantly reduced to Fe^{2+}, as eq. 1 shows. Obviously, this:

$$FeS + H_2S + 2Fe^{3+} \rightarrow FeS_2 + 2Fe^{2+} + 2H^+ \tag{1}$$

reaction can go on only if a considerable supply of Fe^{3+} is being provided to the sediment volume where the pyrite is being produced. Perhaps this is the hydrothermal fluid being provided from the submarine springs. The small proportion of the FeS_2 that has a high ^{34}S content — strongly positive $\delta^{34}S$ — probably did come from a hydrothermal fluid, in which the pyrite would have partly oxidized sulfur in comparison with the non-ferrous sulfides that would have none.)

The authors point out that the $\delta^{34}S$ value in pyrite at Meggen ranges between +16 and +20‰, and they think, therefore, that all of the pyrite sulfur came from submarine hydrothermal springs. From this, it follows to them that the conditions of pyrite precipitation at Meggen were appreciably different than those at Rammelsberg. At Meggen, the barites have a narrow $\delta^{34}S$ spread around +23‰, which they believe to have been that of the sulfate in Devonian sea water.

Field (1966) concludes that, if no large difference exists between the $^{32}S/^{34}S$ ratios in associated sulfides and sulfates, the sulfates probably were formed by supergene processes. If the spread between ratios is appreciably wider, then the sulfates probably were formed by hypogene processes. (From these results, it would appear that, when sulfide ions are oxidized in a near-surface environment, the change in $\delta^{34}S$ from the sulfide to the sulfate is slight. This may be true for all reduction reactions involving sulfate, whether at low temperatures or high. Since little reduction of sulfate sulfur to sulfide sulfur takes place in hypogene solutions, it is impossible to say whether fractionation by this process is more important at high temperatures than it is at low. From the appreciable spread between the $\delta^{34}S$ of sulfides and sulfates formed at high to moderate temperatures, it would appear that considerable fractionation takes place during oxidation in such temperature ranges. If such is the case — and Field's results suggest that it must be — then an appreciable fractionation of sulfides can take place in hydrothermal solutions if the conditions are proper for appreciable oxidation of sulfide ions to the sulfur in sulfate ions to take place. This would seem to have important implications for the spread of $\delta^{34}S$ values in certain hydrothermal deposits. If oxidation of sulfur can take place impressively in a hydrothermal fluid, the spread of $\delta^{34}S$ values in the sulfides between early and late-formed sulfides may be in the same high range as that between sulfides and sulfates in such deposits.)

1966. Massive sulfide deposits

Kinkel (1966) holds that deposits of massive pyrite and pyrrhotite—pyrite occur mainly in volcanic rocks and particularly in the upper pyroclastic parts of submarine volcanic piles. He considers that many of these deposits appear to have been formed in the closing stages of submarine volcanism (or replaced rocks so formed) and were emplaced before these beds were covered by overlying sedimentary deposits. He points out that these deposits are mainly peneconcordant and stratabound, but the sulfides in them were formed by deposition in porous, partly unconsolidated, material and by replacement of flows and pyroclastic beds. He does not consider that they were deposited syngenetically, as the term is commonly used to mean contemporaneously with the enclosing non-sulfide material, since, in some instances, the sulfides follow structurally controlled channelways and show replacement textures with their host rocks. Kinkel points out that large deposits of massive pyrite and pyrite—sulfur have been formed in the recent past and are being formed now around fumaroles and volcanic hot springs in Japan and Taiwan. On the contrary, he thinks that the older, but undeformed, massive pyrite deposits of Cyprus appear to have been formed by hydrothermal exhalations but under submarine conditions. He says that less positive evidence can be obtained from such massive pyrite deposits as those of the Huelva area of Spain or Shasta County, California. He, nevertheless, thinks that these older deposits were formed in the same manner as those of Cyprus. He considers that the material in volcanic emanations is in both molecularly and colloidally dispersed states where the emanations enter a near-surface submarine environment. (Why these materials should not be ionically dispersed, granted the solutions are under sufficient confining pressure, is not clear from Kinkel.) He believes that colloids should flocculate rapidly and of course quickly precipitate when they encounter sea water either in the sea or in sea-water soaked pyroclastic rocks and sediments. He considers that the most reasonable method for forming massive sulfide deposits under these conditions is a filtration of the flocculated and precipitated material, with some accompanying replacement. This filtration would occur where diffusable solutes, sols, coagulates, or precipitates would be forced through a sediment that acted as a membrane.

He concludes that the localization of many of these ore bodies in the vicinity of shear zones may indicate no more than that subsequent deformation of the rock–sulfide system was localized in the more reactive or mobile zones created by the presence of the sulfides. He believes that the colloids are stabilized by excess S^{2-} and HS^- ions and are precipitated when these ions are lost on encounter with sea water. (This suggested mechanism for the formation of massive deposits needs further study before it can be accepted or, as is more likely, rejected.)

1966. Sulfur-isotope ratios

Stanton and Rafter (1966) studied the sulfur-isotope ratios in the Australian deposits of Mount Isa, Captain's Flat, Rosebery, Broken Hill, and Dugald River.

(1) They found that the mean $\delta^{34}S$ for single deposits was always greater (more positive) than for meteoritic sulfur.

(2) In any single deposit, the range in $\delta^{34}S$ was narrow and usually did not exceed 8‰.

(3) Dispersion of $\delta^{34}S$ values, although small, tends to increase with increase of the value of the mean. They consider that this relationship may be fortuitous and well may be due to the small number of samples from each deposit.

(4) The sulfur-isotope ratios are not systematically related to those of the lead isotopes.

(5) From these four facts, they infer that the variation in $^{32}S/^{34}S$ ratios is due largely to fractionation and not to contamination.

(6) On the data available to them, they do not think that fractionation was induced by biological activity and, therefore, the sulfides probably were not produced by bacterial action so far as they could say when the paper was written.

(7) For any single deposit the intensity of the fractionation process was not great, perhaps indicating substantial constancy of conditions of extraction (of sulfur?) and deposition.

(8) If the relationship in (3) is real (and they are not certain this is so — more data must be obtained) the source material had a $^{32}S/^{34}S$ ratio lighter than that of meteorites or similar to it.

(9) If (3) and (8) are correct, the $^{32}S/^{34}S$ ratios in the metallic sulfides would have been induced by enrichment of the heavier isotope in the metallic sulfide phase or by direct combination with heavy sulfur.

If the metallic elements are transported as sulfur-metal complexes, two ways are known — to me — by which the excess sulfur can be removed. One is to oxidize the sulfur to the sulfate ion, thus preferentially removing heavy sulfur from the metal-sulfur complex. The other is to remove the excess sulfur by combination with hydrogen ion, and the removal of the lighter isotope would be favored. The first is achieved by favorable Eh conditions, the second by favorable pH. Since cupric copper is readily reduced under the usual Eh conditions of hydrothermal fluids, sulfides containing cuprous ion must be lower in ^{34}S than associated sphalerite and galena in which no change in the valence state of the metal occurs. Since all the sulfides from these deposits are described by the authors as "mixed sulfides", it is difficult to apply the criterion just discussed but, if the "mixed sulfides" were predominantly sphalerite and galena, then the shift toward heavier sulfur in the metal sulfides would be explained.)

1966. Syngenesis, epigenesis, and metamorphism[1]

Vokes (1966) considers the Bleikvassli sphalerite—galena—pyrite deposit in northern Norway is in the central section of the 1500 km long Scandinavian mountain chain and is confined to a nappe in that structure. The ore zone is at the structural border between an overlying schist and marble division to the west and an underlying division of schists, gneisses, some amphibolites, and fewer marbles to the east. The rocks are early Paleozoic in age and are part of the Nordland facies. Early studies by the author have shown that the ore body and the surrounding rocks have been involved in the Caledonian orogeny and have been subjected to at least two metamorphic events. The wall rocks in the southern part of the ore body are micaceous schists, gneisses, and quartzites, with inter-layered graphite- and sulfide-bearing varieties. To the north, the ore body passes along strike into a body of fine-grained microcline-quartz-mica gneiss. The walls of the ore body are mainly concordant to the schists and gneisses but, in the microcline gneiss, the conformity of ore and wall rocks appears to have resulted from tectonic schisting in the gneiss that is parallel to the strike of the ore body. Vokes believes that this geology can be explained either by: (1) syngenetic or diagenetic premetamorphic emplacement of the sulfides; or (2) syntectonic, epigenetic emplacement. He concludes that the evidence, as known at present, favors choice 1. He goes on to say, however, that he does not consider it necessary that all stratabound ore bodies in metamorphic rocks must, of necessity, have been formed syngenetically. He points out that the geologic history of the Bleikvassli deposit has been quite complex; he does not yet know how to distinguish between a sulfide mass that underwent both metamorphisms (as a syngenetic deposit must have done) or only one of them (as an epigenetic deposit must have done). He suggests a more open attitude on the part of investigators of this type of deposit and a realization that, if it is proved that one or several deposits were formed in a certain way, it does not follow that all "similar" deposits also were formed in that way.

1966—1968. Syngenesis and diagenesis

W.C. White and Wright (1966) express a somewhat different opinion as to the origin of the White Pine copper deposits in Michigan than they did in 1954. To review, the copper zone at White Pine is at the base of the Nonesuch shale, with the main copper mineral being chalcocite. Above the copper zone, the characteristic mineral of the remaining 400—600 ft of the Nonesuch is pyrite. The boundary between these two zones regionally

---

[1] Editor's note: see chapters on metamorphism, e.g. by Mookherjee (Vol. 4), Both/Rutland (Vol. 4), and Vokes (Vol. 6).

transgresses both the stratigraphic layering and the facies gradations, these last being indicative of ancient environmental boundaries. The top of the cupriferous zone is marked by the upward sequence chalcocite—bornite—chalcopyrite—pyrite and by the (secondary?) cadmium sulfide — greenockite. This zoning does not appear to them now (1966) to be syngenetic; instead the regional configuration of the cupriferous zone suggests that the copper was deposited after the deposition of the entire 50 ft. sequence of beds. They do, however, consider that Wise was correct when he suggested that the sulfur for the copper sulfides was provided by syngenetic pyrite. They point out that the locus of maximum copper deposition generally coincides with the area in which the underlying Copper Harbor conglomerate was thinnest; this wedging down of the Copper Harbor may have acted to concentrate a flow of copper-bearing solutions moving outward from the center of the Keweenawan basin.

Amstutz and Park (1967) argue that the stylolites in the southern Illinois fluorspar deposits were developed during the diagenetic stage of the limestones in which they are now located. They also believe that the fluorite and the sulfides are part of the debris that accumulated on the stylolite surfaces. Granted that these statements are facts, then the fluorite and the sulfides must have been deposited at the same time as the sediments, or at least early in their diagenesis, if the sulfides were to have been present during the formation of the diagenetic stylolites. They illustrate stylolites developed in the limestone that fills an erosion channel, and they consider that the erosion that produced the channel took place while the carbonate material still was mobile, being at least partly unconsolidated, and at a time when the stylolites just had begun to form. (The evidence presented may indicate that the sulfides were introduced while stylolites still were forming but does not necessarily mean that either sulfides or stylolites were developed diagenetically.)

Jackson and Beales (1967) suggest that Mississippi Valley-type ores may have been concentrated during the late stages of diagenesis. They believe that, during the normal evolution of a sedimentary basin, an ore fluid generated by sedimentary compaction will form over time. They think that gas- and oil-pools are developed during early diagenesis and metallic mineralization with later diagenetic and metamorphic processes. (Thus, what these authors are talking about is not, in my opinion, a syngenetic process and, as such, should not be included here but is inserted anyway because the editor thinks it should be included.) Nevertheless, the authors consider that the principals that govern the location of this type of ore deposit are sedimentary ones. (This may be true in the sense that the authors would consider their scheme of ore formation as a — late — part of the sedimentary cycle.) They believe that the ores concentrate in the carbonate portion of the sedimentary rock sequence because hydrogen sulfide tends to form and accumulate in carbonate reservoirs and subsequently acts as a precipitant for metallic sulfides. (The flaw in this reasoning is that the hydrogen sulfide, which is almost certainly dissolved in water, is limited in amount by the open space available, the temperature, and the pressure prevailing in the system. At a depth of 500 ft., for example, the confining pressure on the system containing hydrogen sulfide would be about 40 bars and the temperature would range around 100°C. In such a situation, the solubility of H_2S in water is about 1.2 moles/1 or 52.8 g of hydrogen sulfide/liter. If a liter of open space, essentially 1000 cm^3, were filled with sphalerite, it would contain 4000 g of ZnS, which would, in turn, contain about 1333 g of sulfur. From the data just given, it is obvious that the water saturated with H_2S in the openings in the limestone would contain, under the conditions set up, only enough sulfide ion to fill about 3.6% of the open space available. Even this much ZnS could be produced only if the solutions bringing in the zinc ions did not dilute the hydrogen sulfide concentration already present in the openings. Thus, if the open space were to be filled, as much of it is at Pine Point, at least 30 times as much hydrogen sulfide would have to be supplied to the system as be held there at any given moment. The possibility that such quantities of hydrogen sulfide could be generated in solid carbonate rock is negligible. Thus, the authors' statement that "ore bodies are found where regional permeable porous trends, acting as a 'plumbing system' for escaping basinal connate waters that carry ore metals, traverse local H_2S-bearing carbonate reservoirs" seems not to be soundly based unless they add an explanation of how a continuing supply of hydrogen sulfide is being produced at just the right time.)

Jackson and Beales, nevertheless, argue that the geochemical data are consistent with derivation of ore metals from sedimentary basins. Base-metal ions may be sparsely disseminated in argillaceous sediments or deposited as sulfides in concentrations approaching ore tenor in black shale. Carbonate rocks in general, they point out, have low concentrations of base metals. They believe that connate water can derive the necessary metals by desorption, probably mainly from clays or from organic material. They think that burial to critical depths will control the reorganization of clay minerals, the concentration of the connate brines, the release of metals, the breakdown of organic complexes, and the expulsion of connate fluids. These fluids the authors consider to be a type of hydrothermal fluid but would prefer to designate them as "stratafugic". (Another flaw in the authors' thesis is that, under almost any conditions that can be imagined for these stratafugic ore fluids, galena, at Pine Point at least, always would precipitate before sphalerite. At Pine Point much of the galena is late, so a more complex model must be put forward to explain the collection, transport, and precipitation of the Pine Point ores and of any other Mississippi Valley-type ores that contain sphalerite and galena in not widely different amounts.)

Roberts (1967) points out that the formation of a sedimentary ore deposit requires an adequate source of materials, the gathering of these materials by solution or other processes, their transportation to the site of accumulation, and their deposition in that sedimentary basin. Roberts believes that, in most instances, metamorphic energy is needed to consolidate the dispersed phase into an economically workable deposit. The author thinks that weathering could possibly supply a sufficient concentration of metals from such rocks as amphibolites (rich in copper) or from previously existing ore bodies (either in solution or as particles) to form ore bodies. The weathering of the generality of rocks, however, would have to be supplemented by some other means if such a unique event as the formation of an ore body is to be achieved. Thus, the environment must supply enough metals to a correct site of deposition if an ore body is to develop. He says that the source of metals may be volcanoes, hot springs, or other orifices that connect the depths to the surface (he should also emphasize that these hot springs or volcanoes or fissure exhalations must occur underwater if the metallic material brought to the surface by these agencies is to be preserved). He seems on less firm ground when he insists on the need for metamorphic energy to concentrate less than ore quantities of minerals into workable ore deposits. He does not think, however, that sedimentation alone can form ore bodies — some additional form of post-depositional concentration probably is necessary. He then discusses the problems of transporting metals in solution, as vapors or gases, or by processes of recrystallization (remobilization?). He holds that two principal sources for water for the transport of ore materials are connate water and meteoric water; because connate waters normally hold far more dissolved material than meteoric, he considers connate waters as far more likely to be ore carriers. He goes on to discuss the role of sulfide synthesis in the provision of the data needed to understand ore deposition, but this is beyond the realm of this study. He also considers low-temperature sulfide synthesis and the minerals that have been produced by it, plus some mention of the work that has been done on remobilization of such materials.

The problem of the manner of formation of the White Pine copper deposit has received more attention in the literature than any other possibly syngenetic deposit in North America, probably even more than the massive pyrite-rich deposits of New Brunswick, the only other contender for the honor. In 1967, Hamilton pointed out that most of the copper in the deposit occurs in lenses in the upper 7 ft. of the chloritic facies of the Copper Harbor conglomerate. This facies is 1—20 ft. thick and overlies the 80—100 ft.-thick lower hematitic phase of that formation. He believes that the copper normally is as rims around interstitial carbonaceous material. These associated materials usually are found along cross laminae, but exact correspondence between the materials and the sedimentary structure is not universal. He considers that the irregular distribution of the copper along the laminae is related to differences in sedimentary porosity and that, in turn, results from variations in sorting. He believes that the copper was deposited by adsorption on the carbonaceous material, and that this phenomenon was most likely to take place where the pH was between 3 and 8, with an acid pH being the more favorable. He does not believe that any appreciable replacement of carbonaceous material by copper occurred. In addition, the copper could not have replaced a chloritic rim around the carbonaceous matter, because copper rims are wider than the chlorite rims where there is no copper. He rules out the possibility that pyrite could have been the material replaced by copper, saying that, if pyrite had been replaced, the mineral formed would have been an iron-copper sulfide. (This does not necessarily follow because the copper mineral doing the replacing would have been determined by — principally — the Eh and pH conditions of the ore fluid — whatever its ultimate source.) Assuming that the copper was deposited by adsorption on the carbonaceous matter, Hamilton (1967) is forced to the conclusion that the copper was deposited in open space while such open space still existed in the sediment. He gets the organic material to enter the (apparently unconsolidated) sand by migration from the overlying mud and silt. The organic matter, along with "reducing waters" caused the formation of chlorite and the alteration of ilmenite—magnetite grains to anatase. Copper ions then were adsorbed on the surface of the carbonaceous material and this, with other secondary minerals, filled what open space still remained in the sandstone. Later faulting resulted in some remobilization of copper and carbonaceous material. Hamilton (1967) found no evidence as to where the copper came from but he favors the hypothesis proposed by White and Wright (1966) that the copper was deposited by connate water driven out of the large volume of sediments in the Lake Superior basin area (Lake Superior did not, of course, then exist). This water, he thinks, moved from the sandstone into the overlying shale. He adds that the problem needs further study. (This last statement is belaboring the obvious because his explanation of the development of the native copper in the Copper Harbor Formation requires that this copper came from the less-than-consolidated mud and then, accompanied by carbonaceous matter originally in the mud, was forced down into the sand — also still unconsolidated. Surely, a simpler and more straightforward explanation should be possible. Obviously, from Hamilton's concept, the copper ores, both native and sulfide, are diagenetic and not syngenetic.)

Boyle and Lynch (1968) studied the zinc, copper, cadmium, and lead contents of such marine organisms as clams and oysters and found a remarkable enrichment of zinc and cadmium in these two varieties; some oysters also concentrate copper where it is available. They suggest that, during sedimentation, the soft parts of these metal-secreting organisms undergo bacterial humification and putrifaction. During these processes, the sulfur is split out of the sulfo-protein complexes that are present in the form of H_2S; at the same time, the heavy metals are liberated. If this is accomplished under oxidizing conditions, the sulfur is converted to sulfate and the metals added to sea water in soluble form. If the area of sulfur and metal separation is such that sulfur is not oxidized, the H_2S binds with the metals as colloidal sulfides, and these are precipitated with the carbonate sediments. Later, the constituents of these sulfides probably are mobilized by brines of other diagenetic solutions and are moved into available porous zones or structurally prepared sites. Thus, they think that the stages of lead—zinc deposit formation in carbonate rocks is:

(1) Initial concentration of heavy metals and sulfur from sea water by marine organisms, the process being a complex biochemical and physiological one.

(2) A further concentration takes place during humification and putrifaction, the last event in that process being the precipitation of heavy metals by H_2S.

(3) The final concentration of sulfides by brines or other collecting agencies in porous zones or structurally favorable areas.

(It probably is necessary to do no more than ask: "where does the lead come from?")

Greenwood (1968) reports that the Eastport (Silurian) Shale in Maine contains calc—silicate concretions of equigranular epidote and quartz, with minor chlorite and calcite. The valves of linguloid brachiopods are common in the concretions and the surrounding siltstone; these valves are now clear calcite. Most of the concretions contain chalcopyrite, typically replacing whole shell fragments but also as isolated grains, as are lesser amounts of galena and pyrite. The sulfides compose, Greenwood thinks, about 0.1% of the concretions. He believes that the distribution of these sulfides strongly suggests that they formed with the sediment. He points out that the reducing environment necessary for the precipitation of sulfides probably prevailed only in the immediate vicinity of the rotting organisms. He considers these observations as supporting the efficacy of biochemical and diagenetic processes in the accumulation of base-metal sulfides.

1968. Uranium in ancient conglomerates

Gross' (1968) work on the auriferous conglomerates at Jacobina in Brazil shows that the Late Precambrian conglomerates and associated quartzites can be traced along strike for more than 50 miles in the Serra de Jacobina. The section contains 28 conglomerate beds thicker than 0.5 m, and these rocks make up 145 m of a total formation thickness of 555 m. Essentially all of the conglomerate contains a little gold in the matrix between pebbles; economic concentrations of gold occur in a foreset system of cross-beds. Economic concentrations also are at the tops of the conglomerate beds and in conglomerates that come in contact with dikes and sills of ultramafic intrusives. In fractures in some of the conglomerate pebbles or boulders, plates of gold occur. Some high-grade gold is in close association with stringers of pitchblende. Gross interprets these facts to confirm a placer formation of the original gold in the formation and its later remobilization under the influence of folding, metamorphism, and intrusion. Except that the gold in the matrix is reported by Gross to be very fine grained, he does not describe its character, so it is difficult to say if it is typical placer gold or has shapes that mean it either was introduced by solutions external to the conglomerate or was redistributed from its original placer form.

1968. Diagenesis

Park and Amstutz (1968) return to the problems of the genesis of the fluorspar deposits of southern Illinois, this time considering the genetic meaning of what have been called collapse structures that usually contain broken pieces of dolomite, limestone, shale, sandstone, fluorite, and sphalerite. These, the authors believe, actually to be cut-and-fill channels, some of which are bordered by horizontal stylolites that bend downward into the channels and their thicknesses increase (and their amplitudes decrease) near and along the channel slopes. This suggests to them that the channels were produced while the sediments were still quite unconsolidated. The dark clastic fragments that have been mechanically emplaced into open cracks that narrow downward, Park and Amstutz think, were formed in hydroplastic carbonate mud rather than that they were developed by solution in solidified rocks. They believe that the reduction in thickness of the beds in the district (as much as 37%) probably was caused by the development of stylolites and that much of the stylolitization occurred during diagenesis (see their 1967 paper). The cross-cutting relations of some of the late veinlets in the rocks to the stylolites and the more or less horizontal ore layers convinces them that much of the compaction took place after the ore had been introduced during diagenesis. (The formation of the ores by emplace-ment in fractured, lithified rock does not yet seem to have been ruled out.)

1968. Meteoric water

Roedder (1968) shows that the freezing and homogenization temperatures of primary fluid inclusions from the Pine Point ores show that they were formed from exceedingly saline brines, very free of solid particles, in temperature ranges between 50° and 100°C. The salinity of these inclusions prohibits the ore minerals from having been formed from fresh meteoric or sea water. Roedder considers that the ore fluids were interstratal brines (connate waters), and having been provided with their ore materials by leaching the formations through which they moved. He considers that the ore fluids moved deeply enough beneath the surface to have become quite hot. (This well may be correct and it well may be that these fluids own their high salt content to having been brought, while still quite hot, close enough to the surface that they boiled. No evidence has been brought forward to show that they boiled enough to deposit any alkali halides, but they well could have boiled sufficiently to have attained their high saline content.) Roedder would explain the deposition of the ores by mixing of the ore fluids with small amounts of quite fresh, cold surface waters. He also uses this mixing to explain the minute, regular color-banding in some of the sphalerite.[1]

1968. Ore fluids

D.E. White (1968) pointed out that the generation of (many, but not all) ore deposits requires a hydrous fluid that has four critical characteristics:

(1) A source of the ore constituents that usually are dispersed in some medium — a magma, a sedimentary rock sequence, (a volume of stream, lake, or ocean water, or unlithified mud of any type).

(2) The dissolving of the ore and other constituents to be incorporated in the ore body into a hydrous phase (if they are not already in one).

(3) The migration of the metal-bearing fluid.

(4) The selective precipitation of the ore and other constituents in a favorable environ-ment (by replacement or the filling of open space, if in solid rock, or at the bottom of a confined body of surface water).

---

[1] Editor's note: see in Vol. 2 Chapter 4 by Roedder.

Many geologists, he considers, have explained ore deposits through one of several "end-member" models, for each of which all four critical characteristics are interdependent. He reviews the geology of five ore districts with respect to the source of the water of the ore fluids, the source of the dissolved constituents, the force that drove the migrating fluids, and the possible causes of the selective precipitation. He points out that chemical and isotopic data on the fluids phases are necessary for such evaluations. These data were obtained by him largely by analyses of fluid inclusions in the ore and gangue minerals in three of the districts and of existing thermal fluids (in two of the districts where metal-bearing hydrothermal systems are now active). These districts are:

(1) Providencia, Zacatecas, Mexico that has close magmatic affiliations and a probable magmatic source for its ore fluid and its contained metals and sulfide ions.

(2) The Mississippi-Valley lead—zinc—fluorite—barite type in which he considers connate water to have been dominant in the ore fluid during both sulfide and fluorite stages, with the ore constituents most probably having been scavenged from sedimentary rock sources (he does not consider this last aspect as having been proved).

(3) The Salton Sea geothermal system in which meteoric water probably is dominant in the ore fluid; probably most of the metals and salt in this fluid were derived from the neighboring sediments, but some water and metals may have had a volcanic source.

(4) The Red Sea geothermal system in which sea water is dominant in the ore fluid, with dissolved salts and metals, he thinks, coming from associated evaporites and clastic sediments.

(5) The Nonesuch Shale of the White Pine area in which connate water and copper probably were introduced diagenetically soon after the shale was deposited; the copper in the chalcocite was combined with sulfur provided by earlier-formed biogenic sulfur in (syngenetic?) pyrite.

White concedes that the formation of the ores in none of these districts is completely understood and that differences of opinion about the formation of the ores or ore fluids still promote controversy. If, assuming that these five systems as he described them furnish a sample of other base-metal systems, he draws the following conclusions:

(1) Most ore deposits are formed by complex rather than simple end-member processes; the water, dissolved salts, metal and sulfide ions, and other critical constituents of each deposit well may come from more than one source.

(2) The ore-bearing fluids of the metal deposits are Na—Ca—Cl brines, with the total salt content between less than 5 and as much as 40 weight % of NaCl.

(3) Brines of similar major-element composition may form in at least four different ways: (a) magmatic, probably deuteric (he probably means late-stage fluids immiscible in the mainly silicate magma); (b) connate waters that generally evolved from sea water; (c) the dissolving of evaporites by any dilute water, followed by removal of the needed metals by reaction with sedimentary rocks or perhaps with sediments; and (d) membrane concentration of dilute meteoric water.

(4) In at least three, and possibly all five, districts, the ratio of total dissolved metals to

dissolved sulfides in the ore fluids was very high. White holds that base-metal ore deposits always may form from metal-rich but sulfide-deficient brines, with the metals transported as chloride complexes in the presence of some (but not much) sulfide. Sufficient sulfide to form such deposits is provided in seven ways:

(1) Sulfides in the fluids that selectively precipitate certain sulfides as temperature decreases or pH increases (*sic*) as a consequence of hydrogen metasomatism — in the latter case, the H_2S and HS^- of the fluids are converted to S^{2-}. The order of precipitation of the sulfides is controlled by the relative abundances of the metals, the stabilities of their metal—chloride complexes, the solubilities of individual sulfide minerals and reaction mechanisms.

(2) Sulfate in the ore fluid is reduced to sulfide as organic matter in the environment is oxidized — this mechanism obviously is limited to environments in which organic matter is available. The oxidation—reduction reactions may be inorganic at high temperatures or bacterial at low. White points out that a gap between these two mechanisms may exist between $80°$ and $200°C$, a range too high for any known bacterial reactions and too low for significant rates for the inorganic ones.

(3) Sulfide ions are released by the breakdown of sulfur-bearing hydrocarbons.

(4) Sulfide previously stored in (syngenetic or diagenetic) pyrite recombines to form more stable (*sic*) sulfides.

(5) The metal-bearing fluid mixes with another fluid, produced in several possible ways, that is high in sulfide.

(6) Systems that are sulfide-deficient (as determined by an analysis of an average volume of the fluid) actually may be evolving sulfide at rates that are sufficient to precipitate metals sulfides in large amounts in favorable parts of the systems.

(7) Especially at temperatures above $300°C$, many ore fluids may contain abundant total sulfur, dominantly as SO_2, SO_4^{2-}, or some other intermediate-valence species of sulfur. As temperature decreases, White holds that conversion of such sulfide species to sulfur may occur. (Of these seven suggested mechanisms, (2), (3), and (5) may produce ore sulfides at or near the sea floor; these deposits may be either syngenetic or diagenetic.) White also points out that metal-rich brines, deficient in sulfur, amy never from ore deposits but may remain dispersed in rock pores, may escape into other rock types where the metals may be dispersed by adsorption on mineral surfaces, or may escape and pond in porous rocks in structural or stratigraphic lows, retaining their metal content until the brine is flushed out by erosion or downward penetration of meteoric water. (White does not mention what seems to me to be the very real possibility that such concentrated chloride brines, low in sulfur, may, particularly if they boil, deposit such metal chlorides as $AgCl$ and $PbCl_2$. Deposits of such metal chlorides as primary minerals are unknown.) White indicates that the density of a salt-rich brine normally is higher than that of surface waters; the effects of such density contrasts are evident in Red Sea brine pools, and similar hydrodynamic effects are likely for most syngenetic and early diagenetic base-metal deposits.

1969. Massive sulfide deposits

Anderson (1969) emphasizes that massive sulfide deposits of pyrite and/or pyrhotite, plus various amounts of chalcopyrite, sphalerite, and galena commonly are associated with volcanic rocks that had accumulated in eugeosynclines. These deposits usually, but not always, bear the imprint of later regional metamorphism. Of the deposits Anderson considers in his review, two-thirds are in volcanic rocks, they being divided about equally between silicic and mafic volcanic rocks; the other third occur in tuffaceous sedimentary rocks interbedded with the volcanics. He finds little evidence that compels him to relate these deposits genetically to granitic rocks (magmas?), but sees considerable evidence that supports a genetic relationship to volcanic rocks (the source magma chamber of the volcanics?) that accumulated in a submarine environment. He believes that some of these deposits were formed during or shortly after the emplacement of the volcanics. This process of ore formation may have involved an early deposition of pyrite, later followed by the addition of the base metals. He considers that the base metals and some of the sulfur were transported in heated solutions, probably from different sources but all containing sufficient (moderate to high) concentrations that they may be designated as alkali—chloride brines. He believes that the pyrite and base metals may have been deposited not only during or shortly after the accumulation of the volcanic rocks but also during regional metamorphism or the intrusion of granitic plutons. He thinks that the sources of the base metals may have been early metal concentrations distributed through the volcanic pile, trace quantities of these metals in the silicate minerals of the volcanic rocks, or later crystallizing magmas. He points out the need to discriminate between pyrite formed by biogenic mechanisms and pyrite produced by volcanic emanations. (If deposits of these two types are to be different from each other, the sulfur in the sedimentary ones must have been produced by bacterial action on ocean-water sulfates, whereas the sulfur of the volcanic deposits must have accompanied the base metals in their upward journey from their source magma. If the volcanic emanations entered the sea without a sufficient supply of sulfur to produce large quantities of base metals, sulfide ions most have been available in the sea in large quantities over a short period of time, something probably beyond the capabilities of bacteria, particularly bacteria heated by the influx of large volumes of hot volcanic waters. If such bacteriogenic sulfur were, nevertheless, available, the sulfur in both types of deposits would be much the same.) If the sulfurs in the two deposit types are different in origin, they well may be quite different in their contents of selenium and vanadium, as Anderson believes. Pyrite formed from volcanic emanations will tend to be high in selenium, pyrite from biogenic sulfide will be essentially free of selenium. He is less certain that vanadium is high in syngenetic pyrite since he is not sure that the high-vanadium pyrites are actually syngenetic. (Anderson presents another example of a geologist who, when he cannot explain contradictions in his interpretation of the evidence, says that no simple explanation probably can be applicable to the deposit — or deposits — in question.)

1969. Diagenesis

Berner (1969) presents three diagenetic models, based on layered occurrences of pyrite, to show how iron and sulfur may migrate within recently deposited anaerobic sediments. The two most important factors (according to Berner) affecting such migration are the sharp variation with depth in the content of organic matter (this is assumed to be the same in each of Berner's models) and the original content of that proportion of ferruginous material that is reactive toward H_2S (this is varied from one model to the others). In a sediment low in reactive iron, bacterial generation of H_2S in an organic-rich layer lead to a darkening of the layer and the formation of iron sulfide in adjacent organic-poor sediment for a distance far greater than the thickness of the organic layer. Berner derives equations for the calculation of this distance and the rate of darkening from a small number of parameters that characterize the sediment in question. In a sediment high in reactive iron, an organic-rich layer should be enriched in iron sulfide at its boundaries by the diffusion of dissolved iron and sulfate toward the organic layer.

Berner found that adjacent artificial sediments of very different organic-matter content undergo qualitative changes in the laboratory as predicted by the models. He also considers that the laboratory work demonstrates quantitatively the general validity of the model that deals with low iron-content material.

He believes that the reasoning applied to the layered models can be converted to three dimensions to describe changes in an organic-poor sediment surrounding an organic-rich dead body. This situation often is cited as an example of a reducing microenvironment. In the one-dimensional (layer) or the three-dimensional form, the high-iron model may be the correct mechanism for the formation of many pyrite concretions, layers, or pyritized fossils.

In some sediments, iron-sulfide Liesegang bands may form by inter-diffusion of iron and sulfide, if two organic-rich layers or greatly differing iron contents are adjacent to each other or if dissolved sulfide escapes from an organic-rich region into an organic-poor one before it (the sulfide) is fixed by inward diffusing dissolved iron.

Thin pyrite layers may form by the upward diffusion of Fe^{2+} from underlying organic-rich sediment to an oxygenated sediment—water interface. The accumulated hydrous ferric oxide so formed may, upon burial, be converted to pyrite by reaction with H_2S from an adjacent organic-rich layer. (Only the presence of Fe^{3+} in the hydrous ferric oxide makes this reaction possible.)

(These results apply only to iron sulfides formed syngenetically, but they suggest how other sulfides can so develop.)

1969. Ore brines

In 1969, Degens and Ross[1] edited and published a volume entitled "Hot Brines and Recent Heavy Metal Deposits in the Red Sea". It is obvious that space will not permit the summarization of all 50 papers, not all of which are of direct interest to the economic

---

[1] Editor's note: see also their chapter in Vol. 4.

geologist interested in the problems of the syngenetic formation of ore deposits. Two are, in my opninion, most important for the economic geologist. The first of these is by Craig (1969) and the second by Emery et al. (1969).

Craig (1969) considers that the geochemical, chemical, and physical data on the Red Sea brines, available when he wrote, are consistent with a steady-state model in which sea water circulates downward through evaporite sediments and flows northward, driven by density differences between columns of brine and columns of sea water and heated by the local geothermal gradient. The isotopic composition of the water and the dissolved argon concentration indicate to the author that the waters were quite warm, relative to sea water, and were originally developed near the surface. Further, the deuterium and ^{18}O concentrations suggest to him that the source water came from probably about 800 km south of the depressions (on the sea floor), near or on the southern sill (of the Red Sea) where water of 38.2‰ salinity is abundant. He believes that the brine must flow at least 400 km in some few thousand years; he considers that this indicates fissure flow, probably in basalts, along the central rift. He considers that the lack of an ^{18}O shift in the water shows that the temperatures are not much higher than 100°C. He thinks that the enrichment pattern of salts in the brine is consistent with an evaporitic source, probably a halite—sylvite zone. The concentration of certain constituents, however, are depleted as compared with original sea water so he suggests that selective membrane filtration may have occurred. (Other explanations seem possible but are not considered by Craig.) He believes that the flow of present brine was begun about 12,000 years ago when the water in the sea basin rose to the level of the southern sill. He thinks that, in glacial times, the sea level fell and brine flow temporarily may have increased, causing the brine to overflow its present level in the depression and possibly to precipitate large amounts of heavy metals by mixing with sea water. In correspondence with the glacial cycles, there may have been several cycles of heavy-metal precipitation. He points out that sulfur and carbon isotope data show that the sulfides precipitating from the brine could not have been developed by sulfate reduction. He thinks that, in the Atlantis deep, sulfides probably are precipitated by H_2S from the surrounding sediments. He sees no evidence that requires contributions to the brines from volcanic or magmatic sources, but he is not certain as to the source of iron, manganese, and certain trace metals. He accounts for metal enrichment by one of two processes:

(1) Water passing through evaporites approaches saturation in major components early but will continue to extract trace metals in which it is unsaturated.

(2) Recycling by steady-state precipitation in stacked convection cells increases the concentration in the lowest layer by large factors, even orders of magnitude, relative to the actual fluid entering the depressions. The original fluid must be sampled before enrichment factors can be designated as other than "apparent".

Craig points out that further data are needed on:

(1) Detailed salinity, temperature, dissolved oxygen, and trace element profiles across the brine—sea water transition regions.

(2) Dissolved carbonate and ^{13}C and ^{14}C profiles from normal sea water continuously through the brines.

(3) Chlorinity and sulfate measurements taken simultaneously on interstitial waters both in the brine depressions and in normal sediments outside the depressions; these should be correlated with ^{34}S, ^{13}C, ^{18}O, and ^{14}C measurements on the sediments.

Emery et al. (1969) pose a number of questions about the Red Sea brines that must be answered if the deposits and their implications to economic geology are to be understood:

(1) What characteristics of the regional and local geology are so unusual as to indicate how this brine and the deposits from it were developed?

(2) Are the chemical, mineralogical and physical properties of the deposit due to igneous or sedimentary processes or both?

(3) When did the deposit begin to form, and is it still forming?

(4) Are similar environments now present or likely to occur elsewhere in the world, and in what kinds of areas should they be looked for?

(5) What is the probable economic value of the Red Sea deposits?

(6) What are the legal implications of the deposit in respect to its exploitation.

They then discuss in order the geological and geophysical setting, the water and organisms involved, the sediments developed, and the economic and legal implications. Under the first heading, they conclude that the pulling apart of Arabia and Africa to form the Red Sea began about 25 m.y. ago (Miocene time) and continues to the present. The layer of sediment in the area is thinnest along the axial belt and is underlain by a broad ridge of dense rock, probably basalt; this basalt filled the region left by the pulling apart of what was once a continuous granitic shield. The region does not have many earthquakes but considerable secondary deformation has occurred. They believe that the inflow of basalt probably was accompanied by the injection of hot brines through the floor of the Red Sea. Later, this activity lessened and the brine and sediment cooled somewhat; as heating again increased, the temperature of both materials again went up.

Emery et al. believe that the δ^{34}S values, which ranged around +29‰ for dissolved sulfate in all brines and interstitial waters, indicate not only the absence of extreme evaporation in the formation of the brines but also the absence of extensive bacterial fractionation in the brines after they emerged on the floor of the Red Sea. The δ^{34}S of the metallic sulfides of between +3 to +11‰ suggests to them that the sulfide was produced by the reduction of organic matter in the shales at high temperatures; it was not, they think, done by bacteria. They believe that the high temperature, salinity, and concentration of heavy metals in the brines are formidable obstacles to life processes. The only samples they found that produced positive evidence of viable bacteria were those of 22°C water and water at the interface between 22° water and 44°C brine. The sediments contain hard parts of many small pelagic organisms, but these must have lived in near-surface waters, and their remains sank to the depths only after they had died. The authors found that seven facies of sediments could be distinguished by color, texture, and composition. These are: detrital, iron-montmorillonite, goethite-amorphous, sulfide, mangano-siderite, anhydrite, and manganite. They point out that the metalliferous deposits of the Red Sea contain the same minerals that are characteristic of shallow, moderate, and deep vein deposits, but these minerals are not confined to veins but are disseminated in a widespread blanket of sediments. This, they say, is exactly what should be expected of metal-bearing solutions discharged from igneous rocks through deep veins into a deep body of ocean water. They believe that the metals in these solutions obtained the sulfide ion needed to form sulfides from such ions in the brines, the ions having been formed by bacterial reduction of sulfate in the original ocean water. (How they reconcile this statement with their failure to find bacteria in any but the lowest temperature waters is not clear.) They go on to compare the Red Sea deposits with the Kuroko deposits in Japan. They consider that the Kuroko deposits show the lower, third dimension of deposits of the Red Sea type, with deposits similar to the lower parts of the Kuroko ore bodies probably lying beneath the Red Sea sulfide-rich sediments. They suspect that most deposits of the Red Sea type are only transitory. Finally, they mention the economic and legal implications of attempting to mine the Red Sea deposits.

1969. Sea water depths

Jones (1969) considers the possibility of using pillow lavas as indicators of the depth of sea water overlying the sea floor on which the pillow lavas were extruded. He reports that a prevalent variety of pillow lava in Iceland (an interglacial basalt) shows a systematic vertical variation in vesicularity and vesicle size from which he thinks can be inferred the depth of sea water at the site of emplacement at the time of eruption. He also has observed a subsidiary variety of pillow lava that shows no such variation; he believes that this lack of regular variation resulted from partial and varied degassification at a subaerial vent before the pillows formed. From study of an Ordovician sequence of pillow lavas in Wales, he argues that there was a progressive shallowing of the depth of water in which the basalts were emplaced, these depths having been neritic to bathyal. He quotes Moore, with approval, as saying that it is the vesicularity of the pillow at the point at which the lava ceased to flow that can be correlated with the depth of overlying sea water. The vesicularity does not, necessarily, measure the depth of eruption but only that of emplacement. As would be expected, Jones reports that the abundance and size of macrovesicles increases as the lava sequence is followed upward. Further, concentric zonation of vesicles in the more vesicular upper halves of the pillow appears and is intensified, the higher the position of the pillows in the lava sequence. Conspicuously zoned pillows are dominant at depths of less than 350 m and rare below 450 m. On the contrary, faintly zoned or unzoned pillows predominate below 450 m and are rare above 350 m.

1969. Massive sulfide deposits

Lusk (1969) has investigated the distribution of zinc, lead, silver, and copper in (what he calls) a typical volcanic-type stratiform deposit (Heath Steele B-1 ore body) in New Brunswick by studying assay data. He has been able to delineate base-metal zoning parallel to the plane of mineralogical layering (lateral zoning) and perpendicular to it (across-layer zoning). The across-layer zoning he characterizes as being enriched in lead and zinc, with a complimentary decrease in copper, as the deposit is followed stratigraphically upward. He finds a similar relationship for similar deposits in other parts of the area. He characterizes the lateral zoning by similar distribution patterns for zinc, lead, and silver and essentially a reverse pattern for copper. Within given strata, maximum concentrations of zinc, lead, and silver occur at two principal centers, the axes of which project through the deposit approximately perpendicularly to the layering and to the bounding surface of the deposit. Massive sulfides in the vicinity of both centers are underlain by volcanic fragmental or intrusive bodies.

Lusk attributes both types of zoning to the introduction of volcanic exhalations on the basis of consistent relationships between the character of across-layer zoning and stratigraphy on the one hand and the geometrical features of the lateral zoning on the other. The across-layer zoning sequence is similar to the normal zoning sequence in epigenetic hydrothermal vein-deposits and, therefore, may have been formed by a similar metal fractionation process. He does not consider, however, that the lateral zoning away from the two centers is the same as the zoning sequence in typical vein deposits and cannot be accounted for by currently suggested mechanism for the production of zoning. On the contrary, he thinks that the base-metal zoning in this deposit can be explained by a volcanic-exhalative origin. He believes that the consistent relationship between stratigraphy and the character of the across-layer zoning indicates that this deposit (and probably others like it) formed prior to the major folding of the enclosing rocks. He cannot point to a lack of alteration effects above the ores to support such a method for its formation, but this lack, he thinks, may be due to regional metamorphism (not a completely clear explanation). He suggests, however, that the following features of the deposit are compatible with a volcanic-exhalative origin:

(1) The deposit is lenticular in form and concordant with the enclosing rocks.

(2) A parallelism exists within the (volcanic) metasediments between the mineralogical layering and the stratification.

(3) The center axes continue through the ore body.

(4) The center axes are essentially perpendicular to the mineralogical layering.

(5) "Porphyry" occurs below the centers.

(It is difficult for me to see that these five features could not equally well be explained by the introduction of hydrothermal fluids into the metasedimentary rocks after they had been folded. Such fluids do not need to travel directly upward or horizontally, but instead tend to follow bedding and structures, whatever their attitude may be.)

Lusk and Crocket (1969) obtained sulfur-isotope ratios for co-existing sulfides from the Heath Steele B-1 ore body in New Brunswick and for five other stratiform deposits in that area. Isotopic fractionation between given sulfide pairs were quite constant throughout the area. The mean pyrite—sphalerite, sphalerite—galena, and pyrite—galena fractionations were approximately 1, 2, and 3‰, respectively. The authors argue that these consistencies, together with the similarity of the silicate mineralogies in the sulfide deposits and their wall country rocks on a regional basis, show that the sulfur-isotopic equilibrium was closely approached at a fairly uniform temperature. They suggest that the stratiform deposits of the district underwent low-rank regional metamorphism that generated the fractionations they observed and caused localized isotopic homogenization with respect to the three minerals studied. They say that this implies that the fractionation values should be independent of the bulk sulfur-isotope composition. They tested this concept with a simple linear regression model. The hypothesis is that $\Delta \delta^{34}S$ is independent of $\delta^{34}S$ by determining whether β included zero at the 95% confidence limit. This concept was proved to hold for all regression curves, so the hypothesis that $\Delta \delta^{34}S$ is independent of $\delta^{34}S$ was accepted by them. They used the same model to test the dependency of fractionation values on sample position within the ore body, using $\Delta \delta^{34}S$ and the sampling unit as variables. They found the fractionation values to be independent of sample location. Thus, they hold that sulfur-isotope fractionation was established under conditions of a uniform and constant temperature regime that was sufficiently high to permit extensive local exchange of sulfur isotopes between the constituent minerals of the B-1 ore body. They point out, however, that such consistent mineralogical isotope fractionation might be produced during the deposition of sulfides from hot ore-forming solutions or in a long period of heating during regional metamorphism. Such heating might have caused the probably (they say) originally fine-grained minerals to be coarsened, and such recrystallization might have allowed breaking of metal-sulfur bonds to permit exchange of sulfur isotopes and the attainment of (essential) isotopic equilibrium. They admit that this condition of equilibrium might have been attained during precipitation from a hot ore-forming fluid, but argue that the temperature at which quenching took place would have had to have been uniform over the entire B-1 ore body. They think that the introduction of such a hot ore fluid would have set up a temperature gradient between the country rock and the ore body and probably within the ore zone itself. If this were true (and it is not certainly so by any means), the fractionation is not compatible with deposition from a hydrothermal fluid. From this, they conclude that the deposit was formed syngenetically with the rocks in which it is contained and that the essential isotopic equilibrium was achieved during regional metamorphism. (No consideration is given to the possibility, a very real one, I think, that fractionation to the small scale achieved in the Heath Steele B-1 ore body might well have been attained in a hydrothermal—fluid system of ore deposition.)

1969. Metamorphism of ore bodies

Vokes (1969) reviews the effects of metamorphism on ore deposits and says that metamorphism of sulfide and other ores has now been accepted by the majority of ore geologists. (I would have said that all geologists accept the concept that ore deposits can be metamorphosed, the only question on which there is debate is "how many, how much, and at what stage in the life of the deposit?") Vokes points out that most ore deposits considered to be metamorphosed belong to the stratabound type of usually dominantly pyritic ores. He believes that many deposits of this type have been subjected to orogenic deformation and recrystallization that have obscured their original characters to a greater or lesser degree. He argues that many apparently epigenetic features can be produced by metamorphic recrystallization, deformation, and (re)mobilization. These effects may completely dominate the primary, possibly non-magmatic features of such deposits. (Where such effects are shown by vein deposits or ore masses in igneous rocks, geologists generally — but not universally — are willing to accept a magmatic–hydrothermal mechanism of formation for them. Where metamorphosed deposits are conformable to bedding in volcanic or sedimentary rocks, by far the greater proportion of present-day geologists think the original deposit must have been formed syngenetically with their hosts. All that actually is certain about a metamorphosed deposit is, that it was so treated at some stage in its history; because a conformable deposit was metamorphosed does not mean it must have been syngenetically formed. It would be remarkable if all hydrothermal deposits were in, or spatially related to, veins.)

Vokes lists and discusses the types of metamorphism that can affect ores (and other rocks) and groups these into: (1) metamorphism of local extent; and (2) metamorphism of regional dimensions.

Local metamorphism includes: (1) contact metamorphism; and (2) cataclastic metamorphism (dislocational or mechanical metamorphism).

Metamorphism of regional extent includes: (1) the regional dynamothermal type; and (2) the regional burial type. The most pronounced effects of contact metamorphism are in the vicinity of intrusive igneous rock and of cataclastic metamorphism where directed pressures are most intense. The major effect of high temperature contact metamorphism is the recrystallization of sulfides, producing new minerals, new (coarser) textures or both. Vokes considers that, at extremely high temperatures, partial melting (and movement) of sulfides may occur. In sulfides, the most spectacular results of cataclastic metamorphism are in the more brittle sulfides — pyrite and arsenopyrite. Vokes believes that recrystallization results in characteristic crystal shapes and intergrowths that depend on the specific minerals ability to assume euhedral shapes. Such textural relations as are produced are not evidence of paragenetic sequence. The apparent paragenetic "sequence" produced by metamorphism are, in fact, crystoblastic sequences that result from mineral tendencies to achieve minimum interfacial free energies of, and equilibria among, the minerals developed. Variations in growth matrices, impurities, non-stoichiometry, and relative orientation effect the interfacial energies so that different crystoblastic series may result.

Vokes considers that progressive regional metamorphism leads to an increase in grain size, and the increase may be linked directly with the metamorphic grade of the enclosing rocks. The fabrics caused by deformation depend on the relative plasticity of the ore mass at the temperature produced. Little is known as to possible relief fabrics in metamorphosed sulfide ores, mainly because of the ease with which original textures are destroyed by metamorphism. (Metamorphic textures, to some extent, may have been inherited from the replacement of earlier metamorphosed minerals, sulfides or wall-rock minerals.)

Vokes believes that mobilization of sulfides, especially galena, chalcopyrite, and pyrrhotite, may be quite common, particularly where the metamorphism is of high grade. Movements so produced normally have no effects over more than a few millimeters, although they may be more impressive when flow is to crest and troughs of folds.

The older deposits are, the greater the chance they have been metamorphosed and remetamorphosed. Retrograde metamorphism of sulfides Vokes considers to be weak.

None of this material demonstrates that stratabound deposits *must* have been syngenetically formed, as Vokes clearly states.

1969–1970. Syngenesis and diagenesis

Zimmermann (1969a) considers that the barite and other minerals associated with it in the Upper Mississippi-Valley deposits near Shullsburg, Wisconsin, were produced sedimentarily and diagenetically. This work is an extension of that done by Amstutz and Park (1967) on the fluorite ores of southern Illinois. The author believes that the structure and bedding relationships of the contorted sphalerite- and calcite-bearing limestone layers show that the ore minerals were present in the sediment during deformation of the unconsolidated material. Zimmermann (as do Amstutz and Park, 1967) refers to the probitive value of "geopetal"[1] features of the ore minerals in the rocks as to their syngentic formation. Zimmermann argues that the congruency of the ore-mineral layers with contorted limestone beds, especially in the lower part of the Quimbys Mill Member, indicates that the ores were interlayered with the sediments before these were lithified; (he says before deformation, by which I presume he means deformation of unconsolidated material). He says that the sphalerite, galena, pyrite, marcasite, and barite ores and the calcite are so closely associated with the contorted limestone beds that they must have accumu-

---

[1] Geopetal is defined by the "Glossary of Geology" as any rock feature that indicates the relation of top to bottom at the time of formation of the rock, that is, a fabric or internal structure or organization that indicates the original orientation of a stratified rock, such as cross-bedding or grains on a boundary surface. Since there seems to be no doubt as to which is top and which is bottom in the various members of the stratigraphic sequence in the Upper Mississippi-Valley district, the use of the term "geopetal" seems to be strained here.

lated in them early in diagenesis. The ore materials probably, he thinks, were brought into the sediments by drainage from areas at higher elevations. The author believes that it is much more difficult to apply exo-epigenetic theories of emplacement from hydrothermal solutions (apparently even in the broad sense), and to relate these to the sediment—ore relationships. (I suppose that Zimmermann would consider the ores in the pitches and flats as having been placed there by remobilization at some time after the sediments had been consolidated, the ores having been extracted from the sediments and moved into these structures.)

Zimmermann (1969b) describes the barite deposits of the Shoshone Range in Nevada as containing rhythmic layering of several types. In discussing these deposits, he explains why geopetal features are important in determining the nature of their formation. He says that top—bottom features always show some indication of diagenesis in their fabric because, even when formed by syngenetic processes, they went through the entire period in which the diagenetic processes took place. If top—bottom features were developed during diagenesis, at least some evidence of diagenesis will remain in their fabrics. Top—bottom features arising epigenetically through solution of carbonate, although Zimmermann says such have been reported elsewhere, were not seen in the Nevada barite. Stylolite seams, however, occur throughout the barite beds. (Since the likelihood of solution on the scale necessary to develop stylolites in barite is small, the presence of stylolites in barite would suggest that the barite had replaced a carbonate layer in which stylolites had previously developed.) Zimmermann says that the three most readily observable features that often show top—bottom relations are as follows.

(1) Shale layers bend around barite nodules. (These seem as reasonably to be barite replacements of carbonate-nodule remnants in rather broad stylolites.)

(2) Clay seams with iron stains and grains grade upwards into barite with fine-grained quartz. (Again, these seem as easily explained as replacement features.)

(3) Clay seams grade laterally into stylolites. (I am not certain what this means, particularly since no figure shows this relationship.)

At any event, Zimmermann holds that rhythmic layering and top—bottom features (easily explained otherwise) are similar features to those known in unmineralized sedimentary rocks. He considers that these features indicate a sedimentary formation for the barite and that the original deposition was followed by compaction, dissolution, and other diagenetic processes. (It seems to me that there is no doubt but that the rocks show abundant sedimentary features, developed during sedimentation or diagenesis, but they may mean no more than that they were retained during the replacement of much of the stratigraphic sequence by barrite.)

Berner (1970) determined experimentally that pyrite can be synthesized at neutral pH in concentrated sulfide solution and in natural sediments by the reaction between precipitated FeS and native S at 65°C. He considers that similar reactions would occur at sedimentary temperatures, but several years would be required to complete the reaction. The synthetic pyrite so produced has a framboidal texture.

Berner considers the major steps in the formation of sedimentary pyrite to be: (1) bacterial sulfate reduction; (2) reaction of H_2S with iron minerals (apparently not iron ions) to form iron monosulfides; and (3) the reaction of iron monosulfide with elemental sulfur to produce pyrite. If this is true, the presence of elemental sulfur would be essential for, and a limiting factor in, the formation of pyrite. (S^0 can change FeS to FeS_2 by simple addition in the reaction: $FeS + S^0 \rightarrow FeS_2$ in which $S_2^{2-} = (S^{2-}-S^0)$) Berner's concept that S^0 reacts with FeS to form pyrite is somewhat clarified by his saying that if the sediments are overlain by aerobic waters, oxygen will react with H_2S and FeS to form elemental sulfur and, then presumably, pyrite. (He does not give an equation for this reaction.)

Berner goes on to say that, in sediments overlain by anaerobic waters, (euxinic basins), the production of elemental sulfur is (even) less readily explained. In the Black Sea, FeS completely transforms to FeS_2, and Berner can only explain the formation of FeS_2 by suggesting that the sulfur is added from above or some unknown organic reaction is responsible. (These are weak reeds on which to lean the reaction, granted such reeds grow at the bottom of the Black Sea. Perhaps calling on Fe^{3+} to be present is no worse as the following possible reaction shows:

$$2Fe^{3+} + FeS + S^{2-} = FeS_2 + 2Fe^{2+})$$

1970. Ore fluids

J.S. Brown (1970) has tabulated the conclusions of a considerable number of prominent contemporary workers on the Mississippi Valley-type of ore deposits. His work shows that North American geologists increasingly favor an epigenetic formation in which the ore fluid was dominantly connate-marine, with minor additions probably coming from deeper sources. European opinion, however, is divided almost equally between syngenesis—diagenesis and magmatic—epigenetic mechanisms of formation. He adds, however, that many French geologists prefer a hydratogene concept that invokes deeply circulating meteoric waters. European workers have not devoted much study to fluid inclusions. He considers that one theme runs through most of the work on this type of deposit, namely the probability that a factor of considerable importance in the precipitation of ore minerals in deposits of this type has been the mixing of unlike solutions, either magmatic and connate, connate and meteoric, or whatever (Brown's term). He concludes that large amounts of metals are carried in solution as chloride complexes, sulfides complexes or possibly otherwise (again Brown's phrase). He considers it obvious that, for such a solution saturated under particular conditions, any change in conditions (P, T, Eh, pH, and, Brown adds, ect.) must either increase or decrease its metal-carrying capacity, and a decrease will result in precipitation (I know of no one who would argue against this statement). He thinks that the mixing of two or more solutions of different sources and different characters will produce a precipitate so hybrid as to provide a possible explanation for (among other things) the isotopic variability of lead, and possibly to some degree of sulfur, as is typical of the Mississippi Valley lead—zinc deposits. (It should be noted that the European deposits of the Mississippi-Valley type, so far as minerals, host rocks, relations to structures, and trace elements are concerned, are similar to those in the central United States, but they have normal-type lead isotope ratios and not the "J" type of those in that part of the United States. What differences in sources or varieties of solutions is responsible for this difference?) Brown, however, argues that the term Mississippi Valley-type should be reserved for those containing undatable, futuristically anomalous leads. Thus, he excludes from this class all European deposits as well as those of North Africa, that meet all the other requirements for the Mississippi-Valley type. The only acception to this barring of such deposits outside North America would be Laisvall and similar deposits in the Caledonides of Scandinavia. His definition also probably would keep out the deposits of east and central Tennessee, Virginia, and Pennsylvania, and those of Pine Point. Brown would subdivide these low-temperature lead-bearing deposits into three categories:

(1) Normal lead types, for example the English Pennines, in which the lead isotopes might be considered as given ages somewhere near reality.

(2) B-type lead of the Alpine deposits in which the lead isotope ratios appear to give ages older than the time of deposition.

(3) J-type lead of Brown's true Mississippi-Valley type, the isotopic ratios of which suggest that they have yet to be deposited, in the central United States.

(Such a classification shows that much more work is needed on deposits of this type and on the meaning of the lead-isotope ratios associated with them.)

1970. Kuroko deposits[1]

Although the Kuroko deposits of northern Honshu have been discussed mainly by Japanese geologists, the papers from the *1970 Symposium of I.A.G.O.D. in Tokyo and Kyoto* and the field trips before and after the Symposium have resulted in considerable interest by North American geologists in this type of ore. Until the papers in preparation by Hiroshi Ohmoto of the Penn State faculty are published, the most appropriate paper to summarize, in my opinion, is that by *Matsukuma and Horikoshi (1970)* and the book edited by *Ishihara et al. (1974)*.

Kuroko-type ores are composed of four ore types:

(1) Kuroko (or black) ores proper, mainly of sphalerite, galena, tetrahedrite, and barite.

(2) Oko (or yellow) ores, dominantly cupriferous pyrite.

(3) Keiko (or siliceous) ores, cupriferous, silica-rich ores, either disseminated or in stockwork veinlets.

---

[1] Editor's note: see also the chapters by Sangster and Scott (Vol. 6), and by Bernard and Samama (this Vol.).

(4) Sekkoko ores, anhydrite, gypsum, and pyrite.

These ores are closely associated in space and time of formation with Tertiary volcanic rocks and may consist of one or more or all of the four Kuroko-type ores listed above. These deposits are, essentially, stratabound, and most geologists who have studied them consider that they are, in part at least, syngenetic with the rocks that contain them.

Whether the ores were merely emplaced in the volcanic rocks after they were lithified or were deposited on the sea floor at the same time that they (the pyroclastics) were accumulating, the mode of volcanic eruption and the mechanisms of emplacement and sedimentation of pyroclastic materials in the sea is of vital importance in the story of ore formation. Most of these pyroclastic sediments were deposited by turbidity currents that probably were initiated by the submarine eruptions. The first step in the accumulation of a particular phase of pyroclastics was the development of massive tuff breccias, rich in lithic fragments; this was followed by deposition of well-stratified tuffs and intercalated mudstones. Lava flows and domes were formed during the pyroclastic sedimentation. None of the flows was very large, all being under 1 km^2 in area and not more than a few hundred meters in diameter. The dome lavas were brecciated by their rapid cooling on contact with sea water. On the sides of many of the domes, steam explosions often occurred, yet the dimensions of the breccias so produced were only on the order of 30 × 30 × 10 m. Ore mineralization is spatially, if not genetically, related to these domes where they occur. The amount of fracturing in the domes (due to steam explosions occurring on extrusion into sea water) was great enough to provide excellent channelways up which hydrothermal fluids or volcanic exhalations could move to, or toward, the sea floor. Not all such domes are brecciated and not all Kuroko deposits, moreover, are spatially associated with such domes, many deposits having been found in pyroclastic flows with which no domes were developed.

If the Kuroko ores were formed syngenetically, by sedimentation from igneous emanations that reached the sea floor, they must be of essentially the same age as the rocks in which they are found, that is, late Middle Miocene or Late Tertiary. If, on the other hand, the ores were emplaced by replacement and open-space filling, the ore fluids from which they were deposited must have come from the same general magma-chamber source as the lavas of which so much of the Miocene stratigraphic column is composed. This being the case, the ores must have been introduced, at the latest, into the area during the Dewa disturbance at the end of the Miocene, so they would also be, granted this mode or origin is the correct one, of Late Tertiary Age.

Since most Japanese geologists believe that the Kuroko-type deposits formed syngenetically, from volcanic exhalations that reached the sea floor, the tendency to emphasize their stratabound character is irresistible. Nevertheless, such characteristics do exist and should be presented. The ores all exhibit the same general stratigraphic succession to a greater or lesser degree, although some of the members of the idealized section listed below are lacking in individual deposits. The members of the succession are, from bottom to top:

(1) Footwall, normally silicified rhyolite and pyroclastic rocks that are in part veined by sulfides and in part contain disseminated sulfides.

(2) Sekkoko zone, anhydrite, gypsum, pyrite ore.

(3) Keiko zone, chalcopyrite—pyrite—bearing stockwork and disseminated ores.

(4) Oko zone, chalcopyrite—pyrite ore.

(5) Kuroko zone, barite, sphalerite, galena, tetrahedrite (main copper mineral), and silver-bearing sulfosalts.

(6) Ferruginous quartz zone, mainly hematite, quartz, and some pyrite.

(7) Hanging wall, volcanic and sedimentary formation.

The arrangement of the ores within the ore-bearing horizon is not nearly so regular as this tabulation would suggest. Normally, the various ore types are contained in a gypsum zone that only in places is rich enough in gypsum to be classed a mineable gypsum deposit, but some of the Kuroko ores may lie directly on the footwall complex. In other deposits, however, the ores may be contained anywhere within the black mudstone layer, mainly at the bottom but also as far up as its top. In other areas, where the formation enclosing the ores has been more highly deformed, probably largely due to sea-floor slumping but also to some extent to the tectonic effects of the Dewa disturbance, the bodies are much more irregular. Nevertheless, the pattern is such that the ores can reasonably be called stratabound.

Of the actual ores, the metallic one ordinarily lowest in the sequence is the Keiko, that is essentially highly silicified rock in which quartz has replaced rhyolite to a large extent and in which some sulfide minerals, mainly pyrite and chalcopyrite, may be disseminated or contained in stockwork-type veinlets. Only when the chalcopyrite is sufficiently abundant for the rock to be mined is the name Keiko actually applied to it. The ores normally are in stockwork-type deposits that underlie the stratiform portion of the Kuroko-type deposit; these Keiko ores generally are found in pyroclastic breccias, the structure of which usually can be decerned even in highly silicified ore. Essentially all geologists familiar with the Keiko ores are convinced that they are hydrothermal and epigenetic, and most believe that they mark the path by which the ore fluids moved to the sea floor where the Kuroko ores proper were syngenetically deposited. Some of the Keiko ore, however, has so completely changed the original rock that it now consists of cryptocrystalline quartz with spotted globular aggregates of fine crystallized pyrite (and some chalcopyrite) in colloform groups. Some of this material may have formed by open-space deposition.

The Keiko ore gradually changes to the Oko type, which is a massive chalcopyrite—pyrite mixture that is richer in sericitic clay than in quartz. Transitional between Keiko and Oko ores are: (1) siliceous Oko; (2) pyritic Oko in which the pyrite is coarse and the chalcopyrite sparse; and (3) loosely solidified Oko, largely lacking quartz and clay minerals. In the other direction, the Oko ores gradually are converted to Kuroko ores; this Oko is compact and fine-grained or sandy and may have a stratified structure that either reflects the structure of the replaced rock or is a primary sedimentary feature. Sphalerite, galena, tetrahedrite, and chalcopyrite, the main constituents of Kuroko ore, are found in this transitional Oko in varied amounts; locally barite also may be present in this ore.

The black appearance of the Kuroko ore is due to the compact aggregates of sphalerite and galena of which it is principally composed. In addition to these two sulfides, Kuroko ores also contain (in decreasing order of abundance) chalcopyrite, pyrite, and tetrahedrite, although the latter mineral is characteristic of the Kuroko ores only and is the source of the silver they contain. Also observed in places have been bornite, electrum, and native silver. In the lower parts of the Kuroko ore bodies, the main minerals are sphalerite, galena, chalcopyrite, and pyrite; this simple assemblage is complicated at higher levels by the addition of tennantite—tetrahedrite and the silver minerals, while chalcopyrite and pyrite decrease.

Barite ore mostly consists of minute barite crystals that usually are well stratified but only loosely solidified. This sandy ore has considerable lateral extension and thickness and conformably overlies the Kuroko ore bodies at several localities.

Thin layers of ferruginous quartz are known to overlie the barite ore in some places; the iron is present as fine flakes of hematite in the quartz. Rarely, this iron-rich quartz occurs in the lower part of an ore body, but in such a situation, the bodies are small and discontinuous.

The gypsum deposits, Sekkoko ore, underlie either Kuroko beds or stratified Oko bodies and correspond in position to Keiko or siliceous Oko ores in normal Kuroko deposits. The gypsum ores are widely distributed in the mineralized areas, and the amounts of gypsum ore generally are much larger than the metallic portions of the deposits. Some gypsum deposits are associated with only small amounts of metallic ores, but all metallic deposits are associated with gypsum ores. Some of the gypsum ore appears to have been formed by the hydration of anhydrite; many gypsum masses contain anhydrite cores, but that all the gypsum was formed by replacement of anhydrite is uncertain. The host rocks of the gypsum ores (Matsukuma and Horikoshi, 1970) are strongly altered pyroclastic sediments (volcanic breccia, tuff breccia, stratified fine tuff and mudstone) that are characterized by sericitization and chloritization. No silicic volcanic masses are known directly adjoining most of these deposits, but nearly all of them are located around small bodies of highly silicified rock; this rock has often been mistaken for altered rhyolite. These silicified masses are quite similar to the silicified rocks in the footwalls of the Kuroko deposits and to the host rocks of the stockwork Keiko ores; in many cases, the silicified bodies with gypsum ores are weakly mineralized with networks of metalliferous barite—quartz veinlets. This would suggest that the silicification was the earlier (and central) part of the gypsum (anhydrite) mineralization. Most of the gypsum is thought by Matsukuma and Horikoshi (1970) to be better designated as a gypsum-bearing sedimentary rock in which the bedding is moderately clear. The alternative explanation, that the gypsum (anhydrite) is the result of replacement of a sedimentary rock has not been disproved. Where the host rock of the gypsum is such pyroclastics as tuff-breccia and pumice tuff, the ore is a mixture of sericitized and chloritized pyroclastics and globular gypsum aggregates less than several centimeters in diameter. Where mudstone is the host rock, the gypsum occurs in 0.1 m thick layers alternating with similar layers of mudstone. Again, whether this is the result of primary sedimentation or replacement is not definitely determined. However, the presence of gypsum veins cutting across the bedding planes of the various host rocks suggests, in these quite unmetamorphosed sediments, that the gypsum (or its parent anhydrite, if it is derived from that mineral) was epigenetically emplaced after the, at least partial, lithification of the host rock.

The lower portions of the gypsum ores normally are associated with metallic ores of the Keiko type, either networks of veinlets or disseminations, while the upper parts are directly beneath boulder-like and bedded Kuroko ores. In other instances, the gypsum ores change gradually upward into pyritic ore and then to Kuroko. In such instances, the gangue mineral of the Kuroko ores may be anhydrite and gypsum instead of barite.

The Kuroko ores are now always bedded but, on occasion, may be irregular (that is, appear to have been introduced epigenetically) or massive. The gypsum deposits are less likely to be massive than to be of tabular or bedded form. Some of the Kuroko deposits will be much more strongly folded than are the associated footwall and hanging wall rocks, suggesting to some that the ores deformed by pre-solidification submarine sliding or were more easily folded in the Dewa disturbance than were the unmineralized adjacent rocks and to others that the pyroclastic materials were more readily folded than the more massive bordering rocks and provided, therefore, better channels for the ascending ore fluids.

In some deposits, the complete Kuroko-type succession of ores is lacking, in others it is repeated when pyroclastics are separated by mudstones or tuffaceous sediments.

Unanimous agreement seems to have been reached to the effect that the Keiko and Oko ores (the latter at least in considerable part) were emplaced by the replacement of broken pyroclastics of various types. The funnel-shaped, stockwork character of the Keiko ores in particular, and the zones of silicification around them, strongly argues for this explanation. On the other hand, the commonly bedded Sekkoko (gypsum) and Kuroko ores are more usually thought to be slightly changed primary sediments, most changes being due to submarine slumping, than to be replacements and open-space filling of the host rocks (more highly deformed than those of the hanging and footwalls). There is no question but that the bulk of the banded Kuroko and Sekkoko ores faithfully reproduce the bedding of the sediments of which they are a part, but this could have been caused by replacement as readily as by primary sedimentation.

To all geologists, the gradual transition from one ore type to another requires that the various ore types have been formed as parts of a single process. The only question is as to where the rock—sea-water interface was during that process. Those who favor a sedimentary origin for the gypsum and black ores believe that it lay immediately above the funnel-shaped channelways in and around which the Keiko and Oko ores were formed. Those who favor replacement and open-space filling of previously deposited and somewhat diagenetically deformed pyroclastic sediments consider that the ores were introduced in the Late Miocene, after the Dewa disturbance, and that the ore fluids did not encounter the sea until they had deposited almost their entire loads.

1970. Uranium deposits in various rock types

Robertson and Douglas (1970) are firmly convinced that 90% of the western world's uranium deposits are in sedimentary rocks. 60% of these reserves are in Proterozoic quartz-pebble conglomerates (since they include the Rand uranium in this 60%, they must use "quartz-pebble" to include the uranium in the algal mats). They say (observing the caveat just given) that these are all similar in appearance, mode of distribution, and manner of formation. The conglomerates contain uranium minerals and/or gold and abundant pyrite. If they are sedimentary, the crust of Proterozoic time must have contained plentiful quartz and much gold, uranium minerals, and pyrite. These materials must have been eroded, transported, and detritally deposited before the earth had acquired an oxygen-rich atmosphere. (No statements are made as to the existence — or lack of it — of banded iron formations of as old or older age than the mineralized conglomerates.)

They say that about 30% of the western world's uranium deposits were formed epigenetically in arkosic sedimentary rocks distributed through the entire geologic column. They believe that deposits are particularly prominent in the semi-arid parts of the United States.

They describe these deposits as being in lenticular bodies of permeable, stream-channel, sedimentary rocks and as having been precipitated from moving ground water where plant debris or other organic matter had accumulated. The uranium in these deposits may have come from volcanic sediments, commonly associated spatially with sedimentary rocks containing the ores or it may have been leached from exposed granite surfaces or from granitic debris incorporated in appropriately located sediments.

The authors mention that a significant part of the western world's uranium resources are in black shales and lignites. These low-grade uranium-containing rocks are not discussed as to source and manner of accumulation of the uranium they contain. By-product uranium from phosphorites also is mentioned as a potential source of that radioactive element. Finally, they say that reworked sediments may provide the uranium in pegmatites and veins.

(This paper is outlined mainly because, in 1970, the Canadian Institute of Mining and Metallurgy thought it worth publishing.)

1971. Diagenesis

Although *Berner's (1971)* entire book is a necessity to anyone interested in the production of ore deposits by sedimentary processes, Chapter 6 — diagenetic processes — and

Chapter 10 — diagenesis of iron minerals — are of the highest importance. The material presented in these two chapters is too long and the explanations too detailed to be more than summarized here. Chapter 6 is designed to demonstrate how the general problem of diagenesis can be approached from an analytical standpoint and to discuss in detail several important diagenetic processes. These are: (1) compaction; (2) cementation; (3) diffusion in sediments; (4) mineral segregation; (5) layer formation; (6) concretion formation; and (7) Donnan equilibrium. The purpose of Chapter 10 is to discuss the relative stability, under sedimentary conditions of the principal iron mineral produced in that environment — hematite, goethite, glauconite, and pyrite, including the original minerals from which it is derived: mackinawite ($Fe_{1+x}S$) and greigite (Fe_3S_4). As Berner points out, since iron has two oxidation states, stability must be a function of the redox state of the system, which can be expressed in terms of Eh. In addition, he shows that inclusion of several different elements in the composition of these minerals requires the use of numerous additional parameters such as pH and dissolved CO_2. These relationships usually are represented by Eh—pH diagrams for constant total dissolved carbon, sulfur, and silica, and several diagrams normally are required to describe equilibrium between each mineral pair.

A disadvantage of the application of Eh—pH diagrams to sediments is that, in most sediments, particularly the marine variety, the pH is roughly the same. Further, redox equilibrium among all dissolved species, a necesssary assumption in constructing such diagrams for iron minerals, is not approached closely in natural waters. Further, since bacterial activity is not always predictable and since neither are the concentration and distribution of dissolved carbon and sulfur a function of this activity, Berner considers it best to consider these species as independent variables and not dependent, as is assumed in an Eh—pH diagram. Because of this situation, Berner adopts an alternate form of presentation of relative stability. He chooses the variables Eh, P_{CO_2}, and pS^{2-}, the last being the negative logarithm of the activity of sulfide ion, a parameter that can be measured directly with an electrode. For marine sediments, he assumes a constant $a_{Ca^{2+}}$; thus the pH is proportional directly to log P_{CO_2}. Because of the limited range of pH in marine sediments, P_{CO_2} either is held constant or varied over three orders of magnitude ($10^{1.0}$–10^{-4}). These assumptions Berner considers reasonable approximations for most marine sediments. He then presents equations for equilibria between various pairs — siderite—hematite, magnetite—hematite (magnetite because of its common presence in sediments as a detrital mineral), siderite—magnetite, siderite—pyrite, magnetite—pyrite, pyrrhotite—pyrite (pyrrhotite is occasionally authigenic in sediments), hematite—pyrite, pyrrhotite—magnetite, and pyrrhotite—siderite. Glauconite is not included in his diagrams because of a lack of thermodynamic data. Goethite, mackinawite, and greigite also are omitted because of their instability relative to hematite, pyrite, and pyrrhotite. After additional discussion of these diagrams and presenting two examples, Berner goes on to discuss the formation of red beds, of siderite, and of glauconite. Then he considers sources of iron and sulfur, the factors limiting pyrite formation and the mechanism of its formation. In this last, he changes the position he held in his 1970 paper, believing that the principal phases that form first at room temperature and neutral pH, by reaction of H_2S or HS^- with fine-grained goethite or dissolved ferrous iron, are mackinawite and greigite. Such black iron sulfides are common in recent sediments. Their very fine size results from rapid nucleation, due to a high degree of supersaturation during precipitation. These iron sulfides, because of their fine size, are adsorped on larger grains of other materials, making the sediment seem far darker than it really would be without the iron sulfide coatings. Berner considers that the following two reactions proceed with a free energy change of less then 0.

$$FeS \text{ (mackinawite)} \rightarrow FeS \text{ (pyrrhotite)}$$

$$Fe_3S_4 \rightarrow 2FeS \text{ (pyrrhotite)} + FeS_2 \text{ (pyrite)}$$

He admits that the mechanism for the conversion of the black sulfides to pyrite is not well established. He assumes that S° is important in them, without explaining how the S° is produced (in an environment where sulfur is being, and has been, reduced). He gives two possible reactions:

$$FeS + S^0 \rightarrow FeS_2$$

$$Fe_3S_4 + 2S^0 \rightarrow 3FeS_2$$

The second equation might be more clearly expressed as:

$$2S^0 + Fe^2 Fe^3 S_4 \rightarrow 3FeS_2$$

Berner, however, goes on to report that experimental studies have shown that pyrite will form at low temperatures and neutral pH by reaction of H_2S with monosulfides and elemental sulfur. He reports that, if elemental sulfur is absent, he knows of little or no evidence (if he knows of a little evidence, he should not say "no"; if he knows of no evidence, he should not say "little") for pyrite formation under these conditions. (The entire problem obviously needs much more study.)

1971. Massive sulfide deposits — classification

Gilmour (1971) again advocated the adoption of classifications[1] of mineral deposits that are based on observable features rather than on hypothetical mechanism of genesis. His original work on this problem (published in 1962) based his scheme on the mineralogy of the deposits and the composition and tectonic setting of the host rocks. Originally, Gilmour separated occurrences of zinc—lead (copper) deposits in eugeosynclinal sedimentary and volcanic rocks from copper (zinc) deposits in accumulations of rhyolitic pyroclastic material. He later came to the conclusion that the similarities and differences of these two groups could be related and reconciled by regarding them as separate members in a more or less continuous linear series. These relationships are shown in his table 1

---

[1] Editor's note: see in this Vol. Chapter 4 by Gilmour, where he updates and expands his concepts.

(which cannot be reproduced here), and the class of massive sulfide deposits (a series, following Gilmour) is given in his table 2. In table 1, Gilmour puts across the top of his classification, from left to right, the terms — stage of development, sedimentation (or erosion), igneous activity, dominant magma, and type of occurrence and examples. Along the left margin, in the tectonic setting column, from top to bottom, he places "craton" and below that "orogenic belt" that he divides, in turn, into "geosynclinal phase" below and "orogenic phase" above. Under "geosynclinal phase" are, from bottom to top, "eu-" and "mio-". Under the former are early, middle, and late, under mio- is the phrase "or marginal". Under "orogenic phase" is "syn- or late", below, and "post-", above. Within each of the 30 subdivisions (or pigeon-holes) so created are given statements that apply to the various combinations of horizontal and vertical divisions, for example, the "igneous activity" and "middle eugeosynclinal phase" subdivision box contains the phrase — "basic and ultrabasic conformable (Alpine) intrusions". Obviously, the whole classification must be studied to show how well it describes the various conditions that obtain in and around various types of copper deposits. Further study, however, would be needed to make certain that a continuous series is formed by the deposits listed from bottom to top in the "type of occurrences and examples" column. The bottom box of this column is strata-bound massive pyritic sulfide deposits, and it is expanded in Gilmour's table 2. The left hand margin of table 2 contains, from bottom to top, the subdivisions "commonly associated minerals", "principal recoverable minerals", "principal sulfides", "typical original textures", and "characteristic host rocks". Across the top are given twenty examples that fit with the various subdivisions created by the statements opposite the phrases in the left hand column.

In the two diagrams (a and b) given in fig. 1, Gilmour attempts to show how the various types of deposits are related in space (a) and time (b). In short, the diagrams in fig. 1, convert Gilmour's classification from a descriptive into a genetic one. He believes that the contribution of volcanism to the host rocks of these deposits becomes more obvious as it is studied. He also thinks that a soundly conceived, non-genetic classification may be found to reflect fundamental genetic relationships as he attempts to do in figs. 1a and 1b. He concludes by saying that, broadly speaking, variations (differences?) in pH and availability of some of the constituents of massive sulfides may be ascribed to distance from the volcanic center and variations (ranges?) in Eh to the depth of the depositional site. (In other words, he keeps getting closer and closer to a Lindgren-type classification in which a much greater emphasis is placed on volcanic activity and its products than on magmatic activity and its products.)

1971. Syngenesis

Mills et al. (1971) point out that the barite beds of Stevens County, in northeast Washington state, occur as distinct beds that are from 1 to 45 ft thick, are interbedded with quartz siltstones, argillites, and limestones that may range in age from Ordovician to Carboniferous. They consider the six bedded deposits they studied to be similar in many respects to other bedded (and, they say, presumably marine sedimentary) deposits in Nevada and Arkansas. These similarities are:

(1) Conformity with the bedding of enclosing rocks of the same type as are present in the Washington deposits.

(2) Thin beds (less than 1.0–10 mm).

(3) Very fine grain size (0.02–0.60 mm).

(4) Mosaic texture.

(5) Essentially monomineralic character.

(6) (But containing) disseminated euhedral pyrite and amorphous organic matter.

(7) Fetid odor when crushed.

(8) Lack of control by fractures or other potential channels in ingress (for ore fluids?).

(9) Lack of wall-rock alteration.

(10) Lack of zoning with respect to granitic intrusives.

(11) The barite contains no relics of former host minerals that have not been completely replaced.

They report that the uranium content of the bedded barite ranges between 3 ppm and 17 ppm; this contrasts with the uranium content of hydrothermal veins in Stevens County in which the uranium content usually is less than 3 ppm. They consider that these results are compatible with work in which the uranium content of marine barite nodules has been compared with vein barites. Both the conformable barite layers and the enclosing rocks have two generations of folds, the early one is isoclinal and recumbent and the later is upright, with accompanying slip or crenulation cleavage. They believe that post-folding recrystallization is demonstrated by the loss of undulatory extinction, coarsening of grain size in areas of low pressure and in purer beds, and the presence of 120° triple junction points. They believe that the beds were deposited in the Middle–Late Paleozoic, that the folding took place in the Late Jurassic or Early Cretaceous, and that the recrystallization may be related to the Cretaceous granite batholiths of the district. They consider that the presence of folds and cleavage in the barite, that are similar to those in the enclosing rocks, indicates a pre-folding origin for the barite. Although they report that ocean waters normally are undersaturated with respect to $BaSO_4$, they consider that favorable micro-environments may develop locally around decaying organic matter in which barite can be developed. (The total amount of Washington State barite is so large that the environment can hardly be designated as an infinite series of micro-environments.) They believe that this possibility of an infinite number of micro-environments is a probability in the Washington deposits because of the universal presence of organic matter, pyrite, and a fetid odor when crushed. They quote favorably a statement that barium could have been contributed to sea water by the decay of barium-bearing organisms. (Surely this last is hardly in the class with the concepts they already have put forward.)

1971—1972. Massive sulfide deposits[1]

Sinclair (1971) considers that the massive and disseminated sulfides of the No.5 zone at the Horne mine differ markedly from the mineable massive sulfides bodies of that mine. Little geologic work has been done on this body because it has been too low grade to be mined. Sinclair reports that the No.5 zone sulfides seem to be conformably enclosed in their host rocks; although the zone has been developed, it has remained essentially untouched by mine workings. Its host rocks are Early Precambrian, steeply dipping, tabular masses of interbedded tuffs, rhyolite tuff breccias, and rhyolite breccias; it is north of, and stratigraphically above, the Lower H and the other economic bodies of the Horne mine. Pyrite is the dominant sulfide, but the zone contains scattered masses of sphalerite, chalcopyrite, and gold mineralization that are of economic grade but are too small to justify mining. It reaches no nearer the surface than the 19th level, but it extends down dip for 7000 ft.; it has a maximum strike length of 3000 ft., an average thickness of 100 ft., and a maximum thickness of 450 ft. The enclosing volcanics are thin, interfingering units in a complex stratigraphic arrangement. Sinclair thinks that they are the result of pyroclastic activity, followed by submarine slumps on the flanks of an active volcano. These rocks have been strongly silicified, sericitized, and chloritized and have a strong foliation parallel to the bedding. In addition to the sulfides, the sulfide bodies contain quartz, sericite, and muscovite. Altered breccia fragments, up to a foot in diameter are distributed through the zone's metallic-mineral concentration. Individual sulfides masses, containing up to 9 million tons, are elongated parallel to the dip of the volcanics and have horizontal cross-sections that range from oval to thinly tabular. Sinclair believes that the present relationships in the No.5 zone came from post-depositional slumping of the sulfides, along with the breccias, on the flanks of a volcano. He thinks that the sulfides were developed from volcanic exhalations from submarine solfataras or fumaroles. As sulfide muds, these underwent slumping, and successive slumps would account for the overlapping nature of the various elongated masses. He considers it probable that intermixing with breccias produced in the same process resulted in the isolation of smaller blocks and caused sulfides to be disseminated in the host rocks. He conceives of the alteration of the host rocks as having been caused by continued volcanic exhalative activity after sulfide deposition had ceased. (These last two thoughts seem to be the weakest parts of his argument.)

Cataclastic textures have been well developed in the pyrite, particularly along shear zones; recrystallization in the ores, he suggests, was caused by regional metamorphism. These events, he thinks, occurred in the Kenoran orogeny, about 2.5 b.y. ago. Sphalerite is the next most abundant sulfide after pyrite and shows twin-gliding or translational gliding as evidence of metamorphism, but much of the sphalerite is in small stringers and irregular masses of less certain origin. Chalcopyrite, pyrrhotite, and galena do not add anything to determining whether the ores were introduced syngenetically or epigenetical-

---

[1] Editor's note: see also in Vol. 6 Sangster and Scott's Chapter 5.

ly. The amount of cobalt in the pyrite averages 24 ppm. In the adjacent H ore-bodies, the cobalt content of pyrite averages 1000 ppm, and this amount decreases only slightly with depth. In the No.5, selenium ranges between 33 and 88 ppm, whereas the H pyrite ranges between 390 and 1000 ppm in that element. These differences indicate to Sinclair that the H and No.5 zones had different methods of formation. (They simply may have come from different sources.)

Sinclair believes that the sulfide fragments in the adjacent breccias probably were derived from the No.5 mass beneath, suggesting that the sulfides were emplaced before brecciation. Sinclair finds that the mineralogy, texture, and composition of the sulfide fragments are so similar to these characteristics in the massive and disseminated sulfides bodies in the adjoining No.5 zone that they probably were formed by the same process. He considers that the silicate—oxide rims around these sulfide fragments were formed by the reactions of the pyrite with a silicate liquid such as volcanic activity would produce rather than that they are the result of wall-rock alteration. (Probably the relations of sulfides to each other and to their host rocks could be essentially as well be explained by hydrothermal activity, but Sinclair illustrates the present-day trend to interpret every fact by a syngenetic—metamorphic explanation, which may be the right thing to do, if popularity is any criterion.)

Anderson and Nash, of whom Anderson once was convinced of a classic hydrothermal mechanism for the formation of the massive sulfide deposits of Jerome, argue (1972) that the facts known about these deposits can be interpreted to show that the sulfides were syngenetically deposited with the volcanic rocks in which they are enclosed. These deposits, composed largely of pyrite, chalcopyrite, and sphalerite, are in concordant, stratabound lenses in Precambrian massive quartz-bearing crystal tuffs and overlying bedded tuffaceous rocks. The crystal tuffs were emplaced as submarine pyroclastic flows. They believe that the sulfides were discharged into a submarine basin in hydrothermal brines. After deposition and lithification, the volcanic-sulfide complex was folded and metamorphosed; two periods of folding can be recognized. During the second folding, the older folds were deformed along vertical axes, and they believe that some of the chalcopyrite in the major lens migrated downward to form shoots of intersecting veins in the crystal tuff and chloritized tuff (locally known as black schist). Fluid inclusions in both the volaniclastic and vein quartz appear to the authors to be related in formation to deformation and metamorphism and, therefore, give no direct evidence on the nature of the primary fluids that they presume introduced the ore metals into the volcanic rocks that were accumulating on the sea floor. Liquid, CO_2-bearing inclusions are restricted to rocks and veins near the United Verde deposit and to associated chalcopyrite ore sheets. Fluids at the time of the deposition of these materials were CO_2-rich, were at temperatures of about 235°C, and were under confining pressures of 700 bar or less. They do not apply the term "volcanic-exhalative" to these deposits because they do not want to imply that the ores were transported in the vapor phase.

Bain (1973) in his discussion of Anderson and Nash's paper, argues that 95% of the sulfides were formed initially in the same cycle of events that produced the black chlorite schist and that the host rock of both was a subaerial accumulation and not a submarine pyroclastic flow. He believes that metamorphism transformed an upper member of the underlying Cleopatra quartz porphyry and forced magnesium and iron from that member upward to produce, successively, black chlorite schist, massive sulfides, and chert locally against the stratified tuffaceous sediments. Most of the copper mineralization was late and is located along, and adjacent to, axial plane-shears of 10's of feet of displacement through the sequence. He considers that the pyrite body away from the shears and fractures assays less than 0.4% copper and generally less than 0.25%. Copper may be as much as 0.60% in the pyrite mass near the unproductive black chlorite schist, increasing outward from a trace in the center of the massive sulfide body. He says that the copper mineralization begins where weakly mineralized fracture zones intersect nearly north—south shears. He holds that the pattern of copper mineralization (and sulfidization) is strictly tectonic, is on a pitching anticline involving the siliceous body, and is along preore small faults where the chalcopyrite and bornite are of ore grade. The mineralization is stratigraphic only as far as metasomatism is limited to a "Purple porphyry" tuff host and in that rock only where tectonic structures were present also. Anderson and Nash accept Bain's observations but still believe that their model better explains the sequence of mineralizing events at Jerome.

1972. Sea-floor Fe—Mn nodules[1]

Although the problems of genesis of submarine iron-manganese deposits and the sources of their metal content are somewhat peripheral to the subject of this paper, it seems proper to include a summary of one recent paper on this subject.

Bonatti et al. (1972) point out that such iron—manganese deposits are highly varied in their physical and chemical properties, and they emphasize that no one method of formation can account for all deposits of this class. They attempt a classification of such deposits by dividing them into four categories.

The first of these is the hydrogenous; such deposits result from slow precipitation of iron and manganese principally, from sea water in an oxidizing environment. These precipitates occur as microparticles dispersed in sediments, as nodules and concretions on abyssal plains, and as solid pavements on topographic highs.

The second category is hydrothermal; such deposits result from precipitation from hydrothermal solutions that reach the sea floor in areas of volcanism and/or high heat flow. This second type of precipitate occurs along active ocean ridges and rifts and associated with central submarine volcanoes.

The third category is halmyrolitic; such deposits result from the submarine weathering of basaltic debris. These are important, the authors claim, in limited areas of the Pacific Ocean.

The fourth category is diagenetic; such deposits result from remobilization of manga-

---

[1] Editor's note: see also in Vol. 7 Chapter 7 by Glasby and Read.

nese, and other minor elements, in the sediment column. This type of precipitate occurs in the hemipelagic regions of the oceans and in marginal and enclosed seas.

Bonatti and his colleagues report that the iron and manganese contents of sea water have been estimated as about 0.001 and 0.01, respectively. The iron probably occurs as colloidal ferric hydroxide and the manganese in ionic solutions as Mn^{2+}. Precipitation of these materials is favored by the presence of solid nuclei on the sea floor, with the adsorption of amorphous ferric hydroxide on such nuclei being essential for the formation of manganese oxides, with Mn^{2+} ions adsorbed on the hydrous iron oxide coating on the nuclei. This material is there gradually oxidized to 10Å manganites or δMnO_2. These mineral species act as catalists for further precipitation of manganese. The rate of precipitation is on the order of 1 mm/10^3 years. If the simultaneous accumulation of the associated sediments is rapid in relation to that of iron and manganese, these elements are dispersed as microparticles in the dominant sediment. Where, however, the sedimentation rate is similar to, or slower than, that of the iron and manganese, nodules, slabs, or continuous pavements of the iron—manganese minerals will form.

In hydrogenous deposits, the Mn/Fe ratio ranges from 5 to 0.5, and the concentrations of Ni, Co, and Cu are relatively high. In hydrothermal deposits, the Mn/Fe ratio ranges from 1/4 to 1/50, and the contents of Ni, Co, and Cu are much lower than in the hydrogenous type. In the diagenetic type, the Mn/Fe ratio is from 8/1 to 50/1, and the minor metals (Ni, Co, Cu) are higher than in hydrothermal deposits and lower than in hydrogenous. Since uranium is low in sea water and thorium almost lacking, moderately high contents of these elements in the Fe—Mn precipitates indicate a hydrothermal origin for the deposits in question.

Bonatti and his colleagues point out that transitional types of deposits are common, there being some areas at least where the elements involved are being supplied by different means. Thus, they envision hydrothermal—hydrogenous deposits or hydrogenous—diagenetic deposits and any other possible combination. The widest distribution and most abundant development of nodules seems to be of those of the hydrogenous type.

Whereas Anderson and Nash (1972) change an epigenetic deposit into a syngenetic one, *Griffiths et al. (1972)* convert what was thought by such geologists as Borchert and Schneiderhöhn to be a deposit formed syngenetically into one formed in an epigenetic manner. The area of the Ergani—Maden copper deposits in southeastern Turkey is underlain by gently dipping mudstones, interlayered with mafic volcanic rocks and lenticular beds of limestone, mainly of Eocene Age. The layered rocks were intruded and slightly metamorphosed by large masses of gabbro and related mafic rocks, now mainly serpentinized. The copper deposits occupy an east-trending belt about 20 km long, centering near the town of Maden. Near the mines, the main igneous mass has a gently dipping roof and steep sides. Around and south of Maden, the roof of the igneous rocks has barely been removed by erosion, and the gabbro contains many roof pendants of sedimentary and volcanic rocks. In one of these is the largest mine of the area — Ana Yatak. This deposit, in a canoe-shaped pendant of mudstone, is 1 km long and 0.5 km wide; it has a present vertical extent of 170 m. Part of the mudstone is chloritized and partly replaced by ore minerals. The main ore body in this Ana Yatak area is about 550 by 300 by 50 m in maximum dimensions; this massive sulfide body consists principally of pyrite and chalcopyrite, but locally it includes much magnetite or pyrrhotite. The main gangue mineral is chlorite, with less quartz and very little calcite and barite. Under the massive ore is impregnation ore in which veinlets and small irregular masses of sulfides are in chloritized mudstone. Supergene enrichment produced a surface-related zone of chalcocite, covellite, and bornite, above that is a well-developed gossan in many places.

About 1 km northwest of the Ana Yatak mine is the Mihrap Daği mine; it also is near a contact between intrusive and sedimentary rock. It also contains a massive chloritized sulfide ore body, plus some underlying impregnation ore. Griffiths and his colleagues are convinced that the huge ore bodies replaced solid rock in areas where abundant fractures were available for the upward and outward movement of the ore fluids. A sedimentary manner of formation for the deposits, they believe, is clearly incompatible with the demonstrably long time between sedimentation and mineralization; this time span was, they think, at least 8.5 m.y. They found no major compositional layering in the ore, but they do not believe that the primary ores were homogeneous. The amounts of the various minerals differ widely from one area of the mines to another.

1972. Syngenesis

Hallberg (1972) points out that the mechanism for metal-sulfide accumulation in aquatic sediments involves biological, chemical, and physical processes, and the series of events through which metals pass before they are fixed in sediments as stable minerals is most complex. The redox-state and the pH constitute the chemical controls that regulate the mineral equilibria and reaction rates. Physical processes, including temperature that controls all chemical and biological processes, concern the transport of ions, gases, and solid particles during deposition, diagenesis, and metamorphism of the sediment, plus adsorption of metals on solid particles at any stage. Biological processes are the biochemical accumulation of those metals provided by living organisms and the metabolic production and expulsion of ions and gases, at least some of which act to precipitate or dissolve metal-bearing minerals. In addition to attempting to describe the formation of sedimentary sulfides by the processes listed above, Hallberg also tries to express it in energy symbols. He diagrams, in his fig. 1b, the energy circuit system he envisions as being applicable to the cycle of heavy metals in sea water. This diagram is too complicated to reproduce here but must be studied by anyone hoping to understand what Hallberg is trying to accomplish. He identifies the heavy-metal sources to the sea as dissolved and particulate metal-bearing matter discharged from rivers and dropped from the atmosphere (his Q_1), submarine brines (his Q_2), and exhalations from submarine volcanoes (his Q_3). No new energy is generated in the system by the storage of these materials in the sea, and some work is necessary to move these materials in and out of storage. He illustrates the energy input as being from the sun (his Q_4) and from the interior of the earth (his Q_5); these two act as work gates in the energy circuit system. The total amount of metals contributed to the sea is dependent not only on the concentration of the discharged materials but also on their flow rate. The total amount of metals per unit time added to the sea is:

$$k_1 Q_1 Q_4 + k_2 Q_2 Q_5 + k_3 Q_3 Q_5$$

When the discharged material reaches the sea, the anomalously high concentrations of base metals, compared with sea water are diluted. He points out that the volume of the sea acts as a divisive work gate, in contrast to Q_4 and Q_5 that had a multiplying effect. The work of one flow controlling the conductivity of the second may be more complex than simply divisive or multiplicative. The local character of the volume of sea water involved must be considered, for example, if the area is shallow and has a limited circulation; in such a water volume, the metal concentration normally is higher than in a deep sea with good circulation. Brines also give rise to horizontal density-controlled layers, with metalliferous brines below and sea water above. In any study of metal-bearing mineral precipitation, only the brine below the density-controlled barrier has to be considered. The metals exist in two principal forms, dissolved and adsorbed on particles. Hallberg, therefore, divides the metal concentration into two "tanks", one with dissolved metal (his Q_8) and one with particles (his Q_9). The clay on which particles are adsorbed in his Q_7. The flow (his J_1) between the energy storage volumes (Q_8) and (Q_9) is proportional to the difference in the forward and backward force on a single pathway (as in an electrical field) or:

$$J_1 = k(Q_8 - Q_9)$$

In addition to clay, algae and microorganisms absorb dissolved metals, whereas multicellular heterotrophs (his Q_{10}) also absorb particulate metals, especially where they are associated with organic matter. Photoautotrophic organisms (his Q_{11}) receive wave energy, such as light. The living organisms may have a dilution effect on the metals because they store them in their cells. On the contrary, the organisms may concentrate metals in the sediments, if these organisms reach the sediment before they decompose.

Hallberg points out that the biological factor has different enrichment effects, depending on where in the sea such processes take place. In the near-shore environment, the multiplying factor should be greater for some metals than it can be in the open sea; near shore, the benthic algae will be of great importance in metal concentration, whereas at depths of 1000 m, for example, their effect on the metals will be negligible. Some feedback of metals to the sea is provided by organisms that decompose before they join the accumulated sediments on the sea floor. Hallberg assigns the designations J_2, J_3, J_4, and J_5 to these processes, with the feedback being dependent on the size of the organism, small organisms normally decomposing faster than large, and large falling more rapidly through sea water than small. Some metals are more biophilic than others, so various metals are accumulated to different extents in biological systems. With time, however, all metals will deposit on the bottom of the sea in one form or another. The complete diagram of Hallberg's energy circuit system, then, can be used to demonstrate qualitatively the complex mechanism of heavy metal accumulation in sediments. He says that it also can be used quantitatively by using the equations he has derived and combined from the energy diagram.

Hallberg says that, after sedimentation, the sedimentary debris will be attacked by chemical and biochemical processes at the surface of the sediment. These processes differ widely if oxygen is present or absent. The metals in sediments, therefore, exist in only one of two types of environments at a given time. This separation is controlled by the redox state (Eh) of the sediment, and the Eh is governed by: (1) the amount of reducing agents, mainly H_2S; and (2) the amount of oxidizing agents, mainly O_2. The various sulfide solubility products are the major factors in determining the content of the precipitated sulfides. In the presence of oxygen, the metal composition of the products formed will depend on the chemical stability of the authigenic minerals in an environment in which they can be oxidized. Thus, solubility products for such materials as oxides, carbonates, oxyhydroxides, sulfates, and phosphates will determine the metal-bearing mineral composition of the sediment being formed. As the material deposited is more and more deeply buried, it will become lower and lower in oxygen, resulting in a gradual sulfidization of the oxidized layer and even causing some compounds to be redissolved. In this environment, certain intermediate compounds may act as metal chelates and may diffuse upward more rapidly than H_2S, thus causing the chelate to be withdrawn from the reducing zone. This movement of materials as chelates may be repeated one or more times. The final result, however, is the concentration of the chelate above the sulfides. Hallberg quotes Berner (1969) to show that the migration of iron and sulfur can be shown in three models:

(1) A low-metal content gives high H_2S diffusion rates.[1]

(2) A high-metal content gives low H_2S diffusion rates but high rates for chelated Fe^{2+}.

(3) An intermediate metal content causes both H_2S and Fe^{2+} to diffuse, thus producing liesegang structures.

Thus, sulfide mineral formation in a sediment can take place either: (1) straight into a sulfide-rich sediment; or (2) reaching such a sediment only after having passed through an intermediate oxide stage in an oxidized layer. In diagenesis, the sulfide minerals formed in both ways mentioned above will be changed into more stable minerals. Hallberg claims that trace elements, such as Cu, Zn, and Mo and the ratios of these elements should be investigated as possible indicators of chemical conditions in the bottom water during the formation of the sediment in question.

[1] Editor's note: for a detailed discussion on diffusion, see in Vol. 2 Chapter 2 by Duursma and Hoede.

1972. Massive sulfide deposits (and porphyry coppers)

Hutchinson and Hodder (1972) attempt to demonstrate a possible tectonic and genetic relationship between porphyry copper and massive sulfide deposits. They argue that massive sulfide deposits lie conformably above submarine volcanic rocks and below siliceous and manganiferous iron formation. These massive bodies, they consider, may be zones, having a barren pyritic top that grades downward into pyrite with sphalerite (and locally galena) and then into a chalcopyrite-rich base. Such deposits are brecciated in many places, as are the lavas beneath them, and these breccias have pyrite and chalcopyrite as coatings on joints and fractures and as dispersed grains. The lavas also are propylitized, argillitized, and silicified. The authors believe that the massive deposits were formed on the sea floor by fluids escaping through the lavas during period of exhalative or explosive volcanism. They consider that this process may be comparable to deposition of metal sulfides from such hot brines as those accumulated in the depth of the Red Sea.

They hold that some porphyry copper deposits resemble the altered, sulfide-impregnated lavas that lie beneath bodies of pyritic massive sulfides. They point out that, in porphyry copper deposits, pyritic and chalcopyrite occur as veinlets and dispersed grains in intensely fractured tuffs, flows, and subvolcanic intrusions *or other host rocks* (my italics) characterized by propylitic, argillic, and silicic alterations. They believe that metal zoning is present; sphalerite and galena locally occur in manganiferous veins, peripheral to a shell of pyritic veinlets that encloses a central mineable zone of chalcopyrite-bearing veinlets. This arrangement of sulfides, they contend, seems to have been formed by hydrothermal fluids escaping from solidifying magma. They believe that these fluids may be comparable to the Salton Sea brines.

They consider the important differences between massive sulfide and porphyry copper deposits to be:

(1) Massive deposits normally are in differentiated, mafic to felsic, subaqueous volcanics, whereas porphyry coppers mainly are associated with shallow plutons and subaerial volcanic rocks, of intermediate to felsic composition.

(2) Calcic-sodic metasomatism is a concomitant of massive sulfide bodies, whereas potassic metasomatism is characteristic of porphyry coppers.

(3) Ages of massive sulfide deposits range from Archean (Keewatin) to Tertiary, whereas, they say, porphyry coppers are Triassic—Tertiary.

These similarities and differences suggest to Hutchinson and Hodder that massive-sulfide and porphyry-copper deposits are related tectonically and metallogenically. Massive deposits were generated during the early stages of crustal evolution and were repeatedly derived during early stages of younger individual tectonic cycles. In contrast, they hold that porphyry coppers developed during the later stages of crustal evolution — later in both terms of absolute age and relative age within younger individual tectonic cycles, specifically porphyry coppers appear to these authors to be involved events of the middle and late stages of orogenic belts. The massive sulfide deposits, however, are affiliated with early magmatic activity in eugeosynclines, whereas the porphyry coppers formed in late-magmatic, predominantly sub-volcanic settings at the edges of continental plates. They believe that transitional types have been developed in continental margins, where they flank eugeosynclinal sequences. (The authors, thus, appear to believe that a porphyry-copper deposit is not developed below a massive sulfide deposit as part of a continuous sequence of geologic events.)

Lusk (1972) believes that the work done on the massive stratiform sulfide deposits of northern New Brunswick[1] has demonstrated that they are sedimentary in their primary manner of formation and that their sulfur is genetically linked with contemporaneous marine sulfate. He is uncertain, however, if this connection is biologic or abiologic. He considers two possible alternatives: (1) bacterial reduction of marine sulfate; and (2) the mixing of volcanic-exhalative, hot, sulfur- and metal-bearing fluids from an igneous source with contemporaneous Ordovician sulfate. He draws attention to the similarity between $\delta^{34}S$ values for deposits associated with given lithologic units. He believes that the sulfur isotope ratios and variences of $\delta^{34}S$ values in the deposits are best accounted for by assuming the deposits were formed by a submarine volcanic mixing process, which, when combined with a time-dependent variation in marine or connate sulfate composition, should yield the time-independent parallels observed between ore sulfur and marine sulfate of corresponding age. The acceptance of his explanation of the similarity between mean $\delta^{34}S$ values for deposits of given groups and differences between groups in terms of this mechanism depends on a number of possibilities. These include: possible variation in the composition of connate sulfate between areas, between-area variation in source magma depths and/or temperatures, and the degree of cooling in connecting submarine volcanic vent systems. Except for expected differences between mean $\delta^{34}S$ values, the ranges in sulfur-isotope variation shown by the bulk of the sulfur in the individual ore deposits of the district (as well as by average sulfur in all of the ore deposits) compare favorably with the wider ranges shown by some igneous—hydrothermal ore deposits. These relationships, Lusk thinks, are consistent with a volcanic-exhalative mechanism of formation for the ore sulfur. He also considers that both the sulfur-isotope and base-metal zoning patterns across the B-1 ore body are compatible with a common generative process that yielded an over-all temperature decrease during ore formation and that was characterized by short-term variability. In addition, he believes that a volcanic-exhalative formation for the ores offers reasonable explanations for several important features of the deposits. These are:

(1) Reversals in over-all sulfur-isotope trends between individual deposits.

(2) The enrichment and/or depletion in ^{34}S of barite sulfur relative to contemporaneous marine sulfate.

(3) The frequent abundance of pyrrhotite in copper-rich ores toward the stratigraphic bottoms of the ore deposits.

On the contrary, Lusk thinks that a biogenic source of the sulfur is favored by the

---

[1] Editor's note: see in Vol. 5 Chapter 3 by Ruitenberg.

parallels existing between mean compositions for ore sulfur and marine sulfate of corresponding age. This opinion is, he suggests, reinforced by recent work that yielded sulfate-sulfide fractionations similar to those between ore sulfides and contemporaneous marine sulfate.

Still other features Lusk finds ambiguous. Bacterial generation of ore-sulfide sulfur, with means and narrow ranges of isotopic variation comparable with those shown by the bulk of the sulfur in individual deposits, would require rapid reduction rates. These would have to be essentially constant within and between widely separated ore-forming environments. He believes that this requirement is too stringent to envisage in view of the strong dependency of this complex fractionating mechanism on many factors, at least some of which presumably varied within a given area. If the effects of local volcanism were important in producing differences between ores and sedimentary sulfides from equivalent horizons, then differences between reducing conditions that yielded deposits in different areas could reasonably be as large as differences that produced the deposits in given areas.

Nevertheless, Lusk favors the volcanic-exhalative mechanism of ore formation in the New Brunswick massive sulfide deposits because it satisfactorily explains most of the features of the deposits of the area. (I always thought that if the hypothesis did not explain all of the observed facts, it was back to the drawing board.)

1972. Sulfur-isotope ratios

Mauger (1972) conducted a sulfur-isotope study of certain of the generally conformable massive sulfide lenses of the Ducktown mine. In the mines he studied, the sulfur isotopes showed a mean of +4‰ and a range of 0—10‰. In a detailed study of one Calloway mine section, two ore types, characterized by different FeS ratios were shown to exhibit rough stratigraphic conformity with the contacts (assay boundaries) of the ore masses. The distributions of copper, zinc, and magnetite also are roughly consistent with the stratigraphic contacts. Of the two ore types, the high-pyrite ore showed lower $\delta^{34}S$ mean values for pyrite and pyrrhotite than did the high pyrrhotite ore. Thus, he believes that gross compositional, internal compositional, and isotopic characteristics are congruent in a stratigraphic sense. He considers that the ores have been metamorphosed at load and fluid pressures in excess of 5 kbar and at temperatures from 540 to 690°C, based on the presence of kyanite and staurolite, the absence of cordierite, and the iron content of the sphalerite. He suggests that pyrite—pyrrhotite isotopic differences and the iron content of pyrrhotite show equilibrium temperatures of 275°C. Although he thinks that local isotopic equilibrium was attained, he identified variations that he believes probably relate to stratigraphic position. He regards the compositional and isotopic patterns as having survived, in a modified way, from initial sedimentary depositional patterns. He believes that the Ducktown ores were formed from the muds deposited from metal-rich brine pools of the Red Sea type, the metals, at least, having come from an ancient submarine hot-spring vent. He considers that the isotopic characteristics of the lead and strontium and the positive δ^{34}S-values for the sulfides all are compatible with a model in which compaction waters are expelled from sediments in a deep basin. He thinks that the metals were leached from the Red Sea type sediments by the rising compaction waters and were deposited where these waters issued from the top of the sedimentary pile as submarine springs. He admits that pre-metamorphic, epigenetic sulfide deposition in an Ocoe carbonate unit remains as a possibility. He does not, however, think that initial sulfide accumulation during metamorphism is supported by the various isotopic data. (This seems to me another example of the results of the failure to admit that isotopic fractionation can occur in hydrothermal fluids. How anyone, who has seen the retrograde metamorphism of the high-temperature silicate minerals around the present ores, can claim that they were not introduced after metamorphism or during its last stages is incomprehensible to me.)

1972. Sulfate-reducing bacteria

Trudinger et al. (1972) have considered the physical and chemical requirements of sulfate-reducing bacteria, and the general picture that they draw of an environment in which such organisms can function is, at least in relation to ore genesis, that of a basin conducive to the establishment and maintenance of anoxic-reducing conditions in which the pH ranges from slightly acid to mildly alkaline. This basin would need an adequate supply of organic matter (ultimately produced by photosynthesis) and a complex eco-system to degrade the organic matter. Beyond this, they think there seem to be few geochemical factors that would prevent participation of sulfate-reducing bacteria in the formation of sulfide minerals, but it is possible that certain of these factors could combine to render the process ineffective; these are: high heavy-metal concentrations, high temperature, pressure, and salinity, limited organic matter, and lack of oxygen. No one of these factors, however, do they think as discouraging the necessary bacterial life. They believe that sulfate-reducing bacteria could have been active as far back as the time of formation of the most ancient stratiform sulfide ores.

They compare present-day rates of sulfate reduction and carbon production and deduce rates of ore formation that suggest, quantitatively, that no problem exists that would prevent assigning a biogenic mechanism of formation to the Zambian Copper Belt and Kupferschiefer ores. On the contrary, they think that, if the ore bodies of Mount Isa and McArthur River were laid down at rates close to or exceeding those of the sediments in the Atlantic II Deep, then it would be difficult to account for their sulfide contents solely in terms of biological sulfate reduction. Further, they believe that environments, analogous to those of the Black Sea, could be favorable for biogenic sulfide ore genesis, should a source of metals be available. If deposits such as Mount Isa and McArthur River were to be formed in Black Sea-type situations, they would impose additional restrictions.

(1) Average rates of sulfide production and of primary carbon production (which

controls sulfate reduction) would have had to have been maintained fairly continuously throughout the entire period of sulfide-mineral formation.

(2) The major part of the sulfide generated would have had to have been fixed by metals; this implies that the metals were introduced into the environment in amounts close to, or in excess of, those of sulfide. They rule out an excess of metal being added because of the lack of significant amounts of non-sulfidic base metals in the ore zones.

These authors point out that, although as a first approximation, biological sulfate reduction appears to be, in a quantitative sense, an acceptable mechanism for the generation of the sulfide in stratiform ore deposits, the feasibility of biogenic sulfide ores should not be accepted without question. Much more examination of the process is necessary and more precise data, they point out, must be acquired. (If all geologic papers were as well and carefully written as this one, the profession would be saved many journeys down alleys that are, scientifically speaking, blind.)

1973. Syngenesis versus epigenesis

Derry (1973) discussed four classes of ore deposits — Broken Hill, massive sulfides, lead—zinc deposits in unmetamorphosed sediments, and pitchblende veins. He points out that all of them were, at least by most North American geologists, regarded as epigenetic. He considers that all four of these classes have become to be accepted by geologists (he does not modify this word) in various parts of the world as having been formed at or very near the then-existing surface. He thinks that the first two types were formed by direct precipitation at a rock surface-water interface and are truly syngenetic (though they may have undergone later changes from diagenesis and/or metamorphism). He believes that the lead—zinc deposits, regardless of the source of the metals, may, in some cases, have been formed directly at the interface of rock and water but, in others, may have been deposited in porous zones formed by supergene agencies; these, therefore, he would have had formed close to the contemporaneous surface. He considers the pitchblende, in veins or basins, to be much younger than the enclosing rock, but he thinks that important concentrations were developed at a land surface and before the deposition of the unconformably overlying beds. Now, he says, some (perhaps many) geologists in the United States and the U.S.S.R. do not accept what he has said about these four classes of ore deposits. He does not think that what he says describes a passing craze, rather he finds it impossible to disregard the mounting evidence of near-surface formation for these types of ore deposits. He thinks that the reluctance of American geologists to accept the concepts he discusses is because they were concerned with deposits in veins, at least until the end of the first half of the 20th century. He thinks that the reason for the change in attitude toward ore formation that he represents follows from a greater familiarity of Canadian (and other geologists) with bedded, massive or disseminated deposits that are largely, if not entirely, stratabound. He does, however, suggest that of such men as Lindgren, Bateman, Graton, and McKinstry, at least the first and the last would have eagerly grasped the concepts developed since 1960 to explain the origin of strata-bound deposits. He points to hesitency in challenging older geologists as explaining why American geologists are late in joining the syngenetic or near-surface parade. He also mentions that such men as Kinkel, C.A. Anderson, and Jenks have left the hydrothermal fold and have joined the new generation of men capable of evaluating and accepting evidence no matter where it leads. He mentions that V.I. Smirnov is a traditionalist. (To some extent this may have been true, but certainly his recent work is as modern as anyone could wish.)

His final point is that many of the deposits still regard as resulting from magmatic or hydrothermal processes may well have formed near the surface. (I cannot conceive of anyone arguing with this conclusion, but the major reason for accepting it, I hope, is geologic evidence and not a matter of following the leader. Certainly no one has doubted for many years, if ever, that the Bolivian tin deposits are near-surface deposits (xenothermal in Buddington's sense) yet, I think certainly, were formed by hydrothermal solutions. As so many deposits are removed from the hydrothermal class and become syngenetic or diagenetic or no more than near-surface hydrothermal, it becomes reasonable for Derry to say that the larger majority of ore deposits were formed within 5000 ft. of the then-existing surface. Even such an old moss-back as I am would not quarrel much with this statement, but I cannot but think that the pendulum, no matter how distinguished the people who pushed it, has swung too far toward the surface.)

1973. Ore brines

Hackett and Bischoff (1973) described the Red Sea geothermal system as containing three deeps — Atlantic, Chain, and Discovery — that have higher than normal water-temperatures and salinities. The purpose of their study was to make a detailed facies correlation, succession, and history of hydrothermal deposition in the Atlantic II Deep, using cores collected in 1969. The correlation of the facies was done by petrography, mineralogy, and ^{14}C data, plus color and X-ray data. The sediment—water interface was used as the common datum horizon. Within the sediment section, the uppermost lithified pteropod zone had been dated as 11,000—13,000 years B.P. This pteropod zone probably was laid down directly before the beginning of geothermal activity. Other samples from within this sedimentary sequence are from 20,390 to 25,700 years B.P. Sedimentation rates are estimated to run between 11—63 cm/1000 y, within which range 23—43 cm seem, to the authors, to be more probable. The earliest continuous deposition from the brine was a thin sulfide zone, generally less than 10 cm thick and resting directly on the upper pteropod zone. Higher in the section are two additional sulfide zones in the northern and central parts of the Atlantic II Deep. The lower of these is about 50 cm thick, thickening toward the west-central part of the deep. The upper of these two sulfide zones is about 2 m thick; the two units appear to thin and coalesce in the southern part of the deep. These two sulfide units are separated by an iron-montmorillonite in the north and center of the deep, whereas geothite separates the two units in the southern part. Hematite occurs in significant quantities in five of the sixteen cores studied in detail and magnetite was found in one in coexistence with hematite. Nine of the 16 cores bottomed in basalt, and, in two of them, iron oxides occurred directly above the basalt, since the carbonate facies was missing. The basalt is about the same age as the carbonate facies. The iron-montmorillonite is the uppermost and thickest of the facies, being from 12 to 15 m thick; it probably is continuous throughout the Atlantis II Deep. The brine deposits thin toward the margin of the basin and toward the topographic highs in the central and northern parts. The metal deposits are thickest in the north-central and southwestern parts, where the maximum thickness is about 30 m.

The authors consider that the goethite facies begins to develop with the precipitation of amorphous iron hydroxides at the interface between 56° and 44° brines. When this material reaches the bottom, it is dehydrated to goethite. Further dehydration changes the goethite to hematite to a minor extent, hematite development being largely due to local temperature anomalies. The local occurrence of magnetite is considered by the authors to have been caused by the reaction of Fe^{2+} with hematite, the result being magnetite plus hydrogen ion. No oxidation—reduction takes place. The additional Fe^{2+} required by this reaction probably came from a nearby brine vent. At present, the deposits tested in the sixteen cores are not economic; if they were available at the surface, on dry land, they would be worth $2.5 to $8 billion if they were easily sent to a market. The zinc and copper contents of the wet brines have been determined by analysis (wet is the only condition in which the brine could be moved to the surface by any mining method). The zinc ranges between 0.81—0.009% and the copper between 0.13 and 0.003%; this is not ore at sea level and certainly is not 6000 ft. or so beneath the sea surface. Even dried and salt free, the highest zinc content is 5.85% and the highest copper 0.84%. (The authors say nothing about the zinc and copper sulfides present, or how they were formed, or where the sulfur came from; of course, no one paper can do everything.)

1973. Sulfur isotopes

Hoef's work (1973) on stable isotopes is valuable in its entirety for any economic geologist, but the sections on sulfur and on ore deposits are most directly applicable to his problems. His section on sulfur lists the sulfur isotope abundances:

^{32}S: 95.02%; ^{33}S: 0.75%; ^{34}S: 4.21%; ^{36}S: 0.02%.

Taking the ^{32}S/^{34}S ratio of the Canyon Diablo meteorite as 22.22, he diagrams the range in δ^{34}S‰ for sulfur various types of rocks and deposits. In this diagram, the δ^{34}S in sea water ranges narrowly around +δ^{34}S‰. (Plus δ‰ values indicate sulfur heavier, that is higher, in ^{34}S than the meteoritic standard.) The δ^{34}S values for evaporite sulfur are firmly fixed by analytical determinations because evaporite sulfur is readily identified, and they range between about 26‰ and 8‰. (Because ocean water ranges narrowly around +20‰, it is readily apparent that some of the evaporite sulfur must have come from waters in which the SO_4^{2-} was richer in ^{32}S than present-day marine sulfate.) The $\delta^{34}S$‰ in the sulfur in sedimentary rocks ranges between almost +50‰ and −40‰. (It is certain that a large fraction of this cannot have been derived by the reduction of the sulfur in sea water. Even though ^{32}S is more easily reduced than ^{34}S, the change in ratio from sea water to marine sulfide cannot drop greatly below the +20.00‰ of sea water. The highly negative $\delta^{34}S$ values in some sedimentary sulfide must have come from other sources than sea water.) The sulfur in metamorphic rocks has a narrower range than sedimentary rock sulfur, but the metamorphic variety still has a wide range that is almost entirely more negative than that of present-day ocean water. Granitic rocks contain sulfide sulfur that has a narrower range by much than metamorphic rocks, with some of the values on the negative $\delta^{34}S$ side. (The variations in granitic sulfur probably can be explained by fractionation during igneous processes.) Basaltic rocks contain sulfides that have a still narrower range than granitic ones do, essentially all of it being more negative than $\delta^{34}S$‰ values than Canyon Diablo sulfur (this probably resulting from lesser fractionation in basalts than in granites). Even extraterrestrial meteorites and moon rocks show about as much of a spread in $\delta^{34}S$ values as does granite, although these values are shifted somewhat to the negative side.

Hoefs says that two types of reactions are responsible for sulfur isotope variations:

(1) A kinetic effect during the bacterial reduction of sulfate to lighter H_2S; this, he thinks, gives the largest fractionations in the sulfur cycle.

(2) Various chemical exchange reactions, between sulfate and sulfide on the one hand and between sulfides themselves on the other (these reactions can take place in essentially any environment from magmatic through hydrothermal to metamorphic and sedimentary). In a sulfate reservoir, such as the sea, the net isotope fractionation produced will depend on the isotope effects in whatever steps are necessary to convert sulfate sulfur to sulfide (or to remove excess sulfide sulfur from sulfide-metal complexes). The relative speed with which the sulfur moves through these steps (in whatever type of environment) will determine the extent to which the sulfate reservoir is depleted. Hoefs quotes Sakai to point out that the latter's experimental work shows that the fractionation between sulfate sulfur and sulfide sulfur is greatest at lower temperatures. He quotes Ohmoto as showing the influence of the fugacity of oxygen and pH of ore-forming fluids on sulfur-isotope composition.

Hoefs points out that pyrite normally has the highest $\delta^{34}S$ values of a series of time-associated sulfides, but he does not mention that this is to be expected because, at any given instant, one of the sulfers in S_2^{2-} is S^0 and the other S^{2-}. In such a complex, the $\delta^{34}S$ value should be more positive than in sulfides in which no oxidized sulfur occurs. After pyrite, the sequence of decreasing $\delta^{34}S$ values is sphalerite, chalcopyrite, and galena. The experimental determination of equilibrium constants for sulfur-isotope distribution by experiments have produced results that do not agree well with each other. (Probably this is true because the experiments have not correctly reproduced natural conditions, or because the experimenters have different ideas of what natural conditions are.) Certainly these equilibrium data cannot be used as geothermometers until more consistent data are obtained.

Hoefs states that the generation of an ore deposit involves a hydrous fluid that has at least the following aspects: (1) a source of ore; (2) the dissolving of the ore and other components in the fluid phase; (3) the migration of the ore-bearing fluid; and (4) the selective precipitation of the ore fluid in environments favorable to that process. The major problem for any worker in this field is to explain how large accumulations of nearly insoluble sulfides can be accomplished. How can such accumulation be produced without using gigantic volumes of water? He offers some, as he terms them, vague generalizations in separating sulfides formed from magmatic fluids and sulfides produced biogenically:

(1) Sulfides produced biogenically show much greater spreads in $\delta^{34}S$ values than magmatic or hydrothermal sulfides and have a preferential enrichment in the lighter isotope. This generalization needs modification in some way because there are heavy "sedimentary" sulfides also. (Very possibly the heavy sulfides may have been produced by processes other than sedimentary.)

(2) Sulfides probably from magmatic sources rarely differ in $\delta^{34}S$ composition by more than 5‰. (This may not actually be the case but simply may result from the manner in which magmatic sulfides have been defined.)

(3) Sulfides, collected from two ore deposits, which according to geologic evidence have the same manner of formation, may show quite different ratios. On the other hand, in certain ore provinces, the $\delta^{34}S$ value may be essentially zero. (Again, this well may be because we do not understand the problem.)

(4) Systematic regional and depth variations have been found by detailed studies of certain deposits. Such relationships also have not been satisfactorily explained.

(5) Low-temperature sulfides normally have a wider range in isotope ranges than do higher-temperature sulfides. Also, as a result of metamophism, progressive homogenization of sulfides appears to have taken place.

(6) In many suites of coexisting sulfide minerals, galena contains the most light sulfur; others are heavier. These differences normally are 2—5‰.

(7) In general ^{32}S is associated with lead somewhat enriched with radiogenic isotopes 206, 207, and 208, in larger proportions than with the 204 isotope that is non-radiogenic. The major problem with these generalizations is that they do not accurately reflect the isotopic compositions of the ore fluids. He quotes Ohmoto as saying that the isotopic composition of sulfides may be influenced by: (1) isotopic composition of total sulfur in the fluids; (2) temperature; (3) fugacity of oxygen; and (4) pH of the ore fluids. Further, considerations based on the assumption that both chemical and isotopic equilibrium are established among aqueous sulfur species and precipitating sulfur species may not be correct.

1973. Massive sulfide deposits

Hutchinson (1973) considers that ore bodies that are stratabound, lenticular, massive pyrite mineralizations, containing varied amounts of chalcopyrite, sphalerite, and galena in layered volcanic rocks, are formed by volcanogenic processes. Such deposits, in many instances, are found to be overlain by thin bedded siliceous and iron-rich sedimentary rocks, and they usually are underlain by extensive zones of altered, sulfide-impregnated lava. The volcanogenic processes producing them, he considers, took place subaqueously, the ore material being poured out on the sea floor through fumaroles. He distinguishes three varieties of such deposits by their compositions, relative and absolute ages, and rock associations:

(1) Pyrite–sphalerite–chalcopyrite bodies that are in differentiated, mafic-felsic volcanic rocks.

(2) Pyrite–galena–sphalerite–chalcopyrite bodies that are in more felsic, calc-alkaline volcanic rocks.

(3) Pyrite–chalcopyrite bodies that are in mafic, ophiolitic volcanic rocks.

He believes that the time–tectonic–stratigraphic interrelationships of these varieties can be assigned places in the evolutionary process of crustal development. Deposits of type 1, he thinks, are numerous, important, and best developed in Archean greenstones. This suggests to him that they were formed in a thin proto-crust, possibly by degassing of an as yet poorly differentiated proto-mantle. (This is essentially an exact quotation!) Although such deposits recur in later volcanic successions, they become, he says, scarser and smaller with the passage of time. The Archean, Hutchinson holds, lacks deposits of types 2 and 3. Type 2 comes in in the Proterozoic and type 3 is common in the ophiolites of Phanerozoic orogens that typify an early stage of orogenic activity and probably were generated in oceanic-ridge rift environments during the initial stages of the separation of continental crustal blocks. Both types 1 and 2 reappear in Phanerozoic belts, type 1 in the early stages of subduction along continental margins and type 2 in later, more felsic, calc–alkaline volcanics that mark the somewhat later tectonism. Hutchinson advocates the use of these concepts in directing exploration efforts. He also considers that ore deposits may be the product of evolutionary change, and they might be used as gross-scale indicators for correlation or age-determination purposes. These concepts could particularly be applied to highly metamorphosed Precambrian terranes. (The only paper I know that is remotely comparable to this one is the Book of Genesis.)

1973. Syngenesis versus epigenesis

After a detailed discussion of the changes in thinking about the mechanisms for the formation of ore deposits[1], particularly as developed in Australia, *King (1973)* offers three thoughts on questions as to which there are, as yet, no answers:

(1) A big gap in our knowledge (King says) about the formation of stratiform ore is in

---

[1] See also Chapter 5 by King.

the fields of biology and chemistry. Now that many geologists are connecting major sulfide bodies with particular rock types and rock-forming conditions and a number of rock types are being recognized as chemical or biogenic, little doubt remains in King's mind that ore sulfides are producible biogenically (on a large scale). He points out that Lindgren (1935), 40 years ago, saw that bacterial factors might be important in the formation of some ore sulfides. This long ago, Precambrian life was not well evidenced, but now it is known to have been of great importance. Thus, King expects that biogenesis will come to be considered an important factor in the formation of some Precambrian rocks. If the metals are related in formation to the environments in which were formed the rocks with which they now are associated, geologists cannot afford to exclude from their thinking the biological and chemical affinities of associated rocks and ores.

(2) He thinks that difficulties still remain about the nature of some of the rocks that enclose, and are associated with, ore. Where these are obviously sedimentary or volcanic, no problem exists. Where, however, they are rocks with what are called igneous textures and compositions, they tend, as do the ores, to be given names that have genetic connotations (for example granite?) and, thereafter are regarded as magmatic or igneous because of the names given them (give a dog a bad name!). King says that ore textures once were regarded as original and depositional and as of fundamental genetic importance. Now, these textures are known, he says, to be an expression of the conditions to which the minerals involved have been successively exposed. He would take this thought one step farther and ask if the textures of silicate rocks mean no more genetically than do ore textures? He points out that such rocks as marble are known to have acquired crystalline textures under nonmolten conditions. On the other extreme, lavas are known to crystallize from the molten state. Between is a huge variety of rocks that long has been called igneous and thought to have crystallized from the molten state. If this presumption is wrong, and King says it almost certainly is significantly in error, then we do not know where or how to draw the line (granted it is a line and not a broad zone) between crystalline textures of molten and non-molten manner of formation. He thinks that the geological profession is in for some surprises when this problem is studied in detail.

(3) He believes that, if we outgrow the tacit concept that, in the context of ore, geological events happened at one time, for example, sedimentation and/or extrusion, hardening, folding, granite intrusion, and (he says) etc., he ventures to think that there is a considerable overlapping of these events, even that the formation of granite may provide the ingredients for a contemporaneous stratiform deposit at a higher level. (When did any reputable geologist ever say this was not possible, particularly if the stratiform deposit could have been formed by replacement as well as by syngenetic precipitation on the sea floor?)

1973. Sulfate-reducing bacteria

Rees (1973) uses zero-order kinetics to describe sulfate uptake by the bacterium *Desulfovibrio desulfuricans* in a manner more in keeping with the experimental data than is provided by the use of purely first-order reaction schemes. The isotopic behavior of multistep reaction sequences with zero-order first forward steps can be considered conveniently in terms of three distinct regimes:

(1) *The initial behavior regime;* the first product formed exhibits the sum of the isotopic effects in all the forward steps of the reaction sequence. The extents of reaction employed in laboratory experiments have been too great for this first behavior to be observed.

(2) *The steady-state regime*; following the initial behavior, intermediates in the reaction sequence reach steady-state concentrations. The isotope effects for product and residual reactant approach asymptotically the isotope effects for the product and residual effect of a single step reaction. The rapidity with which the asymptotic form of variation is approached depends on the magnitudes of the steady-state intermediate reservoirs, relative to the original reactant reservoir. The overall isotopic effect in the asymptotic one-step reaction is related to the ratios of backward to forward flows between intermediate reservoirs. The experimental data for the reduction of sulfate to *D. desulfuricans* is consistent with this steady-state approach. The extreme overall isotopic effects observed in the laboratory (-3–$46‰$) correspond to the extremes possible for the expression:

$$\alpha_{ae} = \alpha_{ab} + (\alpha_{bc} - \alpha_{ba})X_b + (\alpha_{cd} - \alpha_{cb})X_b X_c + (\alpha_{de} - \alpha_{dc})X_b X_c X_d$$

X_b = the ratio of backward flow (B–A) to the forward flow (A–B); X_c = the ratio of backward flow (C–B) to forward flow (B–C); X_d = the ratio of backward flow (D–C) to the forward flow (C–D). The lower extreme, which occurs when $X_b = X_c + X_d = 0$, that is when the reduction process is unidirectional, allows the isotope effect in the uptake of sulfate to be assigned the value of $-3‰$. According to the steady-state approach, when the reaction proceeds unidirectionally, the overall isotope effect is that in the *first step*. In contrast, a multi-step unidirectional first-order sequence exhibits the isotope effect in the rate determining or *slowest step*. The other extreme experimental value, $46‰$, occurs when $X_b \sim X_c \sim X_d \sim l$; that is, when internal cycling considerably exceeds forward transport. The normal range of isotope effects in the laboratory (0–$25‰$) is explained in terms of the variation of X_b and X_c between zero and unity. Under normal conditions, X_d is zero, and the reduction in sulfite occurs rapidly so that the back flow from sulfite to APS does not build up. (APS = adenosine–5'–phosphosulfate.) The steady-state approach can be extended to the isotopic effects produced by sulfate-reducing bacteria in natural environments. Taking the specific example of the Black Sea, it is shown that the $50‰$ displacement between sulfate and sulfide is not the result of initial summing of isotope effects in a system undergoing an infinitesimal extent of reaction nor the result of partially catalysed isotope exchange brought about by reoxidation of hydrogen sulfide. It is shown that the $50‰$ displacement must be brought about by the internal flow conditions being such that X_b, X_c, and X_d are all close to unity.

(3) *The terminal regime.* The steady-state regime ends when there is insufficient reac-

tant remaining to maintain the intermediate reservoirs at their steady-state values. When this happens, the system runs down, and the intermediate reservoirs become depleted. Details of the isotopic behavior will differ with circumstances, and, in order to detect the terminal behavior experimentally, it is necessary to take reactions to completion and to monitor the instantaneous isotope composition of the reaction product. No appropriate experimental data have been obtained for sulfate reduction by *D. sulfuricans*, but it is shown that an unusual terminal isotope effect during bacterial nitrate reduction is readily explained in terms of a multistep reaction sequence, with zero-order reactant uptake.

Rickard (1973) has constructed another model describing sulfide production in sediments. In this he obtains maximum metal and sulfide fluxes and relates them to simple physical and chemical factors, such as porosity, rate of sedimentation, and organic carbon concentration. He points out that very few examples of bacteriogenic nonferrous metal sulfide concentrations have been found in recent sediments. On the other hand, the sedimentary formation of pyrite seems to be well established, (even though the mechanism has not been explained satisfactorily, even to the authors publishing on the problem). Rickard has constructed simple mathematical models to describe the sedimentary sulfide-producing environment. He assumes that all metal sulfides are formed at the oxidized—reduced boundary at some depth beneath the surface of the sea; all sulfide formed in the reduced zone diffuses to the oxidized—reduced boundary where it either is oxidized or fixed as metal sulfide. He considers that his work shows that:

(1) Bacteriogenic sulfide production in sediments is sufficient to produce metal-sulfide concentrations of ore grade.

(2) Maximum sulfide fluxes are produced in sediments with: (a) high organic-carbon concentrations; in sediments where the organic-carbon content is such that sulfate is not limiting, the sulfide flux is approximately proportional to the fossil-carbon content, all other factors being equal; and (b) grain size; fast-flowing currents depositing coarse particles increase the oxygen flux and lower the depth of the oxidized—reduced boundary within the sediment; thus, the sedimentation of low-density particles is facilitated in more quiescent conditions.

(3) At least 0.1% dry weight organic carbon is necessary to produce metal sulfide deposits containing more than 1.0% dry-weight metal.

(4) An additional metal source, over and above that normally contained in sea water, is essential for the production of ore grade sulfide concentrations by sedimentary processes. Rickard says that the erosion of continental metal-rich rocks may provide the additionally needed metal, but this source probably is insufficient because of the great dilution of the metals produced by their mixing with nonmetallic erosion products (further, a very real difficulty exists in turning on and off such supplies of metals, as the narrow vertical range of most stratabound deposits requires). He assumes, by a process of elimination, that the extra metals must be supplied from submarine volcanic sources.

(5) Since organic-carbon concentrations, and thus bacterial sulfide fluxes adequate to produce ore-grade metal sulfide deposits, are widespread, the factor controlling synsedi-

mentary sulfide ore deposition should, therefore, be in the association of organic-rich, fine-grained sediments, that is shales, with contemporary volcanic associations. (Rickard appears to assume that the volcanic sources cannot be considered as supplying any, or at most more than a small fraction, of the required sulfur.)

1973. Volcanic-exhalative ores

Ridge (1973), basing his work on data supplied by Haas' 1971 paper, determined that, at 200°C, ore fluids with a 5 wt.% salt (NaCl) content would boil before they reached the sea floor if the depth of sea water was less than 130 m. Further, such solutions at 250°C would boil before they reached the sea floor if the depth of sea water were less than 333 m. At 300°C, this depth would increase to 725 m, and at 330°C, it would be 1081 m. For solutions with a salt content of 20 wt.%, boiling would take place for such a solution at 200°C if the depth of sea water were less than 107 m. At 250°C, this depth would be 290 m, at 300°C, 630 m, and 330°C, 945 m. Thus, if the depth of sea water were less than 300 m and ore-fluid temperature were more than 250°C, under either set of conditions, it is unlikely that an ore fluid could reach the sea floor without having boiled at some distance below that floor. If boiling began in the fractured rock or unlithified sediments below the sea floor, the first materials to be precipitated would be those low-vapor-pressure constituents of the fluid that already were near their limit of solubility; these would be mainly sulfides. If the boiling began at considerable depth beneath the sea floor and the fluids were moving very slowly upward, boiling might be continued long enough for more soluble low-vapor-pressure constituents (mainly NaCl) to be deposited. It seems reasonable to suppose that, in the many times that boiling must have occurred in rising ore fluids in some instances at least, boiling would have been long enough continued to have caused the precipitation of salt in large amounts. The lack of such deposits of salt in association with sulfide deposits suggested that one of two explanations must be true:

(1) Ore fluids never reach the sea floor in seas where the depth is 300 m or less, at temperatures above about 250°C.

(2) Ore fluids never reach the sea floor at all.

The second possibility almost certainly can be dismissed, but the lack of salt deposits associated with low-temperature sulfides (such as those of the Mississippi-Valley type) indicates that non-boiling, low-temperature (less than 250°C) ore fluids probably reach the sea floor. The corollary is that if ore minerals wth fluid inclusions that show filling at 250°C or more and are in rocks that must have been formed in a shallow sea, they can only have been formed by emplacement in solid rock. They would have boiled before they reached the sea floor and would have left their sulfide loads in rocks or sediments below that datum.

If the sea water above the area toward which ore fluids were making their upward journey, however, had depths of slightly less than 1000 m or more, boiling could not have taken place at any temperature below 330°C at least. Thus, it becomes important to be able to estimate the depth of sea water above the sea floor at the time that ores in a given rock volume were being deposited. This depth of sea water can be approximated from the type of sediment in which the ores are found; if it is a shallow-water sediment, reef limestone for example, the possibility of the water having been deep enough to prevent boiling of moderate-temperature (250°C or more) fluids is essentially nil. From this it follows that the sulfide minerals that the 250°-ore fluid contained almost certainly would have been deposited in solid rock and would not have been carried into the sea. If the boiling had been long continued, of course, salt would have been deposited, so the fluids must have gone on toward the surface before salt had begun to precipitate.

If the rocks with which the ores are associated are volcanic, then these probably will contain vesicles, and, from these, the size of the depth of water overlying the sea floor toward which the ore fluids are moving also can be estimated. If the lavas were extruded only shortly before the ore fluids entered the rock volume in question, then data from the vesicles becomes pertinent to the manner of ore formation. If the waters were shallow, 300 m or less, then temperatures of 250° or more would guarantee that boiling would have occurred, and the chances of sulfides being deposited on the sea floor in any appreciable quantity would be low. On the contrary, if the waters were deep, 1000 m or more, the ore fluids would not have boiled, and they could have reached the sea floor with their sulfide and salt loads intact. Thus, any massive sulfide deposits, if they are to be syngenetic, must have been formed at depths of the general order of those of the Red Sea deeps, under which conditions metallic materials could have been added to the sea water. There, great gravity, in respect to sea water, would keep these fluids separate from the sea water above them, at least for an appreciable space of geologic time, time enough probably for considerable precipitation to occur.

The high-saline content of inclusions in the minerals of many telethermal deposits strongly suggests that the ore fluids that deposited them underwent boiling before they reached the surface. The boiling concentrated the salt in these fluids sufficiently to raise the salt content from the neighbourhood of 5 wt.% to that of 20 wt.%. At the same time, the boiling would have forced the precipitation of most of the solutions sulfide load, thus leaving little sulfide to deposit on the sea floor.

1973. Massive sulfide deposits

Sillitoe (1973) considers it the concensus among geologists that stratiform, massive pyritic sulfide deposits in sequences of submarine volcanic rocks were formed by deposition on, or replacement immediately beneath, the sea floor, the most necessary ingredient being submarine volcanic fumaroles or hot springs through which metal-bearing ore fluids could debouch. These deposits, therefore, are an important feature of Phanerozoic terranes, not to mention being abundant in certain Precambrian shield areas.

Sillitoe recognizes four main tectonic environments (within the context of plate tectonics) in which submarine volcanism is concentrated:

(1) Ocean rises (mid-ocean ridges), where volcanism is an essential part of the accretion of new oceanic crust at the trailing edges of outward-spreading lithospheric plates. The products are principally tholeitic basalts as pillow lavas, with minor hyaloclastites.

(2) Ocean-basin margins, such as those of the western Pacific, where volcanism accompanies rather irregular basin spreading above the deeper parts of subduction zones. This spreading leads to the separation of island arcs from the facing continents or to the disruption of island arcs themselves to form interarc basins; again the volcanic products are largely tholeitic basalts.

(3) Continental margins and island arcs along convergent plate margins where volcanism is active above subduction zones and may represent mainly the eruption of material generated by the partial fusion of the upper parts of underthrust slabs of oceanic lithosphere. There the eruptive material is calc–alkaline and consists of basalts, andesites, dacites, and rhyolites, with pyroclastic materials being particularly common.

(4) Intraplate oceanic island chains in main ocean basins, where volcanism may be an expression of the ascent of deep mantle plumes over which lithosperic plates drift. The typical volcanic materials are tholeitic and alkali basalts. Since Sillitoe considers that plate tectonics is firmly established as a valid geological theory, he would extrapolate these observed environments of submarine volcanism back to the beginning of the Paleozoic Era, but he is not certain if a model of this type can be considered to apply to the Precambrian. He proposes that massive sulfide deposits formed in such volcanic sites throughout the Phanerozoic, and probably still are forming with tholeitic basalts at spreading centers in the main oceans and in marginal ocean basins and with calc–alkaline volcanics at island arcs (and possibly also at continental edges) along convergent plate margins.

He considers that a scheme such as he has proposed is potentially useful in directing exploration toward massive sulfide deposits both on the continents and under the oceans. He believes that further study and solutions of existing problems in his model may be made easier by determining the nature and sequence of rock types beneath marginal ocean basins. Further help in exploration would be provided by diagnostic field and chemical criteria for the recognized environments of formation of submarine basalts, which he considers to be the hosts of syngenetic massive sulfide deposits in orogenic belts. Once these problems have been cleared up, Sillitoe proposes to go back to the study of such deposits in the Precambrian. Sillitoe considers that such workers as Gilmour (1971) and Hutchinson and Hodder (1972) have had trouble in their work on this problem because they have continued to apply imprecise geosynclinal terminology to the description and classification of the formational environments of massive sulfide deposits. (I would say, however, that Sillitoe has learned much from Hutchinson.)

1973. Sulfate-reducing bacteria

Sweeney and Kaplan (1973) say that their experiments show that the initial iron-sulfur phase that precipitates in mackinawite is in the sence of $Fe_{1+x}S$ or $FeS_{0.9}$. In the presence of oxygen, this mineral oxidizes to pyrrhotite with a composition $FeS_{1.01} - FeS_{1.05}$; only after this conversion to hexagonal pyrrhotite does it convert to pyrite by reaction with elemental sulfur. The mackinawite also can react with elemental sulfur to produce greigite. (This reaction can be written: $3FeS + S^0 = Fe^2Fe_2^3S_4$, the sulfur being reduced to S^{2-} and two of three Fe^{2+} ions being oxidized to Fe^{3+}.) Greigite is produced at $60\,°C$ and pyrite at $85\,°C$. If any intermediate steps occur, they could not recognize them. Neither greigite spheres nor pyrite framboids could be produced in an environment completely lacking in oxygen. The textures of the pyrite derived from pyrrhotite was framboidal aggregates. They conclude that the spherical forms of the pyrite were due to the properties of the solids and not to gel-like properties of such intermediate sulfides as might have been produced. The authors consider that, with the exception of pyrite framboids covered with surface membranes and pyrite-crystal aggregates of non-spherical form, the pyrite textures observed by them in biologically active marine sediments could have been formed abiotically by the pyritization of iron sulfide. The membranes appear to be of an origin external to the pyrite crystallization. As might be expected, the effects of the enclosure of the pyrite surfaces by membranes is to impede outward growth of the pyrite crystals. Their electron-microscope analysis of framboids and aggregates shows that mineral inclusions (with the possible exception of apatite) are absent. This indicates to them that the iron sulfides form either in a void between detrital particles or in a degraded organic structure. Their analysis for carbon was unsuccessful, but, on the basis of overall morphology, replacement of degraded organic bodies is a common situation.

The authors point out that the abiotically produced pyrite from the Red Sea does not occur in framboids; they think that framboids would have formed, based on the amount of dissolved Fe^{2+}, size of the crystals, and the presence of crystal aggregates, but they did not, probably because of a low pH in the sediment. Thus, the only well-studied modern "analog" to ancient stratiform sulfide deposits apparently has chemical properties that inhibit the formation with framboidal texture.

1973. Massive sulfide deposits

Whitehead (1973) proposes, as a model for the genesis of the Heath Steel ore body in the Bathurst area of New Brunswick, a string of basin-like depressions within a marine environment and containing hot springs or fumaroles. During the initial stages of this fumarolic activity, the manganese was reduced within the sediment associated with fumaroles of sulfide deposition (resulting from fumarolic activity). This manganese was removed in solution. Because of the volcanic activity, the surface-oxidizing layer and sediment no longer existed, and the manganese was carried to the edge of the basin, where Whitehead believes it was deposited, probably as a carbonate. The iron stayed behind, and the manganese was preferentially concentrated at the margin of the basin. As sulfides deposited in the center of the basin, manganese continued to be accumulated, through diagenetic processes, at the surface of the peripheral sediments and well as precipitating out of solution in that region. During the final stage, thin sulfide bands (2—5 m thick) were deposited along the margins of the basin; this outward movement of sulfide deposition appears to have resulted from the sedimentary filling of the basin. After sulfide deposition stopped, Whitehead thinks that the peripheral regions of the basin returned to conditions leading to the precipitation of manganese more readily than the central portion of it. Whitehead's work shows that the highest concentration of manganese is in the iron formations (0.68%), whereas the concentrations in the massive sulfides (0.16%) and in the sedimentary rocks (0.07%) are progressively less. The iron follows the same pattern, although the difference between iron in the iron formation (37.5%) and iron in the massive sulfides (36.5%) is slight. In the sedimentary rock, the iron concentration drops to 6.15%. Whitehead thinks that the marked increase of manganese toward the top of the stratigraphic sequence probably is the result of three factors:

(1) The fractionation of manganese in the late stages of volcanic activity.

(2) The reaction of manganese and its subsequent upward migration during diagenesis.

(3) A reducing environment that maintained manganese in a soluble state until an environment was developed favorable to the precipitation of that metal.

Each of these factors, Whitehead suggests, acts in the same direction with the same result, that is, a concentration of manganese at the top of the sedimentary sequence. Whitehead considers it evident that both lateral and vertical differences are true for Mn/Fe ratios and that these ratios appear to be consistent with marine basin deposition and support the general hypothesis of a volcanic—sedimentary deposition for the sulfides also. He suggests that the vertical differences may be useful in helping to ascertain which direction is up in other regions where other criteria are not available. (It would seem equally possible that the iron formation—sedimentary rock sequence could have developed without the sulfides having been formed during sedimentation. Later introduction of the sulfides, by replacement, would have produced the same pattern that Whitehead considers the result of sedimentary processes only, as long as replacement took place at the same horizon at which primary sulfide sedimentation is supposed to have occurred.)

1974. Sulfate-reducing bacteria

The latest paper (at this writing) on the White Pine copper deposits in the upper peninsula of Michigan is that of *A.C. Brown (1974)*. In it he discusses a possible relationship between two bands of sparsely disseminated lead, zinc, and cadmium sulfides that lie above the main copper ores and those copper ores themselves. These two disseminated sulfide bands are: (1) a narrow, blanket-like zone that lies directly above the cupriferous zone and transects the bedding at gentle to moderate angles; and (2) a strictly strata-bound occurrence of these sulfides in pyritic portions of the No.61 bed (or Stripey horizon); this pyritic bed is some 60 ft. above the base of the Nonesuch shale. The lower of these two bands has been considered to have been epigenetically formed but the upper one has been thought to be syngenetic because of its widespread and conformable distribution. Experimental work by Brown has suggested that zones of certain metal sulfides may be spatially separated if a slow influx of the metals is suddenly interrupted by pulsations of the ore fluid. He considers, therefore, that the Pb–Zn–Cd metals of the No.61 bed originally were concentrated immediately ahead of the front of copper mineralization but were later driven to higher levels by one or more abrupt influxes of cupriferous solution from the underlying strata. These abrupt influxes, he believes, could have been caused by minor tectonic adjustments in the sedimentary basin. Such pulsations would be expected to have carried some copper along with them, and the Pb–Zn–Cd sulfides in the No.61 bed do contain some 200 ppm of copper. Brown thinks that stratabound deposition of these sulfides in the No.61 bed is reasonable because it is the first major carbonate-bearing horizon above the cupriferous zone. The pH of the ore fluids, on reaching this zone, would have been largely and rapidly increased, resulting in a decrease in sulfide solubility and a large proportion of the metals being transported in the fluid being deposited in this bed. (Further integration of this concept with the major model for White Pine ore formation must depend on a better understanding of what the White Pine ore fluid was like and where it came from. Obviously, this paper is not an argument for a syngenetic origin of the White Pine deposits but is included here in an effort to present all sides of the arguments about the method of formation of that deposit.)

1974. Sulfide diffusion

Renfro (1974) considers that stratiform metalliferous deposits underlain by continental red beds or other oxidized strata and overlain by evaporites account for about 30% of the world's copper production. He thinks that the Kupferschiefer of Poland and East Germany and the Copperbelt deposits[1] of Zambia and Zaire were formed in a manner that involved the progressive seaward movement of sabkhas (evaporite flats that form along the subaerial landward margins of regressive seas). Because of their unique position, coastal sabkhas are nourished by a subsurface flow of landward-migrating, low Eh-high pH sea water and by seaward-migrating high Eh-low pH terrestrial water. These sabkhas normally are bordered on their seaward side by intertidal flats and lagoons that are carpeted by leather-like mats of sediment-binding, blue-green algae. Beneath the living algal mat is a fetid ooze consisting of interbedded, decaying algae and detrital sediment. In the landward direction, the sabkhas grade into, and initially rest on, sterile, oxygenated desert sediments.

Renfro argues that coastal sabkhas and their related evaporite facies prograde seaward

[1] Editor's note: see also separate chapters on the Kupferschiefer and the Copperbelt.

across adjacent algal-mat facies. As the sabkhas bury the algal mats, the latter become saturated with hydrogen sulfide of anaerobic-bacterial formation. Concurrently, the trailing edges of the sabkhas are buried by prograding terrigenous desert clastics. As the sabkhas migrate basinward, terrestrial water must pass upward through the buried, strongly reducing algal mat to reach the surface of evaporation.

Renfro goes on to explain that terrestrial-formation water, initially low in pH and high in Eh, can mobilize and transport trace amounts of such elements as copper, silver, lead, and zinc. As terrestrial-formation water passes through the hydrogen-sulfide charged algal mat, its load of metals is deposited (Whitehead says reduced, but only the copper, of the list of metals he gives can be so changed) interstitially as sulfides. The metal deposits that result would be generally conformable to the geometry of the hydrogen-sulfide-bearing host material. Such deposits would, he contends, contain suites of minerals that would be zoned from landward to seaward according to their relative solubilities in reactions with hydrogen sulfide. Deposits so accumulated would be underlain by oxidized continental strata and overlain by dolomite, gypsum, anhydrite, and/or halite. The grade and size of the metal-sulfide deposits formed in this manner would, Whitehead believes, be dependent on: (1) the quantity of available reductant; (2) the duration of the sabkha process; and (3) the quantity and chemistry of the metal-bearing terrestrial water. (This hypothesis is a refreshing change from the numerous biogenic sulfur models for the precipitation of iron sulfides which material, when converted to pyrite, it is hoped, will later pick up sufficient quantities of such metals as will make it mineable by some generally unspecified process but sometimes by the timely introduction of volcanic-exhalations into the area of pyrite formation.)

1974. Volcanic-exhalative deposits

Ridge (1974) first discusses the place of volcanic-exhalative deposits in the modified Lindgren classification (1933) and concludes that they should be assigned to II.A.3 headed "by the introduction of fluid igneous emanations and water-rich fluids". He then goes on to point out that the depth of sea water overlying the portion of the sea floor where deposition may take place will be the final determinant as to whether an ore fluid reaches that datum in a dense enough condition to carry material in true solution without boiling or begins to boil before it reaches that surface. The effect of the confining pressure of the rock through which the ore fluid is passing on the boiling of an ore fluid is, except for highly shattered rock near the surface, greater than that of an equal thickness of sea water. As the solution comes nearer and nearer the sea floor, however, the effect of sea water pressure eventually becomes the more important of the two. The first effect of such boiling would be to begin the precipitation of the least soluble constituents, mainly the sulfides. If continued long enough, essentially all of the material carried in solution would be deposited, only that comparably convertable to the gaseous phase would leave with the water vapor. If such complete boiling happened, the amount of sulfide laid down would be dwarfed by the huge quantities of salts, mainly NaCl, that would be precipitated. If boiling did not continue a sufficient time to begin the deposition of NaCl, the salt concentration of that portion of the ore fluid that remained in the fluid state would be increased, the increase being proportional to the time that boiling continued. Thus, it is reasonable that the higher a fluid inclusion is in salt content, the more likely it is that it was derived from an ore fluid that boiled during some part of its journey toward the land surface or the sea floor.

For many stratabound deposits, it should be possible: (1) to determine the temperature of formation of one or more of the ore-deposit minerals (from fluid-inclusion-filling data or from isotope ratios); and (2) to estimate the depth of sea water in which the sediment or volcanic material containing the ore deposit was laid down. If this can be done, it should be possible to say whether the deposit could have been formed on the sea floor or must have been produced by replacement of solid rock or by the filling of open space in such rock. Any deposit placed, by its temperature and the depth of overlying sea water, below the more appropriate one of the two curves given in Ridge's paper is unlikely to have been formed from an encounter between a volcanic exhalation (hydrothermal fluid) and sea water. This follows because the fluid would have boiled before it could have attained the sea floor and would have lost, at least, much of its sulfide content and, with long-continued boiling, at least some of its salt. On the other hand, if the two criteria place the deposit above the more appropriate curve, then the deposit *may* have been formed by the reaction of a volcanic exhalation with sea water but was not necessarily so formed.

CONCLUSIONS

Although the work discussed in this paper was published during the period 1931 through 1974, about half the 110 papers summarized here was printed in 1966 or later. Thus, if the papers cited are representative of the work being done on the problems of syngenetic–diagenetic formation of ore deposits, as much work has been done in the last 9 years (neglecting the time lag between work done and publication achieved) as in the previous 35 years. The papers here presented either discuss work done in North America of by North Americans or work, though done by non-North Americans, that appeared in publications readily available in North America and had considerable influence on North American thought. Obviously, the papers cited here are only those that, in my opinion, have, or should have, contributed significantly to North American thinking about ore deposits that were, or might have been, produced by syngenetic processes. Anyone else's list would have been more or less different from mine.

It is now considered probable or certain, by most North American geologists, that particular types of ore deposits were formed syngenetically (in the broad sense to include diagenesis). Further, it is understood that these deposits were subjected to the same metamorphic events that affected the ores in which they are enclosed. These types are:

(1) Sulfides disseminated in the rocks containing them, these rocks being mainly shales, but other rock types have been suggested as forming contemporaneously with disseminated sulfides.

(2) Massive sulfides in either volcanic or sedimentary rocks, the sulfides being stratiform with these rocks enclosing them.

(3) Conglomerates containing gold and/or uranium minerals.

Most North American geologists would not include, in this syngenetic category, low-temperature deposits of the Mississippi-Valley type. Many non-North American geologists, however, believe that deposits of this type were formed syngenetically. On this continent, most geologists consider that the deposits in carbonate rocks were formed by saline waters (to which any magmatic contribution was probably no more than small) that scavenged their metals from the rocks through which they moved in their generally long journey toward their locus of deposition. Deposition occurred from these solutions when they reached areas where hydrogen sulfide was available to combine with the metal ions to produce the simple sulfides and gangue minerals of the usual teletethermal deposit. Thus, North American opinion favors an epigenetic origin for teletethermal deposits but considers them, at most, only indirectly connected with magmatic processes.

As a result of the extensive field and laboratory work that has been done on the problem of the production of sulfides under conditions in which the sulfur is produced by bacteria, a variety of concepts have been proposed to explain how the initial ferrous sulfide (probably mackinawite in which the formula probably is $FeS_{0.9}$) is converted to pyrite. Usually, the pyrite sulfide is considered to have been produced by reaction with native sulfur. This reaction, however, may take place in stages, with greigite or pyrrhotite being perhaps only some of the intermediate products. It also is claimed that such reactions require help from oxygen, but just how this help is given is not yet clear. Far less is it clear how this pyrite is converted, at least in part, to the ore sulfides found in such disseminated pyrite deposits. The most favored explanation is that the non-ferrous metals are added during diagenesis, but a far lesser agreement exists as to where the metals come from. It is generally conceded that the generality of country rocks cannot produce metals in sufficient quantities to provide, for example, the copper in the Zambian and Zairean Copper Belt or of that metal in the Polish and German Kupferschiefer. Resort has been had to erosion of terranes in which earlier, perhaps hydrothermal, deposits were in more than usual abundance. Even more favored is the concept that the metals are supplied from volcanic vents or fumaroles and combine with the sulfur they find in the sea water they enter. The usually wider spread of the isotopic ratios in such sulfide deposits and their shift toward more positive $\delta^{34}S$ values than in deposits conceded to be hydrothermal is considered to add evidence in favor of a volcanic or bacterial source for the sulfur.

A special case of copper ores in shales and related rocks is that where native copper makes up a significant part of the ore mineralization. Normally the principal copper mineral in such deposits is chalcocite, but digenite, covellite, and even minor amounts of bornite and chalcopyrite may be present. The two major examples of deposits of this type in the Western Hemisphere are White Pine and Coro Coro in Bolivia. In North America, most geologists consider that the White Pine ores were introduced epigenetically, although the source of the copper, its manner of transportation, and the cause of part of it having been deposited in part as native copper instead of as simple sulfides are matters in dispute. Specifically, the part played by bacteria in causing precipitation has not been explained to the universal satisfaction of those who have studied deposits of this type.

A major difference between White Pine and Coro Coro is that the native copper in the latter deposits is in rocks bleached from red to gray; in the White Pine ores, the native copper is in rocks that differ little, outwardly, at least, from those that contain the chalcocite and other copper sulfides.

A rather firm concensus seems to have been reached that the more massive the sulfides in a given stratum, the more likely it is that the metals came from a volcanic source and the sulfur was brought in with the metals. This follows from the difficulties of envisioning the surrounding land surface being able to provide, in geologically such a short time period, the needed metals or the sea to provide the necessary quantities of sulfide through bacterial action.

Most stratabound sulfide deposits, whether massive or disseminated, show what can be categorized as epigenetic features. Such veinlets or breccia cements can have been given their present relationships to their host rocks by having been emplaced after the host rocks had been lithified and broken. In most instances, in these times, such "epigenetic" features are considered most likely to have resulted from metamorphism that affected the rocks, mainly after lithification but possibly during diagenesis as well. Textures that once were searched carefully for clues as to the order on which the primary sulfides were deposited are now interpreted as providing indications of the crystalloblastic sequence of metamorphic recrystallizations. In short, the concept that the ores and their enclosing rocks were formed at the same time requires a complete reversal in the manner in which rock—ore relationships are studied and interpreted.

As is the case with many of the deposits in the world that contain sulfides, all of those conglomerates that contain gold or uranium minerals are now considered to have been sedimentary in manner of formation, although few geologists question the syngenetic character of the economically valuable metallic minerals. As it is true of most sulfide deposits, nearly all of these conglomerates show evidence of having been affected by metamorphism. Anyone who examines the ores of the Witwatersrand under the microscope will find it difficult to believe that the gold could have been transported and deposited in the forms it now has. Agreement about this "fact" has resulted in the development of the "modified" placer hypothesis for the formation of the Rand gold ores. This hypothesis says that the gold was brought in by the same waters that deposited the non-metallic constituents of the conglomerates. But later metamorphism, mainly the effect of deep burial of the gold-bearing beds, caused the gold to be dissolved and moved short distances before it was redeposited in such relations to the primary conglomerate minerals that it might, in a less enlightened era, have been considered to have been precipitated in open space from water-rich fluids of ultimately magmatic origin. The presence of gold in cracks in quartz pebbles also is thought to have been the result of this remobilization of the gold. The problem of the close association of much of the Rand gold with the algal mats that occur in several places through the gold-bearing sequence on the Rand also must be considered from a genetic point of view. This gold, which normally is very finely divided, is thought by those South African geologists that have studied it, to have been deposited in major part by replacement of the carbon that is all that now remains of the algae. Whether this gold was derived from placer particles and moved into the algal mats as gold remobilized from the conglomerates or was brought in in such fine particles that it was an original constituent of the mats is uncertain. All that is certain to a South African geologist is that it was not brought in by hydrothermal solutions.

In the Blind River conglomerates, which are essentially lacking in gold and in which the uranium is in the probably non-placer mineral, brannerite, the brannerite is thought by most geologists to have been produced by reaction between ilmenite and the primary uranium placer mineral (a reaction called the "Pronto" reaction by Ramdohr). The suggestion, however, has been made that the uranium was brought in by ground waters long after the conglomerate and its attendant ilmenite has been emplaced. No one who has published on the Blind River district in recent years has suggested that the uranium was introduced in a hydrothermal fluid, despite the high probability that uranium-bearing hydrothermal fluids could react with ilmenite to form brannerite.

In the conglomerates of the Serra de Jacobina in Brazil, uranium is present locally in mineable quantities in a mineral better described as pitchblende than uraninite. Free gold also has been mined from these conglomerates. In this area, some disagreement exists between workers as to whether or not the gold and uranium were deposited in placer and somewhat remobilized later or was introduced by hydrothermal solutions. Canadians who have published on the deposit favor a modified placer hypothesis for its genesis.

In contrast to the opinion held by probably a majority of geologists that the gold-uranium ores in such conglomerates as those of Blind River and Serra de Jacobina are syngenetic, at least in their original form, the strataform uranium and uranium—vanadium deposits of the Colorado Plateau are considered to be epigenetic. This belief in epigenesis for the ores of this type, however, does not mean that more than a few geologists consider the elements in the ores came directly from magmatic sources. Instead, the concensus appears to be that the ore materials were leached from the surrounding terranes, principally from volcanic rocks; this follows principally because no primary ores deposits are known within hundreds of miles that could have supplied the ore elements. Although the uranium ores often are in horizons that are predominantly red, these rocks immediately adjacent to the ores are bleached to a grayish color, probably as a result of the iron in the originally red sediments having been converted from ferric to ferrous ion as a concomitant of the process of ore deposition. No agreement exists as to how this reduction of ferric iron is related to the ore-forming process because the uranium almost certainly is carried as U^{6+} if it was leached at the surface and transported to the site of deposition by ground water. Since both the ferric iron and the U^{6+} ions have been reduced to Fe^{2+} and U^{4+}, a reducing agent must have been present. Perhaps this was organic matter, although it seems probable that that material would have been reduced completely to carbon before the ore fluids entered the lithified rocks. The S^{2-} of hydrogen sulfide may have been the reducing agent, being oxidized to S^{6+} in the process, but where the hydrogen sulfide came from has not been established with any certainty. The problem needs further study.

Thus, although much greater emphasis and credence is given to the concept of sulfide deposits being formed by syngenetic—diagenetic process than was true even 10 years ago, the details of the processes still are subject to considerable speculation and argument. Even if these syngenetic concepts were to be proved wrong, and most geologists do not expect them to be, they have served the very real purpose of directing attention to three very important classes of ore deposits and have helped or resulted in the discovery of many new deposits of these three categories. Granted that these concepts are correct, they will do even more to promote the finding of new ore bodies.

Attention has only just begun to be given to the genesis of the iron—manganese-nodule deposits that may be in sufficient abundance on the sea floor to be a large-scale supplement to those obtained from land surfaces. In fact, some students think these nodules are abundant enough and are formed rapidly enough that they may provide an inexhaustible source of the elements they contain. Obviously, work on the problems of their formation is essential to being able to find and to mine them (in the broad sense). Certainly these are mainly syngenetic (and to some extent diagenetic) deposits.

REFERENCES

Amstutz, G.C. and Park, W.C., 1967. Stylolites of diagenetic age and their role in the interpretation of the Southern Illinois fluorspar deposits. *Miner. Deposita*, 2: 44—53.
Anderson, C.A., 1969. Massive sulfide deposits and volcanism. *Econ. Geol.*, 64: 129—148.
Anderson, C.A. and Nash, J.T., 1972. Geology of the massive sulfide deposits at Jerome, Arizona. *Econ. Geol.*, 67: 845—863 (discuss., 1973, 68: 709—711, 711).
Anger, G. et al., 1966. Sulfur isotopes in the Rammelsberg ore deposit (Germany). *Econ. Geol.*, 61: 511—536.
Baas Becking, L.G.M. and Moore, D., 1961. Biogenic sulfides. *Econ. Geol.*, 56: 259—272.
Bain, G.W., 1960. Patterns of ore in layered rocks. *Econ. Geol.*, 55: 695—731.
Bain. G.W., 1973. Geology of the massive sulfide deposits at Jerome, Arizona — a reinterpretation. . *Econ. Geol.*, 68: 709—711.
Bastin, E.S., 1933. The chalcocite and native copper types of ore deposits. *Econ. Geol.*, 28: 407—446.
Bell, J.M., 1931. The genesis of the lead—zinc deposits at Pine Point, Great Slave Lake. *Econ. Geol.*, 31: 611—624.

Berner, R.A., 1969. Migration of iron and sulfur within anaerobic sediments during early diagenesis. *Am. J. Sci.,* 267: 19–42.

Berner, R.A., 1970. Sedimentary pyrite formation. *Am. J. Sci.,* 268: 1–23.

Berner, R.A., 1971a. Diagenetic processes (and) diagenesis of iron minerals. In: *Priciples of Chemical Geology.* McGraw-Hill, New York, N.Y., pp. 86–113, 192–209.

Berner, R.A., 1971b. *Principles of Chemical Sedimentology.* McGraw-Hill, New York, N.Y., 240 pp.

Bonatti, E. et al., 1972. Classification and genesis of submarine iron–manganese deposits: papers from a conference on ferromanganese deposits on the ocean floor. *Nat. Sci. Found., Off. Int. Decade Ocean Explor., Washington, D.C.,* pp. 149–166.

Boyle, R.W., 1965. Origin of the Bathurst–Newcastle sulfide deposits, New Brunswick. *Econ. Geol.,* 60: 1529–1532.

Boyle, R.W. and Lynch, J.J., 1968. Speculations on the source of zinc, cadmium, lead, copper and sulfur in Mississippi Valley and similar types of lead–zinc deposits. *Econ. Geol.,* 63: 421–422.

Brown, A.C., 1974. An epigenetic origin for stratiform Cd–Pb–Zn sulfides in the lower Nonesuch shale, White Pine, Michigan. *Econ. Geol.,* 69: 271–274.

Brown, J.S., 1965. Oceanic lead isotopes and ore genesis. *Econ. Geol.,* 60: 47–68 (discuss., pp. 1083–1084, pp. 1533–1539).

Brown, J.S., 1970. Mississippi-Valley type lead–zinc ores. *Miner. Deposita,* 5: 103–119.

Cheney, E.S. and Jensen, M.L., 1962. Comments on biogenic sulfides. *Econ. Geol.,* 57: 624–637.

Chow, T.J. and Patterson, C.C., 1962. The occurrence and significance of lead isotopes in pelagic sediments. *Geochim. Cosmochim. Acta,* 26: 263–308.

Collins, J.J., 1950. Summary of Kinoshita's Kuroko deposits of Japan. *Econ. Geol.,* 45: 363–376.

Craig, H., 1969. Geochemistry and origin of the Red Sea brines. In E.T. Degens and D.A. Ross (Editors), *Hot Brines and Recent Heavy Metal Deposits in the Red Sea.* Springer, New York, N.Y., pp. 208–242.

Darnley, A.G., 1960. Petrology of some Rhodesian Copperbelt ore-bodies and associated rocks. *Inst. Min. Metall. Trans.,* 69: 137–173 (discuss. pp. 371–398, 540–569, 699–716).

Davidson, C.F., 1957. On the occurrence of uranium in ancient conglomerates. *Econ. Geol.,* 52: 668–692 (discuss, 1958, 53: 489–493; 620–622; 757–759; 887–890; 1048–1049).

Davidson, C.F., 1962. The origin of some stratabound sulfide deposits. *Econ. Geol.,* 57: 265–274.

Davidson, C.F., 1964. Uranium in ancient conglomerates: a review (of a publication in Russian of the same title, published in Moscow by Gosatomizdat in 1963 and edited by V.N. Kotlyar). *Econ. Geol.,* 59: 168–177.

Davidson, C.F., 1965. A possible mode of origin of strata-bound copper ores. *Econ. Geol.,* 60: 942–964.

Davis, G.R., 1954. The origin of the Roan Antelope copper deposit of Northern Rhodesia. *Econ. Geol.,* 49: 575–615 (discuss. 1955, 50: 82–83; 880–884; 1956, 51: 391–392).

Dechow, E., 1960. Geology, sulfur isotopes and the origin of the Heath Steele ore deposits, Newcastle, N.B., Canada. *Econ. Geol.,* 55: 539–556.

Dechow, E. and Jensen, M.L., 1965. Sulfur isotopes and some central African sulfide deposits. *Econ. Geol.,* 60: 894–941 (discuss., 1966, 61: 409–414).

Derry, D.R., 1960. Evidence of the origin of the Blind River uranium deposits. *Econ. Geol.,* 55: 906–927.

Derry, D.R., 1973. Ore deposition and contemporaneous surfaces. *Econ. Geol.,* 68: 1374–1380.

Derry, D.R. et al., 1965. The Northgate base-metal deposit at Tynagh, County Galway, Ireland – a preliminary geologic study. *Econ. Geol.,* 60: 1218–1237 (discuss., 1966, 61: 1443–1451).

Dunham, K.C., 1964. Neptunist concepts in ore genesis. *Econ. Geol.,* 59: 1–21.

Emery, K.O. et al., 1969. Summary of hot brines and heavy metal deposits in the Red Sea. In: E.T. Degens and D.A. Ross, (Editors), *Hot Brines and Recent Heavy Metal Deposits in the Red Sea.* Springer, New York, N.Y., pp. 557–571.

Fersman, A.E., 1922. *Geokhimiia Rossii: Nauchnoe Khimichesko.* Tekhnicheskoe izdatel'stvo, St. Petersburg, ca. 220 pp.

Field, C.W., 1966. Sulfur isotopic method for discriminating between sulfates of hypogene and super-gene origin. *Econ. Geol.*, 61: 1428–1435.

Finch, J.W., Chairman, Editorial Committee, 1933. *Ore Deposits of the Western States.* A.I.M.E., New York, N.Y., 797 pp.

Fischer, R.P., 1937. Sedimentary deposits of copper, vanadium–uranium and silver in the south-western United States. *Econ. Geol.*, 32: 906–951 (discuss., 1938, 34: 458–461; 1939, 35: 113–115).

Fischer, R.P., 1950. Uranium-bearing sandstones of the Colorado Plateau. *Econ. Geol.*, 45: 1–11.

Friedman, G.M., 1959. The Samreid Lake sulfide deposit, Ontario, an example of pyrrhotite–pyrite iron formation. *Econ. Geol.*, 54: 268–284.

Garlick, W.G., 1964. Association of mineralization and algal reef structures on Northern Rhodesian copperbelt, Katanga, and Australia. *Econ. Geol.*, 59: 416–427.

Garlick, W.G. and Brummer, J.J., 1951. The age of the granites of the Northern Rhodesian copperbelt. *Econ. Geol.*, 46: 478–498.

Gary, M. et al., 1972. *Glossary of Geology.* Am. Geol. Inst., Wash., D.C., 805 pp., plus 52 pages of bibliography.

Gavelin, S. et al., 1960. Sulfur isotope fractionation in sulfide mineralization. *Econ. Geol.*, 55: 510–530.

Gill, J.E., 1959. The genesis of massive sulphide deposits. *Can. Inst. Min. Metall.*, 57 (510): 316–355.

Gill, J.E., 1960. Solid diffusion of sulphides and ore formation. In: *21st Int. Geol. Congr.*, 16: 209–217.

Gillson, J.L., 1963. The Northern Rhodesian copperbelt: is it a classic example of syngenetic deposi-tion? *Econ. Geol.*, 58: 375–390.

Gilmour, P., 1971. Strata-bound massive pyritic sulfide deposits – a review. *Econ. Geol.*, 66: 1239–1244.

Gray, A., 1959. The future of mineral exploration: In: *Future of Non-Ferrous Mining in Great Britain and Ireland.* Institute of Mining Metallurgy, London, pp. xiii–xxiv.

Greenwood, R., 1968. Syngenetic sulfides in sediments – additional field evidence. *Econ. Geol.*, 63: 188–189.

Griffiths, W.R. et al., 1972. Massive sulfide copper deposits of the Ergani–Maden area, southeastern Turkey. *Econ. Geol.*, 67: 701–716.

Gross, W.H., 1968. Evidence for a modified placer origin for auriferous conglomarates Canavieriras mine, Jacobina, Brazil. *Econ. Geol.*, 63: 271–276.

Haas Jr., J.L., 1971. The effect of salinity on the maximum thermal gradient of a hydrothermal system at hydrostatic pressure. *Econ. Geol.*, 66: 940–946.

Hackett Jr., J.P. and Bischoff, J.L., 1973. New data on stratigraphy, extent, and geologic history of the Red Sea geothermal deposits. *Econ. Geol.*, 68: 553–564.

Hallberg, R.O., 1972. Sedimentary sulfide mineral formation – an energy circuit system approach. *Miner. Deposita*, 7: 189–201.

Hamilton, S.K., 1967. Copper mineralization in the upper part of the Copper Harbor conglomerate at White Pine, Michigan. *Econ. Geol.*, 62: 885–904; discuss., 1968, 63: 190–191, 294; 1969, 64: 462–464.

Helmqvist, S., 1951. Resa till Lipariska öarna. *Geol. Fören. Stockholm Förh.*, 73, 3(466): p. 473–491.

Hess, F.L., 1933. Uranium, vanadium, radium, gold, silver, and molybdenum sedimentary deposits. In: *Ore Deposits of the Western States.* A.I.M.E., New York, N.Y., pp. 450–481, Lindgren Volume.

Hirst, D.M. and Dunham, K.C., 1963. Chemistry and petrography of the Marl slate of Southeast Durham, England. *Econ. Geol.*, 58: 912–940.

Hoefs, J., 1973. Sulfur (and) ore deposits. In: *Stable Isotope Geochemistry.* Springer, New York, N.Y., pp. 32–38, 65–70.

Hutchinson, R.W., 1965. Genesis of Canadian massive sulphides reconsidered by comparison to Cyprus deposits. *Can. Inst. Min. Metall. Bull.*, 58 (641): 972–986 (discuss., 992–994).

Hutchinson, R.W., 1973. Volcanogenic sulfide deposits and their metallogenic significance. *Econ. Geol.*, 68: 1223–1246.

Hutchinson, R.W. and Hodder, R.W., 1972. Possible tectonic and metallogenic relationships between porphyry copper and massive sulphide deposits. *Can. Inst. Min. Metall. Bull.*, 65 (718): 34–40.

Ishihara, I. et al. (Editors), 1974. Geology of Kuroko Deposits – Soc. Min. Geol.

Jackson, A.A. and Beales, F.W., 1967. An aspect of sedimentary basin evolution: the concentration of Mississippi-Valley type ores during late stages of diagenesis. *Can. Pet. Geol.*, 15: 383–433.

Jones, J.G., 1969. Pillow lavas as depth indicators. *Am. J. Sci.*, 267: 181–195.

Jordaan, J., 1961. Nkana. In: F. Mendelsohn (Editor), *The Geology of the Northern Rhodesian Copperbelt.* MacDonald, London, pp. 297–328.

Joubin, F.R., 1954. Uranium deposits of the Algoma district. *Can. Inst. Min. Metall. Tr.*, 57 (510): 431–437.

Kalliokoski, J., 1965. Metamorphic features in North American massive sulfide deposits. *Econ. Geol.*, 60: 485–505 (discuss., 1539–1540).

Kendall, D.L., 1960. Ore deposits and sedimentary features, Jefferson City mine, Tennessee. *Econ. Geol.*, 55: 985–1003 (discuss., 1961: 56: 1137–1138. 444–446; 1962, 57: 115–118).

King, H.F., 1973. Some Antipodean thoughts about ore. *Econ. Geol.*, 68: 1369–1374.

Kinkel Jr., A.R., 1966. Massive pyritic deposits related to volcanism and possible methods of emplacement. *Econ. Geol.*, 61: 673–694.

Kinoshita, K., 1943. *A Report on Research in Economic Geology.* Univ. Kyushu, in Japanese.

Knight, C.L., 1957. Ore genesis – the source-bed concept. *Econ. Geol.*, 52: 808–817 (discuss., 1958, 53: 339–340, 493–494, 622–625, 890–893; 1959, 54: 745–748, 953–956; 1960, 55: 615–617).

Koeberlin, F.R., 1938. Sedimentary copper, vanadium–uranium, and silver deposits in southwestern United States: *Econ. Geol.*, 33: 458–461.

Lindgren, W., 1933. *Mineral Deposits.* McGraw-Hill, New York, N.Y., 4th ed., 930 pp.

Lindgren, W., 1935. Waters, magmatic and meteoric. *Econ. Geol.*, 30: 463–477.

Love, L.G., 1962. Biogenic primary sulfide of the Permian Kupferschiefer and Marl slate. *Econ. Geol.*, 57: 350–366.

Love, L.G. and Zimmerman, D.O., 1961. Bedded pyrite and micro-organisms from the Mount Isa shale. *Econ. Geol.*, 61: 873–896 (discuss., 1962, 57: 118–119, 265–274, 459–460, 460–462, 624–627).

Lovering, T.S., 1963. Epigenetic, diplogenetic, syngenetic, and lithogenetic deposits. *Econ. Geol.*, 58: 315–331.

Lusk, L., 1969. Base metal zoning in the Heath Steels B-1 ore body, New Brunswick, Canada. *Econ. Geol.*, 64: 509–518.

Lusk, J., 1972. Examination of volcanic-exhalative and biogenic origins for sulfur in the stratiform massive sulfide deposits of New Brunswick. *Econ. Geol.*, 67: 169–183.

Lusk, J. and Croket, J.H., 1969. Sulfur isotope fractionation in coexisting sulfides from the Heath Steels B-1 ore body, New Brunswick, Canada. *Econ. Geol.*, 64: 147–158 (discuss., 829).

Matsukuma, T. and Horikoshi, E., 1970. Kuroko deposits in Japan, a review. In: T. Tatsumi (Editor), *Volcanism and Ore Genesis.* Univ. Tokyo Press, pp. 153–179.

Mauger, R.L., 1972. A sulfur isotope study of the Ducktown, Tennessee district, U.S.A. *Econ. Geol.*, 67: 497–510.

Mills, J.W. et al., 1971. Bedded barite deposits of Stevens County, Washington. *Econ. Geol.*, 66: 1157–1163.

Mendelsohn, F., 1961. *The Geology of the Northern Rhodesian Copperbelt.* MacDonald, London, 523 pp.

Moore, J.G., 1965. Petrology of deep sea basalt near Hawaii. *Am. J. Sci.*, 263: 40–52.

Nielsen, H., 1965. Schwefelisotope im marinen Kreislauf und das δ^{34}S der früheren Meere. *Geol. Rundsch.* 55 (1): 160–172.

Noble, E.A., 1963. Formation of ore deposits by water of compaction. *Econ. Geol.*, 58: 1145–1156 (discuss, 1964, 59: 723–724).

Oftedahl, Ch., 1958. A theory of exhalative–sedimentary ores. *Geol Fören. Stockh. Förh.*, 80, 1(492): 1–19, 1959, Replies and discussion, 81, 1(496): 139–144.

Ohmoto, H., 1972. Systematics of sulfur and carbon isotopes in hydrothermal ore deposits. *Econ. Geol.*, 67: 551–578.

Park, W.C. and Amstutz, G.C., 1968. Primary cut-and-fill channels and gravitational diagenetic features – their role in the interpretation of the southern Illinois fluorspar deposits. *Miner. Deposita*, 3: 66–80.

Ramdohr, P., 1957. Die "Pronto-Reaktion". *Neues Jahrb. Miner. Monatsh.*, 10–11: 217–222.

Rees, C.E., 1973. A steady-state model for sulphur isotope fractionation in bacterial reduction processes. *Geochim. Cosmochim. Acta*, 37: 1141–1162.

Renfro, A.R., 1974. Genesis of evaporite-associated stratiform metalliferous deposits – a sabkha process. *Econ. Geol.*, 69: 33–45.

Rickard, D.T., 1973. Limiting conditions for synedimentary sulfide ore formation. *Econ. Geol.*, 68: 605–617.

Ridge, J.D., 1960. Behavior of sulfur magmatic and hydrothermal conditions and its effect on S^{32}/S^{34} ratios. *Econ. Geol.*, 55: 1330, Abstract.

Ridge, J.D., 1973. Volcanic exhalations and ore deposition on the vicinity of the sea floor. *Miner. Deposita*, 8: 332–348.

Ridge, J.D., 1974. A note on boiling of ascending ore fluids and the position of volcanic-exhalative deposits in the modified Lindgren classification. *Geology*, 2: 287–288.

Roberts, W.M.B., 1967. Sulphide synthesis and ore genesis. *Miner. Deposita*, 2: 188–199.

Robertson, D.S. and Steenland, N.C., 1960. On the Blind River uranium ores and their origin. *Econ. Geol.*, 55: 659–694.

Robertson, D.S. and Douglas, R.F., 1970. Sedimentary uranium deposits. *Can. Inst. Min. Metall. Bull.*, 63: (697): 557–566.

Roedder, E., 1968. Temperature, salinity, and origin of the ore-forming fluids at Pine Point, Northwest Territories, Canada, from fluid inclusion studies. *Econ. Geol.*, 63: 439–450.

Roscoe, S.M., 1956. Geology and uranium deposits, Quirke Lake–Elliot Lake, Blind River area, Ontario. *Geol. Surv. Can., Pap.*, 56-7, 21 pp.

Roscoe, S.M. and Stacey, H.R., 1958. On the geology and radioactive deposits of Blind River regions. In: *Proc. 2nd UN Int. Conf. (Geneva) on Peaceful Uses of Atomic Energy*, 2: 475–483.

Sakai, H., 1957. Fractionation of sulphur isotopes in nature. *Geochim. et Cosmochim. Acta*, 12: 150–169.

Sales, R.H., 1960. Critical remarks on the genesis of ore as applied to future mineral exploration. *Econ. Geol.*, 55: 805–817, 629–636.

Sales, R.H., 1962. Hydrothermal versus syngenetic theories of ore deposition. *Econ. Geol.*, 57: 721–734 (disc., pp. 831–836; 1964, 59: 162–167).

Sawkins, F.J., 1965. Oceanic lead isotopes and ore genesis. *Econ. Geol.*, 60: 1083–1084.

Schuchert, C., 1920. Diagenesis in sedimentation (with discusstion). *Geol. Soc. Am. Bull.*, 31: 425–432.

Sillitoe, R.H., 1973. Environments of formation of volcanogenic massive sulfide deposits. *Econ. Geol.*, 68: 1321–1336.

Sinclair, W.D., 1971. A volcanic origin for the No.5 zone of the Horne mine, Noranda, Quebec. *Econ. Geol.*, 66: 1225–1231 (discuss., 1973, 68: 711–713, 713–714).

Solomon, P.J., 1965. Investigations into sulfide mineralization at Mount Isa, Queensland. *Econ. Geol.*, 60: 737–765 (discuss., 1966, 61: 1158–1161).

Stanton, R.L., 1959. Mineralogical features and possible mode of emplacement of the Brunswick Mining and Smelting ore bodies, Gloucester County, New Brunswick. *Can. Inst. Min. Metall., Trans.*, 62 (570): 337–349.

Stanton, R.L., 1960. General features of conformable "pyritic" orebodies; pt. 3: Field association; pt. II: Mineralogy. *Can. Inst. Min. Metall. Trans.*, 63(573–574) 22–36.

Stanton, R.L. and Rafter, T.A., 1966. The isotopic composition of sulphur in some stratiform lead–zinc sulfide ores. *Miner. Deposita*, 1: 16–29.

Stieff, L.R., et al., 1956. Preliminary age determinations of some uranium ores from the Blind River area, Algoma district, Ontario, Canada. *Geol. Soc. Am. Bull.*, 67: p. 1736, abstract.

Suffel, G.G., 1965. Remarks on some sulphide deposits in volcanic extrusives. *Can. Inst. Min. Metall. Bull.*, 58(642): 1057–1063.

Sweeney, R.E. and Kaplan, I.R., 1973. Pyrite framboid formation: laboratory synthesis and marine sediments. *Econ. Geol.*, 68: 618–634.

Temple, K.L., 1964. Syngenesis of sulfide ores: an evaluation of biochemical aspects. *Econ. Geol.*, 59: 1473–1491.

Temple, K.L. and Le Roux, N.W., 1964a. Syngenesis of sulfide ores: Sulfate-reducing bacteria and copper toxicity. *Econ. Geol.*, 59: 271–278 (discuss., 938–939).

Temple, K.L. and Le Roux, N.W., 1964b. Syngenesis of sulfide ores: desorption of adsorbed metal ions and their precipitation as sulfides. *Econ. Geol.*, 59: 647–655.

Trudinger, P.A. et al., 1972. Biogenic sulfide ores: a feasibility study. *Econ. Geol.*, 67: 1114–1127.

Twenhofel, W.H., 1939. *Principles of Sedimentation.* McGraw-Hill, New York, N.Y., 1st ed., pp. 254–259.

Vokes, F.M., 1966. On the possible modes of origin of the Caledonian sulfide ore deposit at Bleikvassli, Nordland, Norway. *Econ. Geol.*, 61: 1130–1139.

Vokes, F.M., 1969. A review of the metamorphism of sulfide deposits. *Earth Sci. Rev.*, 5: 99–143.

White, C.H., 1942. Notes on the origin of the Mansfeld copper deposits. *Econ. Geol.*, 37: 64–68.

White, D.E., 1968. Environments of generation of some base-metal ore deposits. *Econ. Geol.*, 63: 301–335 (disc., pp. 846–847).

White, W.C. and Wright, J.C., 1954. The White Pine copper deposit, Ontonagon county, Michigan. *Econ. Geol.*, 49: 675–716 (discuss., 1959, 54: 947–951, 1127; 1960, 55: 402–410).

White, W.C. and Wright, J.C., 1966. Sulfide-mineral zoning in the basal Nonesuch shale, northern Michigan. *Econ. Geol.*, 61: 1171–1190.

Whitehead, R.E., 1973. Environment of stratiform sulphide deposition: variation in Mn/Fe ratio in host rocks at Heath Steele mine, New Brunswick, Canada. *Miner. Deposita* 8: 148–160.

Wiese, R.G., 1960. *Petrology of a Copper-bearing Precambrian Shale, White Pine, Michigan.* Thesis, Harvard Univ., Cambridge, Mass.

Zimmermann, R.A., 1969a. Sediment–ore–structure relations in barite and associated ores and sediments in the upper Mississippi Valley zinc–lead district near Shullsberg, Wisconsin. *Miner. Deposita,* 4: 248–259.

Zimmermann, R.A., 1969b. Stratabound barite deposits in Nevada. *Miner. Deposita,* 4: 401–409.

TOPICAL INDEX

Garlick and Brummer, 1951
Davis, 1954
W.C. White and Wright, 1954
Sales, 1960

Copper, deposits in red beds
Lindgren Volume, 1933
Bastin, 1933

Diagenesis
Schuchert, 1920
Davidson, 1962
Hirst and Dunham, 1963
Amstutz and Park, 1967
Hamilton, 1967
Jackson and Beales, 1967

Epigenesis
Bell, 1931
White, 1942
Collins, 1950
Sales, 1960
Sales, 1962
Lovering, 1963

Evaporites, involvement in the ore-forming process
Davidson, 1965
Craig, 1969

Isotopes, lead, fractionation of
J.C. Brown, 1965

Isotopes, sulfur, fractionation of
Dechow, 1960
Gavelin et al., 1960
Ridge, 1960

Isotopes, sulfur, ratios of
Dechow, 1960
Dechow and Jensen, 1965
Nielsen, 1965
Anger et al., 1966
Field, 1966

Lead–zinc deposits in carbonate rocks
Kendall, 1960
Garlick, 1964
Derry et al., 1965

Ore fluids, connate waters
Dunham, 1964
Roedder, 1968

W.C. White and Wright, 1966
Hamilton, 1967
A.C. Brown, 1974
Renfro, 1974

Renfro, 1974

Park and Amstutz, 1968
Berner, 1969
Zimmermann, 1969a
Zimmermann, 1969b
Berner, 1971
Hallberg, 1972

Boyle, 1965
Vokes, 1966
W.C. White and Wright, 1966
Bain (cited in Anderson and Nash, 1972)
Griffiths et al., 1972
A.C. Brown, 1974

Renfro, 1974

J.C. Brown, 1970

Nielsen, 1965
Anger et al., 1966
Hoefs, 1973

Stanton and Rafter, 1966
Lusk and Crocket, 1969
Lusk, 1972
Mauger, 1972
Hoefs, 1973

Jackson and Beales, 1967
Derry, 1973

Brown, 1970

Ore fluids, general
D.E. White, 1968

Ore fluids, magmatic
Lindgren, 1935

Ore fluids, meteoric or ground water
Dunham, 1964

Brown, 1970

Ore fluids, volcanic-exhalative
Oftedahl, 1958
Hirst and Dunham, 1963
Anger et al., 1966
Kinkel, 1966

Matsukuma and Horikoshi, 1970
Ridge, 1973
Ridge, 1974

Ore fluids, water of compaction
Noble, 1963
W.C. White and Wright, 1966

Hamilton, 1967
Jackson and Beales, 1967

Ores, in layered rocks
Bain, 1960

Porphyry copper deposits, relations with massive sulfide deposits
Hutchinson and Hodder, 1972

Sea water, determination of depth of
Jones, 1969

Sulfates, supergene
Field, 1966

Sulfides, massive, classification of
Gilmour, 1972

Sulfides, massive, formation of
Friedman, 1959
Stanton, 1959
Stanton, 1960
Boyle, 1965
Hutchinson, 1965
Kalliokoski, 1965
Suffel, 1965
Kinkel, 1968
Anderson, 1969

Lusk, 1969
Sinclair, 1971
Anderson and Nash, 1972
Hutchinson and Hodder, 1972
Lusk, 1972
Hutchinson, 1973
Sillitoe, 1973
Whitehead, 1973

Sulfides, metamorphism of
Lindgren, 1933
Friedman, 1959
Soloman, 1965
Vokes, 1966
Roberts, 1967

Vokes, 1969
Sinclair, 1971
Hallberg, 1972
Mauger, 1972

Chapter 7

SUMMARY OF THE FRENCH SCHOOL OF STUDIES OF ORES IN SEDIMENTARY AND ASSOCIATED VOLCANIC ROCKS — EPIGENESIS VERSUS SYNGENESIS

A.J. BERNARD and J.C. SAMAMA

INTRODUCTION

In France, as elsewhere in the world, the ores contained in sedimentary rocks, and in some cases volcano-sedimentary piles, used to be explained essentially with respect to the nature of the ore, i.e.:

on the one hand, evaporites, phosphates, carbonaceous matter and heavy detrital minerals, resulted clearly from chemical, organic or mechanical sedimentary processes, which convey the *syngenetic* nature of the geochemical concentration;

on the other hand, sulfides and some metallic oxides (notably iron and manganese) which, by their very nature and no matter their relations to the sedimentary host rock, resulted from hydrothermal processes (sensu lato); these showed the nature of the mineralizations and their frequently discordant appearance; that is to say, they conveyed the *epigenetic* character of the concentration.

It was undoubtedly De Launay (1913) who set down the guidelines of the school of thought outlined here. He did so, however, allowing for many shades of meaning, and sometimes with reservations, with which his disciples did not care to burden themselves. Even some twenty years later in the U.S.A., in Lindgren's (1933) *Mineral Deposits* we could still appreciate the caution with which the author treated the question of sulfide ores in sedimentary rocks[1], and wonder why his successors did not show the same discernment. It is in this way that dogmas are born from over-simplifications, a kind of distorted copy of the thoughts of the originators of concepts.

In fact, the above-mentioned attitude is based on two assumptions:

(a) the hydrothermal genesis of sulfides which, it was thought at that time, could only be synthetized under conditions of high temperature; and

---

[1] One finds in chapter XXII, pp. 379–422: "Deposits formed by concentrations of substances, contained in the surrounding rocks by means of circulating waters", the deposits of the so called "Red-Bed" and "Kupferschiefer" types. In the same way, chapter XXIII, pp. 423–443, is entirely devoted to the "Mississippi-Valley" type: "Lead and zinc deposits in sedimentary rocks, origin independent of igneous activity."

(b) the original minerals and their textural relationships remain unchanged, or, in other words, the preservation of the ores in the state of their primary deposit.

These two propositions derive directly from the hydrothermal "vein" model, where they prove to be mutually entirely coherent in the hypothesis of a deposition by decreasing temperature, in an open space (a mineralization called "in-filling"). The vein structures, both banded and symmetrical, are obviously very significant in a deposition of high-temperature paragenesis and are metastably preserved in their initial state.

The extension of these assumptions related to sulfide impregnations, whether diffuse or massive, in sedimentary rocks raises numerous difficulties; precisely those which led both De Launay and Lindgren to express many reservations about the purely igneous hydrothermal genesis of these ores. One of the most important reservations is the one which raises the issue of the existence of deposits pseudomorphosing the sedimentary structures, and therefore implying, ipso facto, a metasomatic replacement at a low temperature (around 100°C). In a rarely mentioned, yet fundamental article, Dreyer and Garrels (1952) brought out all the contradictions harboured by the concept of a mineralizing solution capable of dissolving the rock in which it is circulating, and at the same time depositing sulfide mineralizations of the same volume. In this article they introduced the problem of the chemistry of hydrothermal solutions (about which, we must acknowledge a posteriori, nobody knew anything while expecting quite a lot). Ten years later, Helgeson (1964) offered the first coherent chemical model of these solutions and since then new light has been shed on the genesis of the peri-plutonic deposits; nevertheless, though we have a better idea of them, the dissolvant and metallizing capacities of the hot chlorinated brines still do not explain the telethermalism which is still the best defined by McKinstry (1955):

"That the source of ore solutions is a magma seems indicated by much circumstantial evidence, to be found particularly in epithermal and mesothermal deposits. Other deposits, especially those classed as telethermal are more difficult to ascribe with confidence to any particular intrusive body, and while there are reasons for believing that their parent solutions come from a magma, their source, whatever it may have been, is so remote that it has no more than an indirect bearing on the behaviour of solution at the site of deposition."

The very ambiguity of such a definition does not, in itself, constitute a priori an admission of failure: telethermalism is a belief, in which we may or may not put our faith, but we cannot justify it without the means to prove its validity.

In short, with the theoretical study of strata-bound sulfide deposits leading to a dead end, the need to return to the facts was felt, and all over the world (cf. Amstutz, 1962) researchers began work on describing these mineralizations in a better way, and especially on placing them more precisely into their geological environment. In France, as early as 1950, the School at Nancy launched the primary revisional studies, at first descriptive then theoretical. The Laboratory of Applied Geology at the Sorbonne took over several years later; finally Routhier (1963), judging by the amount of space he accorded in his treatise to strata-bound deposits, has diffused on a wide scale the ideas which we will here qualify as "sedimentaristic".

At the outset, the verification of the existence of mineralizations very early in the history of their sedimentary country rocks led to the notion of syngenetism. Indeed, it is not so rare to find sulfide impregnations involved in slumping phenomena developed in non-compacted silt (Fig. 1) or mineralized pebbles in intraformational conglomerates reworking the underlying metalliferous bed (Fig. 2). This penecontemporaneity of the

Fig. 1A. Pyritic banded ore involved in early deformation: contorted pyritic seams due to slumping prior to rock lithification. Gradually, increasing deformations lead to intra-formational micro-breccia. Polished section: St-Felix de Paillères, northern flank (cf. Fig. 3).

Fig. 1B. Schalenblende and galena banding deformed in early intra-formational folds. Late crystallization of dolomite and barite (white areas) by in-filling of empty spaces opened by the deformation. Polished section, same scale as 1A: Les Malines, bed no. 1 Alby-Fontbonne (Gard, France).
These two figures clearly show the existence of ore minerals (pyrite, galena, Schalenblende) before the end of lithification. These mineralizations are thus, at least, diagenetic.

Fig. 2A. In a coarse arkosic sandstone, we observe a local concentration of galena limited to a soft pebble of silt. The feature of a soft pebble in a detrital environment is very common, and, in this case, the underlying silt (roof of bed no. 2) has been eroded by a coarser detrital event in which the mineralized pebble occurs. This strongly suggests that the underlying bed was already mineralized before its erosion and before the deposition of the overlying detrital bed.
Polished hand specimen, bed no. 1, Largentière (Ardèche, France).

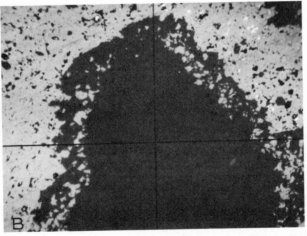

Fig. 2B. Around a detrital grain of quartz (black area), we observe a complex growth of galena closely associated to secondary quartz. Here, secondary quartz as well as galena correspond to a diagenetic feature of crystallization.
Polished section, X50, La Gravouillère, level 266 (Gard, France).

sulfide concentration and the sedimentary deposit of the country rock clearly establishes the syngenetism of the concentration and of the rock. But just as the mineral substances which constitute the rock evolve during lithification and sedimentary epigenesis, and even anchi-metamorphism, the metallic sulfides recrystallize according to their own physicochemical properties. These differences in crystallogenetic evolution often lead, especially at the microscopic level, to structural divergencies such that at the end of the same evolution the ores seem discordant — or more exactly, incongruent — with respect to the preserved sedimentary structures. Their appearance is epigenetic, which is quite normal since these crystals are developed in consolidated rock. Paradoxical though this may appear, however, most of the concentrations syngenetic with their sedimentary country rock have today, after diagenesis and burial, epigenetic characteristics with respect to these same host rocks.

It is most necessary to understand fully that the syngenesis-versus-epigenesis debate is only meaningful when it applies to concentrations and their initial forms, in relation to the country rock, and has no meaning when it is a question of the more or less discordant aspects of the ores with respect to these country rocks. Looking at it in this way, the total congruency of the ore and its host rock establishes a strong leaning towards the syngenetism of the concentration; on the contrary, discordance does not signify anything. In fact we must refer to arguments of a different sort, notably the geochemistry of the deposit and of its immediate and more distant surroundings in order to decide on the syngenetism or epigenetism of the concentration.

Also, in spite of the heading of this chapter, we will not be using the terms syngenesis and epigenesis too much, since they lead to confusion rather than clarity. Concerning the ores' appearance, the purely descriptive terms of congruency and discordance will better replace them; concerning the concentration, the reference to such processes as sedimentary, diagenetic, karstic, or exhalative, will lead to unequivocal information by more specific and more precise genetic models. There remains, nevertheless, the custom (Roukhine, 1953) of separating different stages in the evolution of a sedimentary rock: syngenetic, diagenetic or epigenetic, before anchi-metamorphism. It is within the context of this meaning that we shall go on using the terms syngenesis and epigenesis as related to the evolution of country rock.

At the end of this introduction, it goes without saying that the previous assumptions in favour of generalized hydrothermalism become superfluous, at least where a good number of strata-bound sulfide mineralizations are concerned[1]. They could be replaced by the following propositions:

(a) the genesis of sulfides is not exclusively igneous-hydrothermal; even at normal temperatures the sulfides precipitate each time the conditions of pH-Eh are right for it (cf. Christ and Garrels, 1967); and

(b) like all mineral species, the sulfides and notably the colloidal sulfide gels are sensitive to a crystallogenetic evolution, i.e., crystallization and/or re-crystallization, if pressure and temperature increase.

It would, however, be misleading to set the now formulated assumptions against those

---

[1] In fact this is not always the case: for example, it is probable that the volcanic-exhalative sulfides of the massive pyrito-cupriferous accumulations are precipitated at high temperature. Nevertheless, they change during later burial and the second assumption no longer applies.

on pp. 299–300 and to reduce the crystallochemistry of the sulfides to opposing propositions, thus repeating the error committed by the purists of hydrothermalism. It is the geological history of concentrations which we should translate into terms of solution chemistry and crystallochemistry of mineral species, in order to appreciate the genesis of a sulfide deposit, whether vein-like or strata-bound. The following pages will be devoted to the
latter.

THE CASE OF THE KUPFERSCHIEFER

The existence in Europe of a formation of cupriferous shales lying at the base of the
Upper Permian stage (Zechstein) early attracted the attention of metallogenists to
sulfide strata-bound mineralizations. A brief look at this field will help us to situate the
problems which preoccupied French economic geologists during the last twenty years and
which we will treat later in more detail.

In spite of the extraordinary regularity of an ore horizon 0.50–2 m thick over a
surface of more than $6 \cdot 10^5$ km^2, and offering abnormally heavy metal contents almost
everywhere, these mineralizations were a point of furious controversy between believers
in the sedimentary nature of the deposits and tenants of a metasomatic hydrothermal
origin. It was only very recently that Wedepohl (1971) put an end to these discussions by
proposing, after an exhaustive synthesis of the works on the Kupferschiefer, that these
metallizations should be considered as the prototype of the sedimentary, syngenetic concentrations of sulfides. (Cf. in Vol. 6, Chapter 7 by Jung and Knitzschke.)

Still the problem is not resolved, for even if the trapping of heavy metals in an euxinic
marine area is relatively easy to understand, the origin of these metals is unknown. Three
hypotheses of metalliferous supply synchronous with sedimentation were taken into
consideration:

(1) volcanic (or hydrothermal) exhalation in sea water; (2) pure marine (or
thalassogenetic supply) from sea water which could be normal or only slightly abnormal;
(3) continental, by leaching of normal (or abnormal) rocks which have emerged before
and during deposition of the mineralized beds.

Starting from the fact that the present-day sedimentary phenomena lead at best to
geochemical anomalies and not to ore deposits, some authors (Dunham, 1964) do not
hesitate to introduce, ipso facto, the intervention of a cause other than purely exogenic
(marine[1] or continental) and, more precisely, to resort to an igneous-hydrothermal
source, exhalative or not.

---

[1] The Kupferschiefer is in fact a great geochemical anomaly which is far from being workable in all
cases, even with economic criteria less rigid than those of the occidental economies. One might wonder
in fact if local conditions, a shallow gulf with restricted sea-water circulation to put forward by
Brongersma-Sanders (1966), might not enhance the regional anomaly to the level of an exploitable
concentration. In this two-stage model, i.e. basin-wide anomaly and local, coastal, over-concentrating
environment, it is the general anomaly which remains to be explained.

While remaining aware of the fact that this way of thinking is rather too cursory to take up the complexity of the phenomena considered, we willingly admit (Bernard, 1962, 1973a,b) that the formation of a workable deposit comes from the spatial connection and the contemporaneity of two favourable causes: an exceptional metalliferous supply and a sedimentary trap. One without the other is only likely to lead to the creation of geochemical anomalies, the one of regional extension, the other depending on the size of the trap. Let us look more closely at these propositions.

THE TRAP STRUCTURES

The interest attached in France to the metallogenic role of the black euxinic shales is better understood after this outline of the genetic problems raised by the "Kupferschiefer—black shales geochemically very abnormal because of their high metallic content". In fact, and as Wedepohl (1971, p. 269) has recently pointed out, it is far more a question of a Pb-Zn marly bituminous shale than of a Cu one.

At the outset, this remark drew our attention to the petrographic diversity of the fine-grained (less than 50μ) and bedded rocks, that is to say the shales, and in this instance marly shales[1].

The proportion of carbonates, phyllites and detrital carbonaceous matter may vary widely: Vine and Tourtelot (1970), then Tardy and Vine (1973) have studied this problem and listed statistical correlations which exist between the petrographic constituents and the trace elements. The geochemical spectrum presented in this way allows the permanent sedimentological factors to emerge from these somewhat ubiquitous facies (Krumbein and Sloss, 1963). These are those of the euxinic milieu: quiet-water, reducing environment permitting the conservation of organic matters, and slow or very slow rate of sedimentation.

It is within these permanent sedimentological peculiarities that we are led to look for the causes of the high content of metallic trace elements, so often found in black shales. The role of adsorption was recognized very early (Goldschmidt, 1937) but even today it is still not fully appreciated, notably not in the area concerning the maximum strenght of concentration induced by this phenomenon.

The main factors governing adsorption appear to be well-listed: a heavy-metal amount in the solutions, nature and development of the adsorbent surface, granulometry of the adsorbent matter and duration of fixation. However, characteristics much more difficult to estimate, such as the state of the surfaces, even for the same mineral, also seem to act as determining factors in the effectiveness of the phenomenon.

---

[1] Eisenhut and Kautsch (1954) noted a mean composition of the Kupferschiefer in which there occurred sulfates (gypsum, anhydrite) and a siliceous gel (29%). These components are no longer quoted in later publications (Haranczyk, 1964, 1970; Erzberger et al., 1968; Wedepohl, 1971) which destracts in no way from the pre-evaporitic nature of this sediment.

Lastly, the organic colloids deriving from bacterial fossilization play a special role; bacteria, while alive, are apt to concentrate a good number of heavy elements (Baas Becking, 1959; Turekian, 1969). This concentration factor is often minimized, for the correlation percentage of carbonaceous matter vs. amount of sulfides is usually only positive and significant for a certain range, rather low in the percentage of sulfides. On the other hand, we should consider that carbonaceous matter conserved in the black sulfide-rich shales is only residual: in fact, sulfate-reducing bacteria consume organic matter in order to maintain their metabolism and to liberate CO_2 (see, in Vol. 2 the chapters by Saxby, and Trudinger). The formation of sulfides and the conservation of organic substances appear thus like two antagonistic processes in which it is difficult to decipher, after diagenesis, the role of metallic concentration due to pure mineral adsorption, and that due only to organic activity. The nature of the concentrated metals occasionally permits a clarification of the dilemma, if one is referring to normal sea water, but this argument loses a great deal of its weight in the case of derivation of rich sulfide impregnations from abnormal sea waters.

Unfortunately, the petrographic diversity of "black shales" might lead to equally diverse and, for the moment, little-known diagenetic evolutions. The syngenetic formation of bisulfides of iron and notably of pyrite, though obvious geologically, has only recently been elucidated (Roberts et al., 1969). In the same way, the model of early diagenetic fixation of heavy metals as sulfide has only just been outlined (Hallberg, 1972; Rickard, 1973; Trudinger, Chapter 6, Vol. 2).

Rickard points especially to the probable role of heavy-metal chelates which would have, in the very first phase of diagenesis, an antagonistic action to that of the precipitation of simple sulfides. As chelation increases with the amount of metal, only very reducing environments, characterized by a negative Eh of the water/sediment interface would be able to concentrate easily chelatable metals, for instance copper. Apparently these environments are relatively rare and generally of restricted extent.

The European Kupferschiefer appears finally as characteristic in a very reducing environment where diagenesis essentially consists of a growth of sulfide grains from crystalline germs, without any great movement of heavy metals (centimetric scale). In that way they also constitute a prototype. It is probable indeed that most formations of black shales (cf. samples of Vine and Tourtelot, 1970) were deposited in oxygenized bottom waters, as indicated by the low values of the Cu/Zn ratio and the small amount of Mo (Hallberg, 1972).

It are then finally the products of solubility of the different sulfide species which explain most clearly the zonal dispositions observed from the shorelines in real euxinic environments, as much in Mansfeld as in Rhodesia (Garlick, 1961).

In consequence, the metalliferous black shales form a very special group characterized, from a sedimentological point of view, by very calm environments and by a very low sedimentation though rich in carbonaceous fine detritals favouring the development and the activity of bacterial sulfate-reducing colonies.

In marine areas this brief outline seems sufficient to encourage prospection in such euxinic environments. One may equally well wonder whether similar conditions do not occur in a continental environment, since it is, after all, as much an abundant generation of H_2S in a metalliferous solution which remains (and this will surprise no one) the major cause of the sedimentary fixation of chalcophile metals. Let us examine these two points separately.

The marine traps

One may first of all think, as for European Kupferschiefer, that the metalliferous zones are individualized within vast basins of common bituminous shales, that is to say presenting a water/sediment oxidizing interface and passing rapidly down to reducing horizons. Starting from micro-values in heavy metals due either to mineral adsorption or to organic fixation, the antagonistic action between chelation and sulfide fixation will lead to a certain early diagenetic retention of the chalcophiles. Let us remember at this point that a low percentage of metal favours the diffusion of H_2S and, therefore, sulfide fixation (Berner, 1969): a weak metallic supply would be almost completely precipitated. The high heavy-element content of certain horizons of a normal black-shale formation does not, however, necessarily involve highly reducing facies. Fuchs (1969), studying this problem in the Toarcian on the southern and western Massif Central borders, showed that these Pb-Zn anomalies were always related to a higher proportion of either clastic or soluble terrigenous elements. These host layers thus register an exceptional supply of heavy elements at the site of sedimentation, and not a variation of the redox of the water/sediment interface. As with the bitumens of the Toarcian black shales, the heavy metals seem well fixed, and it is only in special conditions of diagenesis that small-scale lateral circulations of connate waters sometimes induce a mineralogical form, i.e., sphalerite (Bernard and Fuchs, 1964). By the way, the host rock should not be confused with the concept of a "source-bed" (Knight, 1957; Noble, 1963) whose metalliferous connate waters constitute rather a moveable reserve during compaction. When fixed, the micro-amounts of the host-layer preserve in the geologic record of an ordinary euxinic environment, an exceptional supply of heavy metals (the Toarcian one is terrigenous) in the sedimentary basin.

The notion of host layer, even if it does not apply to our research of highly reducing Kupferschiefer environments, at least constitutes an excellent indication of exceptional supply, especially when the metalliferous horizon has a large geographic extent. We will use this argument later. At this time, however, the study of the host layers leads us to make another observation.

When for the same isochronous host-layer, irregularities of the basin sea floor lead to variations in thickness, it is always the least-thick deposit, i.e., the slowest rate of sediment accumulation, which is also the richest in heavy metals. This is a direct consequency of the syngenetic adsorption preferentially developed on the sea-floor heights, sites characterized by a low rate of sedimentation.

This remark joins the conclusions made by one of us (Bernard, 1959) on a regional scale, in the "sous-cévenole" province (St-Felix-de-Paillères, southeast of the Massif Central, France). It is on the tops and on the flanks of the active domes (and in this way generative of lateral changes of the lithological series), that the sulfide strata-bound accumulations develop selectively (Fig. 3). At the start only a simple academic conclusion, this remark has since become a rule of mining prospection, analogous to the so-called "buried-hills" rule, used, in other sedimentary environments, by petroleum geologists (Levorsen, 1955). Nevertheless, this general remark needs here a particular explanation.

Before Lombard (1953), very little was known about the sedimentological causes of a decrease in the rate of marine deposition: the parallel disposition of strata had become such a commonly known fact that no one concerned himself any more with the nature and significance of joints.

Now, in fact these thin horizons or these surfaces which limit the outcropping strata, correspond with decreasing phases of sediment accumulation on the sea floor, or even, when the joints erode the underlying beds, with submarine erosion. Thus, between a thick and homogeneous layer deposited in the basin and an isochronous diastem on the heights, the area of low-rate sedimentation coincides with the so-called wedge of the layer. We are concerned here with the condensation wedge (Fig. 4) which is not the same as the stratigraphic wedges of the shore lines which are called reduction wedges (Grabau, 1906).

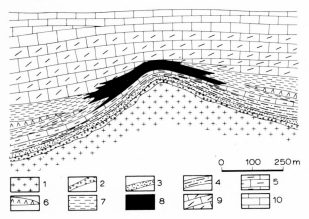

Fig. 3. The St-Felix de Paillères buried hill at the end of the Sinemurian. The condensed metalliferous series of sea-floor heights occurs principally during the Lower Keuper interfingering with the basin evaporitic series, thus illustrating the subsiding nature of Upper Saliferous formations in the basin.
1 = Hercynian basement: granite; *2* = fluvial conglomerate: Bundsandstein; *3* = blanket arkose; *4* = marls and evaporites: Lower Saliferous; *5* = marine dolomite: Muschelkalk; *6* = marls and evaporites: Upper Saliferous; *7* = variegated marls: Upper Keuper and Rhaetian; *8* = bituminous and metalliferous shales and dolomites; *9* = Hettangian dolomites; *10* = Sinemurian limestones.

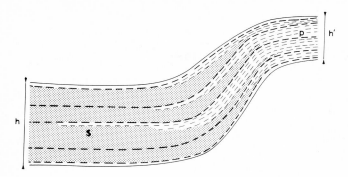

Fig. 4. Sketch of a condensation wedge (after Grabau, 1906). The decrease in thickness of each sandy stratum (S) is accompanied by an enrichment in phyllites (p); h and h' are the thicknesses of the isochronous sedimentary piles, in the basin and over the height, respectively. (Courtesy Geol. Soc. Am.)

The petrographic change which accompanies the condensation of a layer from the basin to the heights consists essentially in the progressive disappearance of the most abundant mineral in the basin deposit. Thus, a calcareous layer will progressively become marly and dolomitic, then argillaceous, before disappearing completely as a simple surface of discontinuity. Everything happens, in fact, as if the argillaceous sedimentation were constant over the whole area (basin and heights): on the heights and on their flanks, it is only an argillaceous sedimentation, whereas it is diluted by carbonates in the basin area. This is only a frequent and simple example of wedging by condensation, developed in a hydrodynamically calm environment, but it is quite obvious that numerous possibilities of lateral changes occur, though they are governed by stratonomical rules which are relatively simple (Bernard, 1959).

Generally speaking, the most probable sedimentation of the heights or their flanks is finely detrital (fine clastics, inherited clays, organic colloids, or, in some cases, carbonate muds). It is, therefore, a question of black shale (argillaceous or marly) deposition and, consequently, of very fine-grained sediments, that is to say with a large specific surface and finally of particularly adsorbent muds.

In other words, if one admits, as we have done previously for phyllites, that the deposition of heavy metals may be the same by surface unity over the whole area of sedimentation, the sediment on the heights will seem to be enriched in relation to the sediment in the basin. The level of apparent concentration C_a can be calculated, as a first approximation, by the ratio of the thicknesses h/h'. In fact, it is here a minimum evaluation which should be multiplied by a coefficient of lateral change T, taking into account both the increasing adsorbent properties of the sediment from the basin towards the heights, and a factor of time of contact between the adsorbent particle and the solution which is greater on the heights than in the basin. One would therefore have:

$$C_a = \frac{h}{h'} \cdot T \qquad \text{(cf. Fig. 4)}$$

We have been able to verify this assumption (Bernard and Samama, 1969) over a wedge of carbonate host-beds, with separate strata (which is not the case for clay formations).

Thus, it is essentially the preferential adsorption of slow sediment accumulation in quiet environments, rather than the redox character of the water/sediment interface, which governs the geochemical anomaly of the heights. It is obvious that the development of a real euxinic environment, very rich in H_2S, could only enhance this metallogenetic trend.

Finally, by going back to monographic descriptions of ore deposits, one could verify that the petrographic and tectonic nature of the heights is not important: epeirogenetic horsts (Ohle and Brown, 1954), biohermal reefs (Callahan, 1964; Monseur and Pel, 1973), early anticlinal folds and diapiric uplifts, all lead equally to geochemical anomalies in heavy metals. If this trap takes place in a basin receiving an exceptional supply – which we will examine later (p. 319) – the levels of concentration thus reached lead to economic deposits. Their genetic history is, however, complicated by the localization itself, on the top or on the flanks of the heights. On the one hand and because of the lateral changes, the host rocks are heterogeneous, on the other hand, the slow lithification of these fine muds deposited on a slope leads to frequent slumpings, either early (convolute beddings) or late (brecciation by slumping). All these factors greatly complicate the diagenetic evolution of the ores. The existence of convolute mineralizations (Fig. 1) implies the syngenetic existence of the concentration before slumping begins. In any case, when dealing with carbonate environments, the common preferential distribution of sulfides in the cement of the slump-breccia poses a delicate problem of mobilization, or rather modification, of the primary concentrations. The reasons put forward for the formation of breccias are not without importance: for some (Launay and Leenhardt, 1959) the slope of the synsedimentary bank is sufficient for a slump when the accumulation surpasses the level of stability of the mud bank. For others (Snyder and Odell, 1958) the differential compaction of the basin facies is a necessary preliminary to the beginning of the slumping[1]. However, in both cases, the slump breccias generally occur on the flanks of structures. Syngenetic, the metal concentration acquires its final aspect during diagenesis of the host rock.

The history of these syngenetic concentrations in the carbonate environment does not stop there: the trend to uplift as illustrated by sea-floor heights sometimes goes on until emersion. Important karstic modifications then intervene: we will discuss these at the same time as the continental traps (p. 317).

However, before leaving the subject of the marine traps, we must consider the Pb-Zn mineralizations which are sometimes baritic (Triassic beds in Upper Silesia, Poland,

---

[1] Here, it is a reef height (Callahan, 1964; Gerdemann and Meyers, 1972): the growth of the calcarenitic height being faster than the depositional rate of sediment accumulation in the basin and – this is the general case – the slumps are due to the only effect of slope thus created.

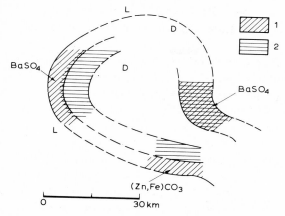

Fig. 5. Sketch of the zoning of Pb-Zn mineralizations on the High-Silesia Triassic basin (Gogolin beds). (After Gruszczyk, 1966; courtesy *Econ. Geol.*)
Plombiferous (*1*) and zinciferous (*2*) ores occur at the limit between limestone (*L*) and dolomite (*D*) fields.

according to Gruszczyk, 1966), and the fluorite mineralizations (Betic Cordilleras, Spain, according to Jacquin, 1970, and Alabert, 1973): these are strata-bound mineralizations systematically located at the limit of calcareous and dolomitic facies in the general environment of a carbonate epicontinental platform. As it happens, the two examples chosen are both Triassic.

Upper Silesia, Poland. On a small scale, the metalliferous dolomites of Upper Silesia constitute, at several levels of Muschelkalk, an outlier of the continental border facies in the basin (Fig. 5). These dolomites which are syngenetic or early diagenetic, are only mineralized near their contact with their calcareous surroundings, when they correspond to marine regressions. A pre-evaporitic environment is involved.

On a larger scale, the sulfide strata-bound accumulations present a clear calcareous footwall and an irregular dolomitic hanging wall. The transition from a well-defined marine environment (shelly limestone) to an environment of syngenetic pene-saline dolomites is marked, in fact, by the mineralized formation which in terms of sedimentology appears essentially as a sulfuretum[1] . Is this so surprising? We do not think so, given that in other environments hydrocarbons occur more particularly in zones of transition between dolomitic plateau facies and relatively deep calcareous facies (Perrodon, 1966).

In any case, Gruszczyk (1966, p. 175) refers to another process of selective deposition: he considers a colloidal contribution of sulfides, apparently terrigenous,

---

[1] According to Galliher (1933) a sulfuretum is an anaerobic environment where sulfate and sulfide ions remain in equilibrium.

which would flocculate in the deposition zone of dolomites. In addition he (p. 175) points out the presence of argillites and bitumens among these colloids (cf. above).

Betic Cordilleras, Spain. In a Middle-Triassic formation the lead- and fluorite-rich strata occur within alternating calcareous and dolomitic horizons generally in association with cherts (Sierra of Gador, Almeria). The dolomitic host-rock is very often fetid and sometimes presents banded facies (alternating light and dark bands: facies locally called "Franciscan") or intraformational breccia facies. The study of mineralizations in this very particular environment led Alabert (1973) to resort to the solution of pre-evaporitic lagoons where magnesian brines, poor in calcium (see p. 323), maintained the fluorine in solution as MgF^+ complex (Cadek and Malkovsky, 1963). It is enough, therefore, to consider a calcium supply, either by percolation of these brines in limestone layers, or by the periodic invasion of sea water, in order to induce the dolomite formation and the fluorite precipitation by destruction of the complex. The establishment of sulphuretums (fetid dolomites, fixation of galena), the frequency of the siliceous occurrences, the localization within the basin of these periods of metalliferous and slowed-down sedimentation, would lead to the choice of the last proposition of this alternative.

We could hardly leave these mineralizations in carbonate environments without at least looking briefly at the diversity of metal to be concentrated there. The sous-cévenole border offers us another two examples of the pre-evaporitic dolomites of Muschelkalk characterized here by their germanic facies: barite, on one hand (Bernard, 1959), and stibnite, on the other (Samama, 1970).

Layers of siliceous barite are interbedded between syngenetic dolomites of Muschelkalk at the hanging wall, and ferrugineous jaspers grading to silty shales at the footwall. This sequential position of metagenetic[1] deposits shows that barite here represents a sediment associated with siliceous colloids which mark the end of the detrital sedimentation of Buntsandstein and the beginning of the carbonate sedimentation of Muschelkalk. Laterally, these baritic layers grade to the sulfate evaporites of the "Lower Saliferous" (cf. Fig. 3 where the barite often occurs between the evaporites of the basin and the sulfides on the heights or their flanks).

Stibnite ores occur on the top of a small sea-floor height in the carbonate Muschelkalk (Samama, 1970). A dolomitic and then euxinic, pre-evaporitic argillaceous series, rich in pyrite, becomes reduced and condensed on the rising height marked by a breccia facies of silicified dolomite-bearing stibnite in its matrix. This is a remarkable development of a geochemical Sb anomaly found within this lithostratigraphic level over a range of several kilometres. Note here that, in opposition to the earlier cases, the richest metalliferous concentration is found at the top and not on the flanks of the height.

---

[1] According to Lombard (1953) a positive sequence of strata begins with clastics, first coarse and then fine, and goes on with chemical sediments: carbonates and evaporites. Commonly, between these two sets of rocks, highly comminuted clastics (ultra-detritals) or chemically precipitated colloids occur (silica, Fe hydroxydes, sulfides). These particular sediments are called metagenetics.

Finally, to reach a conclusion about marine traps, one can correlate strata-bound sulfide concentrations, either a geochemical anomaly or a workable deposit, to very characteristic sedimentary surroundings. These could be defined by slow-rate accumulation of metagenetic sediments (ultra-detritals and chemically precipitated colloids) in quiet and reducing environments. While not having the space here to develop this idea, we should note, however, that the metals with a polyvalent behaviour in the redox fields of the marine realm (Mn, V, Mo, V) could equally be concentrated, in a reduced form, in these highly euxinic surroundings.

Particularly well marked with shales, the chemically reducing nature of a sedimentary environment is not so evident in the petrography of carbonate or siliceous colloids. There is, however, nothing chemically incompatible between the deposition of these particles and the potential redox. We have tried to illustrate this proposition in a carbonate environment. The liaison, however, appears significant even with the siliceous colloids: these are often apparent as petrographic disturbances of the carbonate series, their restricted spatial development leads only very rarely to economic concentrations. One of the finest examples known to us is that of the zinc mineralizations of the Tri-State Grand Fall chert of the Mississippian (U.S.A.), so well described on his time (1906) by Sienbenthal (in Lindgren, 1933, pp. 432—436), that it does not seem necessary to go back over it.[1]

Nevertheless, the character of the more problematic and, no doubt, the rarer coincidence of "reducing environment and carbonate or siliceous colloidal deposit" leads us to a consideration of the corresponding distribution of the heavy-metal traps in these environments. In his treatise, Routhier (1963, p. 347) alludes to the western border district of the Cévennes. There, a Liassic series exhibits, especially at its base (Hettangian), a multitude of barite, sphalerite, galena and sometimes copper-mineral occurrences: either as micro-concretions or in small concordant lenses, in veinlets or sometimes in simple crustifications, these disseminated occurrences nonetheless represent a strong regional geochemical anomaly. In much the same way the "cubic dolomite"[2] (Hettangian), sampled in small quantity outside these occurrences, shows only common low-metal grades. Miners and geochemists are unable to estimate the average percentages of such a formation, or rather, of such a host series. A similar example might be that of the sulfide and chert disseminations of the Upper Mississippi Tri-State as well as the "Metallifero" of the Cambrian in Sardinia (Brusca and Dessau, 1968). Some geologists claim that there is no continuity between the clarke values of the sedimentary rocks and the metal content of mineralized beds (Krauskopf, 1967). We believe, on the contrary, that as a function of the sedimentological effectiveness of marine traps and their spatial distribution, all transitional cases exist to the extent to which one is able to understand and to sample them. Much remains to be done in this field which, until now, one must

---

[1] Cf. also the chapters in Vol. 7 dealing with Mississippi Valley-type ores.
[2] Local term for a dolomitic rock unit which exhibits a well-marked tri-rectangular joint system: its weathered outcrops split in cubic pieces.

admit, has interested neither geochemists nor mining geologists. The former have been preoccupied with the estimation of clarke values and with the mormal cycle of elements, and more often than not the latter have been confronted with a set of occurrences where they could choose only the economic ones and ignore all the others. In both cases the host-rock units are never analysed and described as such.

The continental traps

The continental environment is generally an oxidizing one and it may seem paradoxical to look there for sufficiently reducing surroundings for the sulfide precipitation. Obviously, one first thinks of lakes and, to a lesser degree, of hydromorphic zones in a fluviatile flood-plain. This idea is confirmed by occurrences of siderite and of iron bisulfides (sometimes sulfides of other metals), the so-called coal balls sometimes associated with oxydized sediments and ores (swamp ores), formed in such environments. In any case, it is a question of small and low-grade concentrations suitable only for exploitation by small local operators.

In particular, the phenomenon of supergenous cementation pointed out by De Launay (1897) has a certain economic interest, while casting new light on the behaviour of a chalcophile metal near the surface. After much debate, cementation was recognized as a phenomenon of supergene sulfide formation: historically, the exclusively hydrothermal nature of sulfide formation was attacked for the first time and attention was drawn to the redox of stagnant water-tables. However, it was not until 1953, that the problem of supergene sulfides was treated in general terms by R.M. Garrels: he showed that the heavy-metal sulfides were liable to precipitate each time certain physico-chemical requirements (pH, Eh, composition of the metalliferous solutions) were fulfilled (Garrels, 1953).

Later works (Michard, 1967) have shown that pH-Eh diagrams were perhaps not the best way to describe the sulfide-sulfate fields of stability; in fact, the sulfide-sulfate equilibrium does not exist, the antagonistic oxydo-reduction reactions proceeding with very different rates. In any case, while taking into account the catalytic activity of bacteria in the sulfate-sulfide reduction, equilibrium conditions are fulfilled and one may consider that the pH-Eh diagrams of Garrels present a good approximation of the reality.

Equipped with the above indispensable tools of chemistry, let us analyse two continental traps, the "red-bed" and the paleokarstic models which have been studied by the Economic Geology Laboratory in Nancy, France.

The "Red-Beds" model (Samama, 1968, 1969). The term "Red-Beds" originally characterized the red terrigenous formations of the "Colorado" type. The association of copper concentrations with grey levels within these environments, has led to these copper occurrences themselves being called "Red-Beds" (Tarr, 1910; Rogers, 1916; Finch, 1928), while the frequent association of uranium and vanadium and the uniformity of the group

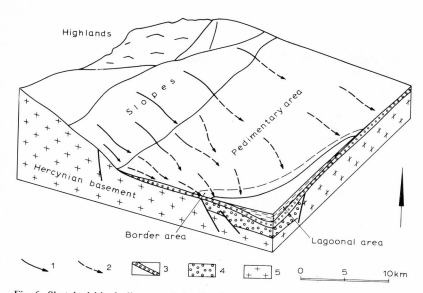

Fig. 6. Sketched block diagram of the distribution of physiographic areas during the Lower Triassic in the Largentière area (Ardèche, France). (After Bernard and Samama, 1970.)
1 = Surface water flow; *2* = underground water flow; *3* = sedimentary spread; *4* = Permian basin; *5* = Hercynian basement.

(Lindgren, 1933) has led to an enlargement of the definition of the type to include these two metals. In the same way, the paleogeographic similarity between this environment and that of Pb-Zn concentrations in arkosic formations has led to an extension of the term to the whole group of metals (Cu, U, V, Pb, Zn), strata-bound occurrences in detrital rocks (siltites to rudites). These concentrations will occur in continental or epi-continental areas although the red color, due to the liberation of iron hydroxides, may then be hardly apparent, or even non-existent.[1]

The Largentière deposit would seem the most suitable introduction to the metallogenic model of the Red-Beds because of the simplicity of the facts and the quality of the information it exhibits. The paleogeography of the Lower Triassic is that of a pediment complex deriving from the Hercynian basement (Fig. 6). The downstream part of this continental complex is composed of a lagoon-like evaporitic basin. From sedimentological indications, it appears that the climate would have been arid and dry with precipitation during a short rainy season. At that time the pediment plain would be flooded and the silico-aluminous detrital materials would spread rapidly without any great differentiation. On the other hand, during the long dry seasons, the continental waters would run underground, being limited to the depression and thus reaching the supersaline evaporitic water of the basin.

---

[1] Editor's note: see in Vol. 7 the chapter by Rackley on U in sandstones, and in Vol. 6 that by Smith on Cu in similar host rocks.

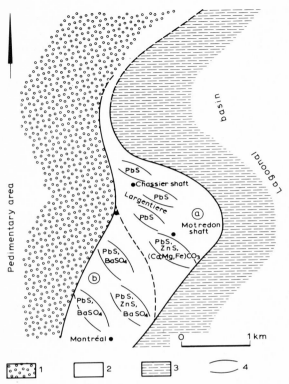

Fig. 7. Semi-theoretical sketch of the distribution of a chemical precipitation area (cementation) at Largentière. (After Bernard and Samama, 1970.)

Cementation of the pedimentary area (*1*) is phyllitic, that of the border area (left in white) siliceous; two fields can be demarcated: northwards (*a*) a very reducing zone (S^{2-} only), southwards (*b*) a less reducing zone (S^{-2} and SO_4^{2-}). Only channels (*4*) are mineralized in economic grades. The influence of the basin saline aquifer (*3*) is illustrated by a carbonate and sulfate cementation.

The most obvious result of these conditions is a zonal cementation in the detrital materials. From continent to basin there is a systematic distribution (Mainguet-Suares and Samama, 1969) of cements forming a sequence: clays, silica, carbonates and sulfates. These cements (or matrix) result from phenomena of internal sedimentation, which according to Shepard and Moore (1955)[1] still belong to sedimentary syngenesis. It is an essential and fundamental fact that the sulfide impregnations (galena and sphalerite) in the cement of arkoses always appear between the argillaceous zone of cementation and the carbonate one. The richly mineralized facies always develop in the same sequential position within

---

[1] Diagenesis will not be considered as initiated before the sediment has taken its definitive place in the erosion–transport–sedimentation cycle. The change or transformation of the particles which intervene before this final immobilization will not be considered as diagenetic but as an influence of the surrounding conditions on the particles.

the permanent channels where phreatic flows preferentially occurred even during dry periods.

To cut short the enumeration of the thorough investigations carried out in this field (Samama, 1968, 1969; Bernard and Samama, 1970), it is at the oscillating boundary between the saline water table and the continental one, that the sulfide and baritic cementations occur (Fig. 7). The saline water impregnates the sediments of the pediment where organic vegetal matter is present, especially around the permanent channels. There the sulfates of marine waters are reduced by anaerobic bacteria and H_2S is produced: from this derives the zonal distribution $PbS-ZnS-BaSO_4$. Once past this Eh barrier, the barium reaches the sulfate and lagoon-like waters and precipitates before the calcium sulfates but after the first carbonates (ankerite). Here, once again, we can see the pre-evaporitic character of sedimentary barite.

Different modalities can of course be considered in the application of this model, here only briefly outlined: variants would be necessary in the case of the deposits considered here ($Pb-Zn-BaSO_4$) and to an even greater extent for deposits of other metals: Cu, U, V (cf. the chapter by Samama on the environment of strata-bound deposits: Chapter 1, Vol. 6).

The paleo-karstic model (Bernard, 1973a). The karstic systems, once having matured (Fig. 8), offer a magnificent frame for metallogenic concentration. If one outlines the phenomenon, according to Cvijic (1918), it appears that the characteristics of the internal waters will induce the karstic sedimentation of several products.

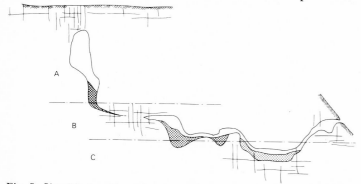

Fig. 8. Simplified section of a mature karst system. (After Cvijic, 1918; courtesy *Rev. Trav. Inst. Geogr. Alpine.*)

A. Percolation zone: meteoric waters percolate vertically through carbonate formations, i.e., through fissures which they widen, thus opening cave sinkholes, gash-veins, etc. Irregular flows, occasionally torrential streams; mechanical erosion largely prevails over chemical leaching. Collapse breccias and coarse detrital sedimentation.

B. Permanent circulation zone: water circulation is mostly horizontal when rock jointing allows it; resulting caves exhibit marked lateral extent: galleries. Free or forced flow, always irregualr, leads to intense chemical erosion, fine detrital sedimentation.

C. General imbibition zone: comprises the country rock below the water table. Very slow circulation, if any, i.e., stagnant waters. It is essentially a zone of ultra-detrital and chemical sedimentation.

We will not give too much space here to the phenomena acting on the formation of the cavities and leading to the collapse breccias to which the intra-karstic mechanical erosion is essentially due (cf. Leleu, 1966b). (Cf., in Vol. 3 Chapter 4 by Zuffardi.)

First of all, clastic material, then ultra-detrital (carbonates and clays) and lastly, chemical material will deposit inside the cavities. It is fundamental to note here the presence of an euxinic micro-milieu coinciding with the so-called imbibition zone of the system (Fig. 8): argillaceous or marly shales rich in carbonaceous matter (notably pollen) are deposited there.

Let us now consider the behaviour of heavy metals (Pb, Zn, for example) in such a karstic system by plotting the solubility of their principal compounds on a Garrels (1953) pH-Eh diagram (Fig. 9). Dissolved in carbonate form in the superficial zone which may be either slightly acid or neutral and oxidizing, (10^{-7} mol/l for $PbCO_3$; 10^{-4} mol/l for $ZnCO_3$; Fig. 9, point A) these metals will be integrally precipitated as sulfides in the zone of imbibition, slightly basic and very reducing (10^{-21} mol/l for PbS; 10^{-15} mol/l for ZnS; point B).

In other words, all Pb and Zn traces contained in the carbonate rocks then dissolved by karstic erosion will be more or less integrally trapped in the zone of imbibition. It should be noted that, at the end of the evolution of the system, this zone will be situated directly under the peneplaned surface.

We could not provide a better example of these Pb-Zn occurrences under an unconformity (Callahan, 1966) than that of the doline (sink-holes locally called circle-ground) of the old Tri-State, so well set out in their time by Lindgren (1933). If, as we have supposed (p. 313) the disseminations of the "sheet-grounds" are synsedimentary, poor but forming an enormous metal stock, we need not look further, in this case, for the

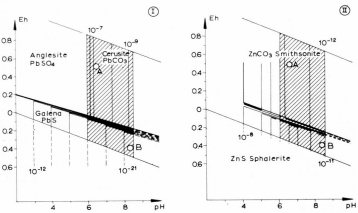

Fig. 9. Diagrams of theoretical stability fields of minerals as functions of pH, Eh and solubility products of various lead (*I*) and zinc (*II*) species. (After Garrels, 1953; courtesy *Geochim. Cosmochim. Acta*.)
The range of pH-Eh variations in karsts is superimposed (shaded area). Travel from *A* to *B* represents the presumed mean karst—water evolution in mature systems.

source of the concentrated metals in the circle-ground and their adjacent "runs" (spaces of brecciation or solution cavities filled by mineralizations).

Here again, numerous peculiarities are liable to act upon the working of a karstic system for the same metal and even more so for other elements (F, PO_4^{3-}; Cu, U). The best thing we can do here is to refer the reader to the original studies (Benz, 1964; Leleu, 1966a,b, 1969; Lagny, 1969; Rouvier, 1971; Padalino et al., 1973).

One last remark needs to be made. As explained before (p. 310) the uplifting trend shown by some synsedimentary sea-floor heights goes on sometimes up to emersion. The karstic erosion of previous synsedimentary concentrations then produces a new phase of super-concentration which will appear, finally, under an unconformity and genuinely epigenetic in relation to the carbonate host rocks. Despite this, it is neither hydro- nor telethermal.

This two-stage model (first synsedimentary strata-bound pre-concentration and then karstic reworking during an immediately subsequent emersion) is very common and it is understandable that many authors have wished to see in it a single type, the Mississippi Valley type. Now it must be appreciated that the genuine synsedimentary type exists (Upper Silesia) and that the genuine karstic type also exists (Salafossa, Dolomites, Italy). In the absence of this essential distinction, pointed out by Callahan (1964), it is obviously very hard to explain these mineralizations as having the same genesis (Brown, 1970), and it is rather meaningless to surmount the difficulty, even by fits and starts, using a model with 127 different possibilities (Laffitte, 1966). We will discuss later the incidence of fluid inclusion results on the credibility of this paleo-karstic model.

It would be appropriate, after this look at continental traps, to examine the metallo-genic role of the paleosurfaces and tackle the problem of metallizations per descensum (Moreau et al., 1966), already initiated with the karst model. Such a generalization would require the insertion of a new chapter in this work. We will simply touch on it together with the question of exceptional metalliferous supplies, which we should now take into consideration.

THE EXCEPTIONAL METALLIFEROUS SUPPLIES

Generally speaking, four types of exceptional metalliferous supplies may be listed for a sedimentary environment:

(1) terrigenous: etymologically engendered by the earth; (2) vulcanogenous: (and hydrothermal); (3) thalassogenous: i.e., pure marine; and (4) fluidogenous: i.e., by non-magmatic deep basinal waters.

Let us examine them a little more closely.

Terrigenous supply

The products of continental erosion are not carried and distributed in bulk to basins of sedimentation, that is a well-known fact. Otherwise it would be difficult to explain how a sialic continent with a granitic composition, for example, could supply the epicontinental sedimentation with only Ca and Mg as often occurs.

In fact, three major differentiating processes intervene between the site of erosion and the final place of sedimentation: the pedological process, the transportation and the sedimentary process itself. Thanks to Erhart (1956)[1] who showed us that the pedological phenomenon literally pilots the other two, it is in fact the physiographic conditions which essentially fix the proportions of resistates, hydrolystates (residual or neoformed) and the soluble salts of the materials which will later be supplied to transport and sedimentation. (Cf., in Vol. 3 the chapter by Lelong et al.)

How do the heavy metals behave in this pedological context? Goldschmidt (1937) offered a very general reply with the idea that ionic potential Z/r (ratio of ionic charge Z to ionic radius r) apparently governs the distribution of trace elements in the different geochemical products of weathering (resistates, hydrolystates, soluble salts). However, on the one hand, the relative proportions of these different products are not the same at the outset, and on the other, the control of the distribution of oligo-elements by their ionic potential is not as rigorous as Goldschmidt hoped. Certain typomorphis elements (chemical or mineral) according to Perel'man (1967) create important distortions with respect to the expected behaviour of the trace elements. For example, iron and aluminum as hydroxides seem to trap copper in certain laterites and bauxites (Gordon and Murata, 1952), whereas this metal ought to be eliminated and leached outside such evolved weathering profiles.

Thus, it appears that the supergene mobility of one element evaluated on the ionic or mineral scale cannot be extrapolated onto the soil-profile scale and even less so onto that of a catchment basin or a continent. In order to sketch this global behaviour, it is better to use the concepts of residual or migrating character taken directly from the analysis of residual elements in weathering profiles and from the analysis of migrating elements in waters percolating through the same weathering zone (Tardy, 1969).

Present data on the global behaviour of oligo-elements in the different climatic types of weathering are still very few: nevertheless for certain elements, it is possible to draw up a table of their tendencies (Fig. 10). One may therefore note that the global pedological behaviour of the metals present in red-beds differs greatly according to the various climates.

The metallogenic implications of these data are quite evident. During a steady weathering period of a given type, the residual accumulation of certain heavy metals will induce a pedological pre-concentration, while the migrating metalliferous phase is dispersed in the sediments of the basin. During a climatic variation, these pre-concentrations could be

---

[1] Little mentioned in English-language journals, the bio-rhexistasis theory of Erhart has a good deal of fame in western Europe, particularly justified by the dimensions and the simplicity of the relationships shown between two disciplines which had followed different paths until then: pedology and sedimentology. Even a condensed presentation of this theory would require at least a chapter of this work and we can do no more here than to refer the reader to the original paper.

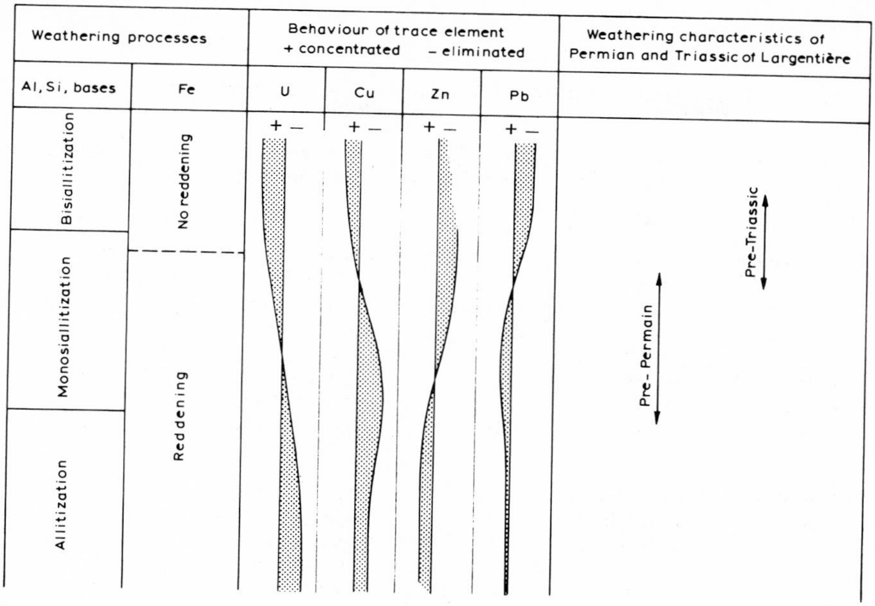

Weathering processes		Behaviour of trace element + concentrated – eliminated				Weathering characteristics of Permian and Triassic of Largentière
Al, Si, bases	Fe	U	Cu	Zn	Pb	
Bisiallitization	No reddening	+ –	+ –	+ –	+ –	Pre-Triassic
Monosiallitization						Pre-Permian
Allitization	Reddening					

Fig. 10. The weathering processes and the corresponding behaviour of heavy metal traces. (After Samama, 1973.)

The weathering zones, characterized by their major chemical and mineralogical processes, are presented as a semi-theoretical succession according to the latitude in the Northern Hemisphere. The behaviour of trace elements is represented by (+) or (–) symbols according to their residual (+) or migrating (–) character during weathering.

either partially or wholly mobilized and thus supplied to the sedimentary traps, whether continental or marine. This is what has been verified during two climatically different periods, the Permian and Triassic, in the same catchment area, the Ardèche border of the Massif Central in France (Samama, 1973).

The pre-Permian weatherings develop essentially in a zone of monosiallization (paleo-profiles study): the pre-concentrations will therefore (cf. Fig. 10) be essentially cupriferous and uraniferous. During Permian time local epeirogenic phenomena and a climatic variation tending towards bisiallization led to an abrupt reworking (rhexistasic according to Erhart's terminology) of this pedological cover. It is, therefore, normal to find in impregnations of red-beds types, small copper and uranium concentrations, as well as a geochemical background high in zinc and low in lead ($Zn/Pb = 1.5$ to 2). Therefore, lead accumulated on the continent during this pre-Triassic weathering (enrichment confirmed by study of the paleo-profiles locally conserved under the Triassic cover) will be in its turn mobilized during the Buntsandstein sedimentation by a new episode, both tectonic (epeirogenesis) and climatic (return to conditions of monosiallization): the Largentière red-bed presents a large excess of lead over zinc ($Pb/Zn = 6.8$) and over copper ($Pb/Cu = 5 \cdot 10^4$).

TABLE I

Comparison of environmental characteristics of uranium, copper and lead ores of sandstone type

Elements	Examples/ types	Sedimentary environment	Possible relations between types	Lithogenetic environments	Post-sedimentary mobilization
U(V)	Colorado/Niger	mainly alluvial (continental)	$\updownarrow$ frequent	hot, hardly arid (strong hydrolysis, liberation of Fe^{3+})	very intense to intense
Cu	Udokan/Ural Permian	deltaic (?) basin border (often evaporitic)	$\updownarrow$ rare	hot, more arid (noted hydrolysis, liberation of Fe^{3+})	medium to low
Pb/Zn	Laisvall/Largentière	pedimentary, evaporitic (?) basin border		hot, semi-arid (absence of hydrolysis, no liberation of Fe^{3+})	low

The geochemistry of continental landscapes and the peculiarities of the immediately overlying detrital formations can be considered as genetically related: it is the continental weathering, changing with time in a given place, which controls the exceptional metalliferous supply of the rhexistasic series (Table I).

At this time it should be noted that the more acidic and richer the basement is in elements at the head of the radiogenic series, the richer in radiogenic isotopes the lead fixed by soils will be. During Buntsandstein sedimentation, lead rich in radiogenic isotopes was abruptly leached out and rapidly precipitated; thus it could not acquire the Triassic composition and indicates a future apparent age (Pb of J type). This suggestion is in fact very close to the model elaborated by Doe and Delavaux (1972) for the mineralizations of Bonneterre Dolomite. According to the evidence, lead solutions must percolate (cf. Fig. 6) through the pedimentary detrital rocks (Lamotte Sandstone) before reaching the marine basin. Higher up in the stratigraphic series, galena precipitated further from the shorelines derived from solutions diluted by sea water and by its lead: it will be slightly less rich in radiogenic istotopes (Brown, 1967).

Once again, the sketch here is of a two-stage model which only introduces very low rates of concentration (or dissemination) as must be the case in the supergene field. Thus, Bubenicek (1970) refers to the iron pre-concentrations, represented by the Toarcian bituminous shales (rich in siderite and pyrite), in order to explain the exceptional metalliferous supply in the Aalenian delta of the Luxembourg Gulf. Just sedimented and hardly compacted, these euxinic shales were brought to emersion by a local epeirogenic uplift in the Ardennes Massif. There the sedimentary trap is due first to precipitation, then to the oolitic accretion of ferric hydroxides in the turbulent zone where continental and marine waters meet. The same trap seems to work with the manganiferous beds near shorelines (Nikopol, Tchiatoura, U.S.S.R., Imini, Morocco); the nature of the continental pre-concentration seems, however, more difficult to elucidate. The explanation most called upon is the weathering of basic volcanic rocks yielding pedological and residual pre-concentrations; these were reworked by marine transgression and gave the final shoreline concentrations. Easily accepted for these oxidized iron and manganese deposits, the two-stage models are taken into consideration, with much difficulty, when other metals are in question.

This explanation of exceptional terrigenous and metalliferous supply by the rhexistasic reworking of pedological pre-concentrations would deserve supplementary information (extending to supplies in marine environments, cf. Fuchs, 1969) as well as other variations: parent-rock either basic or already mineralized. Otherwise, the metallogenic trends of the different types of climate have been defined on peneplains: now, the influence of the physiography, which radically changes the conditions of drainage (see, e.g., the works of Goldich, 1948 on laterite formation), can locally prevail over the purely climatic factor. In spite of these modalities, excellent correlations between the climatic paleozones and the distribution of sulfide strata-bound deposits (Fig. 10) have been recorded (Samama, 1973). The spatial and temporal links so frequently noted, between evaporites and these deposits (Davidson, 1965a,b; Brongersma-Sanders, 1968) simply underline, to our mind, this climatic dependence, though estimated here in a very cursory way; in fact, the pre-evaporite (Red-Beds, Lower Cévennes; Kupferschiefer, Upper Silesia) or intra-evaporite (sous-cévenole border; Fig. 3) characters do not have a very precise climatological significance.

On the other hand, it seems that paleoclimatology, properly handled, offers interesting possibilities for problems until now unsolved, because of too much blind comparison with present-day climates. Thus Guillou (1972) shows clearly why sedimentary magnesites take the place of sulfated calcic evaporites in an atmosphere richer in CO_2 than the present one. In the same way the excellent model of Lepp and Goldich (1964), which explains the Precambrian iron deposits of Lake Superior type (B.H.Q.: banded hematite quartzite) by terrigenous supply, is easily imagined in a non-oxidizing atmosphere (cf. Rutten, 1967; Cloud, 1968). In a warm and humid climate (e.g., conditions of allitization) an intense hydrolysis of the silicates will liberate a considerable stock of silica and iron liable to migration. Indeed, in a reducing atmosphere iron remains in its Fe^{2+} state. It is in sea water, where oxygen of the hydrosphere is liberated, that iron precipitation will take place, at least partially in a tri-valent state.[1]

To what extent might the peculiarities of other Precambrian strata-bound deposits be explained by taking into consideration the corresponding paleo-atmospheres and paleo-climates? Without a doubt the terrigenous supplies were different at that time from what they have been since the Permo-Triassic.[2]

---

[1] See, in Vol. 7 the chapter on iron ores by Eichler.

[2] See, in Vol. 3 the chapter by Veizer.

Perhaps as an end to this survey of the terrigenous supply, it would be useful to recall the geotectonic localization of strata-bound deposits with terrigenous supply. The types considered until now were systematically located in sedimentary cover transgressive over a basement, i.e., a peneplaned mountain range. In detail, it is at the base of the so-called cover series that the mineralizations occur. Briefly, this involves epicontinental border series, where the sea-floor configuration is often induced by tectonic or epeirogenic movements in the very close underlying basement. The unconformity surface which separates the peneplaned basement from its overlying cover generally represents a long hiatus in stratigraphic recording, i.e., a long emersion. This last remark obviously brings us back to the likelihood of pedological pre-concentrations.

In the light of this observation, studying the genesis of a sulfide (Fe and Cu, or, Pb-Zn) impregnation which developed along the shoreline of a shallow epicontinental sea, or one which developed at a depth of more than 2000 m in the axial deep of an oceanic ridge, is not the same. In much the same way, it seems unreasonable to consider the metalliferous deposits of a pre-orogenic submarine formation with acid pyroclastics and the sulfide impregnations of arkoses which are continental, pedimentary and post-orogenic as being red-bed types. Pure academic nuances for some though they may be, these distinctions appear to be essential both from the point of view of the typology of these deposits and from the strictly genetic point of view. We will try to explain this more clearly while briefly examining vulcanogenous supply.

Vulcanogenous supply

Recent works concerning thermal sources and geothermic energy have shown first of all that most of the fluids due to volcanism, notably water, are of meteoric origin, and secondly that the amount of base metals in these waters is in the order of $n \cdot 10^{-9}$ (White et al., 1963). These fluids are probably not igneous-hydrothermal mineralizing solutions capable of exceptional metalliferous supply.

However, it could be a matter of superficial aerial outlet, and one wonders whether in the depths these fluids do not come from large reservoirs of real chloride solutions (Na or Na-Ca; Ellis, 1967). Even if they do not reach the very high salinity of the Salton Sea waters, these brines could constitute an adequate medium for metal transport (Helgeson, 1964). They would deposit their metallic salts long before reaching the surface (White, 1967), at the level of convective circuits developed in the vicinity of extrusions. The close attention which the study of these convective circuits has received during several years, is quite significant (Ellis and Mahon, 1964; Elder, 1966; White et al., 1971); one finds, there, a very attractive metallogenetic model.

We will not linger on this point, which deviates too far from our subject, but as we must point out, references to peri-volcanic, submarine convective circuits, and to their changes of functioning with respect to the depth of discharge are lacking. This is deplorable for, as Anderson (1969) states with great lucidity, the massive sulfide deposits are today generally considered as volcano-sedimentary when they are unambiguously associ-ated with volcanic rocks in a marine environment.

It is, therefore, possible that thermal springs of submarine discharges emit chloride-mineralizing brines; at least this is what we will assume as we deal with the problems posed by the volcano-sedimentary deposits.

Japanese authors (Horikoshi, 1969; Masukuma and Korikoshi, 1970; Tatsumi and Watanabe, 1971; Sato, 1971) now connect the origin of the mineralization of the Kuroko type with the emergence of submarine, hydrothermal springs.which are bearers of metal-liferous sulfides. The Shakanaï mine (Akita District) even shows the feeder structures in the form of stockworks mineralized and developed in silicified zones of volcanic rocks (Fig. 11). One can now appreciate that cooling and/or contamination of the hydro-thermal brines by sea water led to a sulfide and sulfate deposit located very close to the point of discharge, if not above or within the stockwork itself. However, phenomena of slumping, developed on the sloping flanks of the acid volcanic accumulation, would, during diagenesis of metalliferous muds move the sulfide deposits far away from the thermal springs. The scale of these phenomena remains, however, in the order of a hundred meters, if one evaluates it by the dimension of the volcanic protrusions which break up the sedimentary floor.

In the Red Sea (cf. Degens and Ross, 1969, and their Chapter 4 in Vol. 4) or in Cyprus (Hutchinson and Searle, 1971), the ore deposits seem to be located, spatially, very near the place of emergence. The study of the present submarine thermal springs (Honnorez,

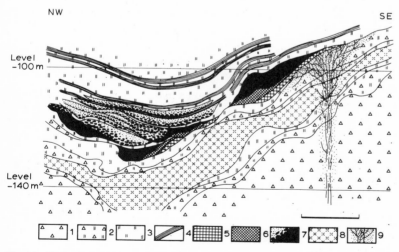

Fig. 11. Geologic profile of the Shakaniï deposit (Akita Prefecture, Japan). (After Kajiwara, 1970.) *1* = Rhyolite volcanic breccia; *2* = tuff breccia; *3* = tuff and lapilli tuff; *4* = mudstone; *5* = pyrite ore; *6* = yellow ore; *7* = black ore (Kuroko); *8* = gypsum; *9* = metallic veinlets.
Black-ore representation recalls the fragmental and graded structure of this massive sulfide mineraliza-tion. (See also in Vol. 6 Chapter 5 by Sangster and Scott.)

1969; Honnorez et al., 1973. Valette, 1973) confirms this point of view. One may wonder in considering these examples, where undeniable spatial and temporal[1] relationships demonstrate the volcano-sedimentary origin of ore, whether the field of action of the exhalations is considerably extended, in space. This is, in fact, the principal metallogenic problem of this exceptional metalliferous supply: we must recall that some geologists see in it the source of most, if not all, synsedimentary deposits (Dunham, 1964. Maucher, 1966).

To discuss this problem here would take us very far from our original purpose. At the moment too little is known about the volcanic emanations[2] (Krauskopf, 1967) to approach the question seriously from this angle. On the other hand, the geotectonic localization of definite volcano-sedimentary ores leads us to draw notable conclusions.

Again, it is the massive sulfide beds which will guide us, and indeed do lead us, to the geosynclinal environments (Anderson, 1969; Watanabe, 1970) and more exactly, to the pre-orogenic events of these environments. Two important types of deposits appear (Bernard and Soler, 1971; Bernard, 1972):

(1) Deposits related to spilite-quartz keratophyric associations in which we are willing to class the kurokos, as well as deposits of the South-Iberian province (Rio Tinto). Developed on sialic flexures which characterize the early evolution of the Cordillera structures (Dewey and Bird, 1970), the quartz keratophyric volcanism seems to localize very precisely the massive sulfide accumulations.

(2) Deposits related to oceanic ridges and, at least, to oceanic crusts. This case is more difficult to approach for it must be envisaged in a dynamic evolution: born in the very early phases of the individualization of a ridge (Red Sea type), these mineralizations move further and further away during its life (Cyprus type). They end up almost inevitably in a zone of subduction (Okuki type, Japan, Watanabe et al., 1970; Besshi type, Japan, Kanehira and Tatsumi, 1970), where they are considerably transformed by metamorphism.

These two genetic groups represent most of the known massive pyritic accumulations, to the exclusion of nickeliferous sulfide deposits whose geotectonic localization and genesis seem very different. Apart from this discrepancy, the two great genetic groups proposed here are identical to those set down by Miller (1960).

Again, we are led to pre-orogenic, geosynclinal environments when considering deposits of iron and manganese oxides (Lahn and Dill in Schneiderhöhn, 1941; Borchert, 1970), of iron carbonate (Krautner, 1970), of cinnabar (Almaden in Saupe, 1973) or of scheelite and stibnite (Maucher, 1965; Maucher and Höll, 1968; Lahusen, 1972). Almost everywhere there is an undeniable connection with volcanic rocks, basic or intermediary and more or less metamorphosed.

By the number of metals likely to be concentrated by volcanic emanation, the vul-

---

[1] And notably the close association of mineralization with volcanic rocks of submarine origin.

[2] Which is true for submarine emanations, but far less true for aerial manifestations.

canogenous supply appears both powerful and eclectic though it is limited to pre-orogenic geosynclinal surroundings. In aerial environments it seems that mineralizations are deposited well below the surface, in the classic subvolcanic epithermal field. There remains the epicontinental marine domain: the direct relationships of mineralizations to submarine volcanism are very rare here, if they exist at all and the probability of the exceptional volcanogenous supply remains very doubtful. The continental reworking of effusive volcanic formations by aerial erosion, which yields to the marine sedimentation a large metal supply, is indeed plausible (cf. oxide of Mn, p. 323); it will, however, only be completely effective if the general, and especially the climatic conditions, characterizing the terrigenous supply, are respected. There can, therefore, be no confusion between the two types of supply. Thus, it seems meaningless today to refer systematically to a vulcanogenous supply, notably in epicontinental sedimentary environments, without being able to produce solid arguments in favour of the proposition.

Thalassogenous supply[1]

It is well known, especially for evaporites and phosphates, that some concentrations derive directly from sea water. Basing her argument on the model of Kazakov (1950), Brongersma-Sanders (1968) campaigns in favour of a purely thalassogenous supply of certain base metals (Cu, Pb, Zn) in the epicontinental marine domain: upwelling currents would be responsible both for this exceptional supply (notably by the bias of planktonic organisms) and for the creation of an arid climate, generative of evaporites, on the neighbouring continent. This model is extremely ingenious and it is difficult to refute several of Brongersma-Sanders' arguments. One may simply ask, as far as heavy metals are concerned, whether the model is likely to lead to anything other than geochemical enrichments and not to ore deposits (see p. 315). The large epicontinental anomalies in clay and/or carbonate environments, might well come from such phenomena which, in such cases, appear to be much more plausible than the igneous-hydrothermal source in the absence of any contemporaneous and close volcanic occurrence.

The thalassogenous source through upwelling seems much more difficult to accept when the host layer exhibits a terrigenous inheritance or when the traps are both small and rich (buried hills). In any case, the thalassogenous supply can explain neither the concentration of red-beds, nor their paleoclimatic distribution, and if, as Wedepohl (1971) suggests, Kupferschiefer are productive only when they transgress mineralized red-beds, then the theory of upwelling loses much of its metallogenic interest. Lastly, and above all, the criteria for recognizing the former zones of upwelling currents are still sadly lacking.

We will only touch upon the problem of the ferromanganese nodules of great oceanic depths. It seems definite that they result from a very general phenomenon (sea or lakes)

---

[1] Cf. p. 319.

needing no exceptional metalliferous source, not even a vulcanogenic one as was often suggested as a final hypothesis. Any immersed object (pebbles, organic debris) is rapidly covered by a film or crust of iron and manganese oxide, whatever the nature of the sea floor and whatever the depth. This process, which is similar to an alteration and which develops at the water/sediment interface, seems to govern by colloidal properties the geochemical behaviour of elements 24 to 29 in the periodic table (Mn, Fe, Ni, Co, Cu). Much remains to be done in this field, but the approaches now seem well understood and, looked at in this way, the metalliferous supply is seen to be certainly purely thalasso-genous. (Cf., in Vol. 7 the chapter by Glasby and Read.)

Fluidogenous supply

Several writers (Pospelov, 1969; Dars et al., 1971; Wolf, 1976), in pointing out analo-gies of localization between sulfide ore bodies and oil deposits, refer to the expulsion of connate waters, outside of euxinic sediments, during compaction, in order to explain some sulfide mineralizations in sedimentary rocks. To their mind, this "fluidogenous" theory offers an alternative to the igneous genesis (magmatic or convective) of the metal-liferous hydrothermal solutions. This theory especially concerns concentrations which are discordant with their immediate surroundings, in the way Mackay (1946) envisaged the impounding structures. Evoked here simply as a reminder, this contribution has but little influence at the level of the synsedimentary formation of the metalliferous concentra-tions.

It is possible to imagine that the emergence of such connate fluids occurs at the time when overlying sediment is deposited. Resurgences of deep circulation waters of meteoric origin, could occur in similar synsedimentary conditions (Pelissonnier, 1966). In fact, for both hypotheses the synsedimentary heavy-metal trapping appears only as one particular case of general fixation which is most often epigenetic (by filling or replacement) in sedi-mentary rocks. Beneath we will examine the probability of such mineralizing solutions by studying the problem of the burial (diagenesis and metamorphism) of synsedimentary mineralized bodies.

At the end of this study of metalliferous traps and supplies it should be remembered that both must coexist in one place for a long time in order that suspensions or fairly diluted solutions could deposit a tonnage large and rich enough to be economic at the present time.

Very often, the sedimentological characters of traps are registered very clearly in the lithological column, and this is obviously very valuable in tactical mining prospection. However, it is much more difficult to measure the quality and duration of metal supplies, i.e., at the stratigraphic level and the quality of the metallogenic province. The most one can do is to define, among several provinces, a certain number of favourable characteristics: the geotectonic framework, the paleoclimatology and the magmatology of sub-marine emissions. It must be emphasized on this point that if the geology of epicontinental environments (either aerial or sub-marine) is well known, this is not the case for sub-marine volcanism (especially acid emissions) and above all, as we have pointed out, for the environment of upwelling currents.

THE POST-SEDIMENTARY TRANSFORMATIONS

The transformations which the synsedimentary concentrations undergo during burial (diagenesis, sedimentary epigenesis, anchi-metamorphism and metamorphism) remain certainly the greatest difficulties to overcome for many economic geologists. This question is more a matter of principle than a real scientific blockage: no one would think of denying the reality of the phenomena of metamorphism when applied to silicates, and their geological implications. Why then is it so difficult to accept that ore minerals and particularly sulfides, are equally sensitive to the P.T. conditions of burial, and from this draw up the geochemical and geological implications?

It is certain that metallogeny has registered considerable delay in this field compared to petrography, and that the Leitmotiv: "epigenetic aspect = hydrothermal contribution" has paralysed the metallogenic discipline in a scandalous and illogical way. Ramdohr (1953) was, to our knowledge, the first to strike a serious blow at the automatism of the system. He attacked the principle itself of epigenetism showing, with mineralogical proof to support him, that ore concentration within the host rock was pre-existant to the final crystallized form. Though little followed, he was not opposed: the breach was made and no honest mind could any longer come to the conclusion "hydrothermal contribution" without stating his case. Thus, progressively, papers balancing facts, interpretations and ideas came out. The transformism of mineralizations still does not appear, twenty years later, as a perfectly coherent and established doctrine, and the exposition of the main directions of research would alone deserve a whole chapter of this volume. We shall, therefore, only refer the reader to the first synoptic papers where he will find concepts and references laid out.

Two rather sketchily defined fields may be marked out: that of the diagenesis of sedimentary concentrations, and that of the metamorphism of sulfide mineralizations.

Diagenesis of sedimentary metalliferous concentrations

This field involves studies, as yet essentially descriptive, on ores in sedimentary rocks lightly affected by burial (epicontinental ore deposits). These rocks are, of course, lithified and sometimes show neoformations of an anchi-metamorphic[1] type, but most of the textures and structures of the orebearing sedimentary rocks are preserved, and it is

---

[1] Anchi-metamorphism covers burial transformations developed before the attainment of greenschist-facies P.T. conditions.

interesting to work out the sequence of crystallization of ores in relation to the stages of the diagenesis of these rocks (diagenetic diagrams: Bernard, 1959; Amstutz et al., 1963; Amstutz and Bubenicek, 1967).

These studies show clearly how ignorant we generally are about the forms of very early crystallization of sulfide species. Studies carried out on pyrite now lead to coherent models (Roberts et al., 1969; Hallberg, 1972; Sweeney and Kaplan, 1973); however, even for this mineral much remains to be done in the field of the growth processes. Considering the open or closed nature of the chemical system where growth occurs acting directly on the isotopic evolution of sulfur, one may ask at present if any interpretation of an isotopic composition of sulfide in sedimentary rock does not rely first of all on the very diagenetic history of the rock. An analytical tool particularly well adapted to this problem is undoubtedly the epitaxic oxidation[1] (Arnold, 1969), which, in revealing the fine structures (sometimes as far as growth spirals), brings to light complex crystallogenetic histories quite unthought of in classic metallography.

In the same way, analysis of the fluid inclusions of authigenic species and a study of the present connate waters will perhaps allow us to bring closer the neoformations and their mother-solutions in coherent models: this is what Bernard (1973a) has attempted for the particularly suitable case of paleokarstic connate waters.

The problem of diagenetic circulations of metalliferous connate waters during the compaction of clayey sediments deserves also to be touched upon briefly: the metallogenic, fluidogenous model, which would result from this generation of metalliferous solutions, would include source rock and reservoir thus linking the formation of sulfide deposits with that of oil (Pospelov, 1969; Dars et al., 1971; Wolf, 1976). In order that heavy metals, fixed in the syngenetic mud as sulfides, may move into compaction fluids or solutions, they must be complexed according to a reversible reaction of the kind:

$$MeS + L^- \rightleftharpoons MeL^- + \frac{1}{n} S_n^{2-} + 2\left(1 - \frac{1}{n}\right)e^-$$

(L = complexant; Me = bivalent heavy metal; n = number of atoms of S in the polysulfides).

The complexing agent may be organic (chelation) or chlorine. Now, as far as it is known, if the phenomena of chelation act under the water/sediment interface (Hallberg, 1972) it is before the syngenetic fixation to which they are opposed. Below a certain Eh level, chelation stops, sulfides are fixed and diagenesis begins. For chlorine, complexation may be envisaged with chloride brines: connate waters are known that are sufficiently rich in salts to operate such sulfide dissolution. But, as Gaida and Von Engelhardt (1963) have shown, these connate waters with a high salinity are precisely those which remain within the compacted sediments after the expulsion of much less saline solutions; these

---

[1] This method puts to profit the orientated oxidation products (epitactic) formed through a mild etching on a pyrite polished surface.

brines, therefore, cannot be considered as a medium for heavy-metal transport for they are strongly tied to the host rock.

Chemically, the fluidogenous model leads to an impasse. It may be wondered if this is not the same with hydrogeology; the theory of metallogenic drainage [slow percolation of vadose waters through metalliferous euxinic shales which would be leached out (Avias, 1972)], seems too far from the solid and classic notions of impounding (Mackay, 1946; Pelissonnier, 1966) to be taken reasonably into consideration. In the same way, it would be a good thing to estimate the geological probability of a large circulating system of vadose waters which, upon being heated and mineralized at depth, would be likely to reappear at the surface as thermal and mineralizing solutions [so-called paleo-insular theory (Pelissonnier, 1961)]. The charge and discharge mechanisms of such systems would merit especially thorough hydrogeological investigations before constituting a base for good metallogenic reflection.

In fact, the association as well as the similarities often noticed between the origins of sulfide and oil deposits must further be explained by convergences in the formation of the source rocks. At a time when oil geologists are discovering the diversity of productive environments and are defining oil as a sediment, lending more and more importance to the traps acting during sedimentation and early diagenesis, these convergences of formation of euxinic environments, both source and host rock, seem to us to limit considerably the role and the extent of migrations (Perrodon, 1966). There exist otherwise important differences between oil and metalliferous euxinic surroundings, if only in the rate of accumulation of organic matter, but that is another problem.

Metamorphism of sulfide mineralizations (particularly, synsedimentary concentrations deposited in a geosynclinal environment)

In fact, because of their tectonic position and their early origin in the orogenic cycle, these concentrations are normally involved in a complex geological history including deformations and metamorphism. Accordingly, the Kuroko and Cyprus deposits are exceptional occurrences because they are so weakly transformed (though the Kurokos were buried, apparently, in the zeolite-facies conditions: Matsukuma and Horikoshi, 1970).

According to the excellent synopsis of McDonald (1967)[1], the possible transformations of sulfides by metamorphism include: modifications of textures, mineral neoformations, differential mobilizations of the constituents, reactions with sulfide minerals and, lastly, and only possibly, the generation of sulfide concentration by metamorphism.

We will not dwell here upon phenomena of recrystallization in situ or without great mobilization of primary concentrations, except to point out that, before reaching the epimetamorphic stage, these concentrations have previously undergone diagenesis and sedimentary epigenesis. It is on deposits already fairly modified by a long sedimentary history that the first anchi-metamorphic transformations take place. From this stage on,

[1] See also the chapters by Mookherjee (Vol. 4) and Sangster and Scott (Vol. 6).

and notably from the acquisition of the cleavages (of fracture, or slaty), the obliteration of synsedimentary characteristics is often considerable and these mineralized bodies will constitute the initial state of the real metamorphic transformations. It is not necessary to add that it is almost impossible to decipher the synsedimentary features when starting from a metamorphosed deposit; to restore the anchi-metamorphic state is often a tour de force. In fact, the basic observation remains the recognition of the ante-schistosity character of mineralizations (Williams, 1962), which cannot be made at once, if one wishes to eliminate all the implications of the metasomatic post-schistosity replacement.

We will try, on the other hand, to estimate the scale of the differential mobilizations of the constituents and, by extension, the probability of metamorphogenous accumulations (Domarev, 1967). After the work of Ramberg (1952) it now seems accepted that essential movements of matter in metamorphic transformations are effected by ionic and/or molecular diffusion according to gradients of chemical potential. According to the dimensions of the openings, these diffusions operate in stationary solvent or by following the mineral surfaces. Only the first of these processes seems geologically effective: but in any case, it explains in a reasonable way the phenomena of lateral secretion, most often of trace elements deriving from source beds or from mineral concentrations themselves (Avias and Bernard, 1956; Boyle, 1959, 1968). Already apparent in sedimentary epigenesis and anchi-metamorphism, they will develop better in the epizone using capillary or super-capillary openings, establishing locally strong gradients of pressure, a large proportion still of aqueous medium, and relatively high temperatures. The mineralizations which result (veins and networks of veinlets, saddle-reefs) have sometimes been called pseudo-hydrothermal.

Diffusions operating along crystal surfaces (surface diffusion) seem much less effective and the differential mobilizations described in the mesozone bring to light more differences in mechanical behaviour in the constituents of sulfide bodies. Under mesozone conditions, the mobilization of oligoelements fixed by silicate structures may be related to yet another process, the destruction of the host structure, which is possible by metamorphic transformation or by a strong leaching, the latter impossible in a stationary solvent.

In other words, if a pyritic formation buried in epizonal conditions is fractured, then the open system thus created will liberate mobile constituents (H_2O, SH_2, CO_2); the pyritic ore body will be affected in its mass (pyrrhotite, magnetite) and in its immediate surroundings (lateral secretions). If the pyritic body remains in closed systems it will reach the mesozone without other modifications than some reworking of its mineral species and internal structures.

CONCLUSION

We come, thus, to the end of our consideration of the development of a metallization in a rock which was sedimentary. The impression which emerges from studies of diagenesis and metamorphism of sulfide bodies is, with a few modifications, that of the conservation of the mineral concentration. As a corollary, this observation considerably restricts the probability of metamorphogenous concentrations deriving from trace amounts (Devore, 1956). Very strong T.P. and chemical potential gradients seem to be necessary in order to mobilize the oligo-elements once these are fixed in their sedimentary host rocks (cf. Barnes, 1959). Two geochemical fields are clearly capable of this: plutonism (sensu lato, i.e., magmatism of basaltic and anatectic origin, including periplutonic convective phenomena: Marmo, 1960; Krauskopf, 1971), and, meteoric weathering. This survey of mineralizations in sedimentary rocks has allowed the reader to appreciate the metallogenic importance of the supergenous cycle; if we have succeeded in doing this, our aim will have been reached.

REFERENCES

Alabert, J., 1973. *La Province plombo-zincifère des Cordillères bétiques (Espagne méridionale). Essai typologique.* Thesis, Nancy, 148 pp.

Amstutz, G.C., 1962. L'origine des gîtes minéraux concordants dans les roches sédimentaires. *Chron. Mines Rech. Min.*, 308: 115–126.

Amstutz, G.C., Ramdohr, P. and El Baz, F., 1963. Diagenetic behaviour of sulphides. In: G.C. Amstutz (Editor), *Sedimentology and Ore Genesis.* Elsevier, Amsterdam, pp. 63–90.

Amstutz, G.C. and Bubenicek, L., 1967. Diagenesis in sedimentary mineral deposits. In: G. Larsen and G.V. Chilingar (Editors), *Diagenesis in Sediments.* Elsevier, Amsterdam, pp.417–475.

Anderson, C.A., 1969. Massive sulphide deposits and volcanism. *Econ. Geol.*, 64: 129–146.

Arnold, M., 1969. *L'Oxydation épitaxique: une Méthode de Résolution des Structures et des Microstructures des Bisulfures de Fer.* Thesis, Nancy, 85 pp.

Avias, J., 1972. Réflexions sur le rôle de l'hydrogéologie et la paléohydrogéologie, dans la genèse, le remaniement et la destruction des gîtes minéraux en général et des gîtes métallifères stratiformes en particulier. *Mém. C.E.R.G.H., Univ. Montpellier,* 2 (4): 1–16.

Avias, J. and Bernard, A.J., 1956. Sur l'origine des gîtes de nickel et de cobalt de la région de Zinkwand-Vöttern (Autriche). *Congr. Géol. Int., Mexico, Sect.* 6: p. 83 and *Sci. Terre,* 11: 375–383.

Baas Becking, L.G.M., 1959. Geology and microbiology. *N.Z. Dept. Sci. Ind. Res., Inform. Bull.,* 22: 48–64.

Barnes, H.L., 1959. The effects of metamorphism on metal distribution near base-metal deposits. *Econ. Geol.,* 54: 919–943.

Benz, J.P., 1964. Le gisement plombo-zincifère d'Arenas (Sardaigne). *Trav. Lab. Sci. Terre, Ecole Mines, Nancy,* 2: 128 pp.

Bernard, A.J., 1959. Contribution à l'étude de la province métallifère sous-cévenole. *Sci. Terre,* 7 (3/4): 123–403 (Thesis, Nancy).

Bernard, A.J., 1962. Notions de métallogénie sédimentaire. *Notes Mém. Serv. Géol. Maroc,* 181: 267–282.

Bernard, A.J., 1972. A propos des limites actuelles de prospection indirecte en mines métalliques. *Ann. Mines Belg.,* 7/8: 1–21.

Bernard, A.J., 1973a. Metallogenic processes of intra-karstic sedimentation. In: G.C. Amstutz and A.J. Bernard (Editors), *Ores in Sediments.* Springer, Berlin, pp. 43–57.

Bernard, A.J., 1973b. Essai de revue des concentrations métallifères dans le cycle sédimentaire. *Geol. Rundsch.,* (in press).

Bernard, A.J. and Fuchs, Y., 1964. Contribution à l'étude de concentrations zincifères diagénétiques: les nodules de Laguépie (Tarn-et-Garonne). *Bull. Soc. Géol. Fr., 7e Sér.,* 6: 707-711.

Bernard, A.J. and Samama, J.C., 1969. La remobilisation: essai de définition suscité par l'exemple d'une anomalie géochimique en plomb. In: E. Mulas (Editor), *Convegno sulla Rimobilizzazione dei Minerali metallici e non metallici.* Cagliari, pp. 19–35.

Bernard, A.J. and Samama, J.C., 1970. A propos du gisement de Largentière (Ardèche). Essai méthodologique sur la prospection des "Red-Beds" plombo-zincifères. *Sci. Terre,* 15 (3): 207–264.

Bernard, A.J. and Soler, E., 1971. Sur la localisation géo-tectonique des amas pyriteux massifs du type Rio Tinto. *Compt. Rend. Acad. Sci.,* 273: 1087–1090.

Berner, R.A., 1969. Migration of iron and sulfur within anaerobic sediments during early diagenesis. *Am. J. Sci.,* 27: 19–42.

Borchert, H., 1970. On the ore-deposition and geochemistry of manganese. *Mineralium Deposita,* 5: 300–314.

Boyle, R.W., 1959. The geochemistry, origin, and role of carbon dioxide, water, sulphur and boron in the Yellowknife gold deposits. NW Territories, Canada. *Econ. Geol.,* 54: 1506–1524.

Boyle, R.W., 1968. The source of metals and gangue elements in epigenetic deposits. *Mineralium Deposita,* 3: 174–177.

Brongersma-Sanders, M., 1966. Metals of Kupferschiefer supplied by normal sea water. Geol. Rundsch., 55: 365–375.

Brongersma-Sanders, M., 1968. On the geographical association of strata-bound ore deposits with evaporites. *Mineralium Deposita,* 3: 286–291.

Brown, J.S., 1967. Isotopic zoning of lead and sulfur in southeast Missouri. In: *Genesis of Stratiform Lead-Zinc-Barite-Fluorite Deposits–Econ. Geol., Monogr.,* 3: 371–377.

Brown, J.S., 1970. Mississipi Valley type lead-zinc ores. A review and sequel to the Behre Symposium. *Mineralium Deposita,* 5: 103–119.

Brusca, C. and Dessau, G., 1968. I giacimenti piombo-zincifero di S. Giovanni (Iglesias) nel quadro della geologia del Cambrico Sardo. *Ind. Minerarid,* Oct-Nov., pp. 1–53.

Bubenicek, L., 1970. *Géologie du Gisement de Fer de Lorraine.* Thesis, Univ. Nancy, 146 pp.

Cadek, J. and Malkovsky, M., 1963. Contribution to the problem of transport of fluorine at low temperature. *Symp. Probl. Postmagmatic Ore Deposition, Prague, 1963–1965,* 2: 407–412.

Callahan, W.H., 1964. Paleogeographic premises for prospecting for strata-bound base metal deposits in carbonate rocks. *Cento Symp. Mining Geol., Ankara,* pp. 191–248.

Callahan, W.H. 1966. Mississippi Valley. Appalachian ore deposits. In: *Genesis of Stratiform Lead-Zinc-Barite-Fluorine Deposits–Econ. Geol., Monogr.,* 3: 14–19.

Christ, C.L. and Garrels, R.M., 1967. *Equilibres des Minéraux et de leurs Solutions aqueuses.* Gauthier-Villars, Paris (translated by R. Wollast), 335 pp.

Cissarz, A., 1957. Lagerstätten des Geosynklinalvulkanismus in den Dinariden und ihre Bedeutung für die geosynklinale Lagerstättenbildung. *Neues Jhrb. Mineral. Abh.,* 91: 485–540.

Cloud, P.E., Jr., 1968. Atmospheric and hydrospheric evolution of the primitive earth. *Science,* 160 (3829): 729–736.

Cvijic, J., 1918. Hydrographie souterraine et évolution morphologique du karst. *Rev. Trav. Inst. Géogr. Alpine,* 6 (4): 56 pp.

Dars, R., Allegre, C.J. and Michard, G., 1971. Sur la formation des gisements stratiformes de plomb-zinc: un modèle de formation proposé d'après les études géochimiques. *Compt. Rend. Acad Sci.,* 273: 1261–1264.

Davidson, C.F., 1965a. A possible mode of origin of stratabound copper ores. *Econ. Geol.,* 60: 942–954.

Davidson, C.F., 1965b. The mode of origin of banket orebodies. *Bull. Inst. Mining Metall.,* 700: 319–338.

Degens, E.T. and Ross, D.A., 1969. *Hot Brines and Heavy-metal Deposits in the Red Sea.* Springer, Berlin, 600 pp.

De Launey, L., 1897. *Sources thermo-minérales. Recherche, Captage et Aménagement.* Béranger, Paris, 636 pp.

De Launey, L., 1913. *Gîtes minéraux et métallifrès, I, II, III*, Béranger, Paris.

Devore, G.W., 1956. Surface chemistry as a chemical control on mineral association. *J. Geol.*, 43: 159–190.

Dewey, J.F. and Bird, J.M., 1970. Mountain belts and the new global tectonics. *J. Geophys. Res.*, 75: 2625–2647.

Doe, B.R. and Delevaux, M.H., 1972. Source of lead in Southeast Missouri galena ores. *Econ. Geol.*, 67: 409–425.

Domarev, V.S., 1967. Metamorphogenic mineralization. *Int. Geol. Rev.*, 9: 1290–1298.

Dreyer, R.M. and Garrels, R.M., 1952. Mechanism of limestone replacement at low temperatures and pressures. *Bull. Geol. Soc. Am.*, 48 (5): 337–357.

Dunham, K.C., 1964. Neptunist concepts in ore genesis. *Econ. Geol.*, 59: 1–21.

Elder, J.W., 1966. Heat and mass transfer in the earth hydrothermal systems. *Bull. N.Z. D.S.I.R.*, 169: 115 pp.

Ellis, A.J., 1967. The geochemistry of some explored geothermal systems. In: H.L. Barnes (Editor), *Geochemistry of Hydrothermal Deposits*. Holt, Rinehart and Winston, New York, N.Y., pp. 465–514.

Ellis, A.J. and Mahon, W.A., 1964. Natural hydrothermal systems and experimental hot-water rock interactions. *Geochim. Cosmochim. Acta*, 28: 1323–1357; 31: 519–538.

Eisenhut, K. and Kautsch, E., 1954. *Handbuch für den Kupferschiefer Bergbau*. Geest und Portig, Leipzig, 335 pp.

Erhart, H., 1956. *La Genèse de Sols en tant que Phénomène géologique*. Masson, Paris, 90 pp.

Erzberger, R., Franz, R., Jung, R., Knitzschke, G., Langer, M., Luge, J. and Rentsch, H., 1968. Lithologie, Paläogeographie und Metallführung des Kupferschiefers in der D.D.R. *Géologie*, 17: 716–791.

Finch, J., 1928. Sedimentary metalliferous deposits of the Red-Beds. *Trans. Am. Inst. Mining. Metall. Eng.*, 76: 378–392.

Fuchs, Y., 1969. *Contribution à l'Etude géologique, géochimique et métallogénique du Détroit de Rodez*. Thesis, Nancy, 257 pp.

Gaida, K.H. and Von Engelhardt, W., 1963. Concentration changes of pore solution during the compaction of clay sediments. *J. Sed. Petrol.*, 33 (4): 919–930.

Galliher, E.W., 1933. The sulfur cycle in sediments. *J. Sed. Petrol.*, 3: 51–63.

Garlick, W.G., 1961. The syngenetic theory. In: F. Mendelsohn (Editor), *The Geology of the Northern Rhodesian Copperbelt*. McDonald, London, pp. 146–165.

Garrels, R.M., 1953. Mineral species as functions of pH and oxidation-reduction potentials, with special reference to the zone of oxidation and secondary enrichment of sulfide ore deposits. *Geochim. Cosmochim. Acta*, 5,4: 153–168.

Gerdemann, P.E. and Meyers, H.E., 1972. Relationship of carbonate facies patterns to ore distribution and to ore genesis in the southeast Missouri lead district. *Econ. Geol.*, 67: 426–433.

Goldich, S.S., 1948. Origin and development of aluminous laterites. *Bull. Geol. Soc. Am.*, 59: 1326.

Goldschmidt, V.M., 1937. The principles of distribution of chemical elements in minerals and rocks. *J. Chem. Soc., Lond.*, 1937: 655–672.

Gordon, M. and Murata, K.J., 1952. Minor elements in Arkansas bauxite. *Econ. Geol.*, 47: 169–179.

Grabau, A.W., 1906. Types of sedimentary overlaps. *Bull. Geol. Soc. Am.*, 4 (17): 567–636.

Gruszczyk, H., 1966. The genesis of the Silesian-Cracow deposits of lead-zinc ores in genesis of stratiform lead-zinc-barite-fluorite deposits. *Econ. Geol., Monogr.*, 3: 169–177.

Guillou, J.J., 1972. La série carbonatée magnésienne et l'évolution de l'hydrosphère. *Compt. Rend. Acad. Sci.*, 274: 2952–2955.

Hallberg, R.O., 1972. Sedimentary sulfide mineral formation. An energy circuit system approach. *Mineralium Deposita*, 7: 189–201.

Haranczyk, C., 1964. Investigation of copper-bearing Zechstein shales from the Wraclaw monocline (Lower Silesia). *Acad. Pol. Sci. Bull., Serv. Géol. Géogr.*, 12 (1): 13–18.

Haranczyk, C., 1970. Zechstein lead-bearing shales in the fore-Sudetian monocline, in Poland. *Econ. Geol.*, 65: 481–495.

Helgeson, H.C., 1964. *Complexing and Hydrothermal Ore Deposition*. Pergamon Press, Oxford, 128 pp.

Honnorez, J., 1969. La formation actuelle d'un gisement sous-marin de sulfures fumerolliens à Vulcano (Mer thyrrénienne). *Mineralium Deposita*, 4: 95–112.

Honnorez, J., Honnorez-Guerstein, B., Valette, J. and Wauschkuhn, A., 1973. Present-day formation of an exhalative sulfide deposit at Vulcano (Tyrrhenian Sea), 2: active crystallization of fumarolic sulfides in the volcanic sediments of the Baia di Levante. In: G.C. Amstutz and A.J. Bernard (Editors), *Ores in Sediments*. Springer, Berlin, pp. 139–166.

Horikoshi, E., 1969. Volcanic activity related to the kuroko-type deposits in the Kosaka district, Japan. *Mineralium Deposita*, 4: 321–345.

Hutchinson, R.W. and Searle, D.L., 1971. Strata-bound pyrite deposits in Cyprus and relations to other sulfide ores. *Soc. Minig Geol. Japan, Spec. Issue*, 37: 198–205.

Jacquin, J.P., 1970. *Contribution à l'Etude géologique et minière de la Sierra de Gador (Almeria, Espagne)*. Thesis, Nantes, 501 pp.

Kajiwara, J., 1970. Syngenetic features of the kuroko ore from the Shakanai mine. In: T. Tatsumi (Editor), *Volcanism and Ore Genesis*. Tokyo Univ. Press, Tokyo, pp. 197–205.

Kanehira, K. and Tatsumi, T., 1970. Bedded cupriferous iron sulphide deposits in Japan, a review. In: T. Tatsumi (Editor), *Volcanism and Ore Genesis*. Tokyo Univ. Press, Tokyo, pp. 51–76.

Kazakov, A.V., 1950. The phosphatic facies. Origin of phosphorites and the geological factors in the formation of phosphorite deposits. *Tr. Nauchn.-Issl. Inst. Udobr. Insektofung.*, 145.

Knight, C.L., 1957. Ore genesis the source-bed concept. *Econ. Geol.*, 52: 808–817.

Krauskopf, K.B., 1967. Source rocks for metal-bearing fluids. In: H.L. Barnes (Editor), *Geochemistry of Hydrothermal Ore Deposits*. Holt, Rinehart and Winston, New York, N.Y., pp. 1–33.

Krautner, H.G., 1970. Die hercynische Geosynklinalbildung in den rumänischen Karpaten und ihre Beziehung zu der hercynischen Metallogenese Mitteleuropas. *Mineralium Deposita*, 5: 323–344.

Krumbein, W.C. and Sloss, L.L., 1963. *Stratigraphy and Sedimentation*. Freeman, San Francisco, Calif., 660 pp.

Laffitte, P., 1966. Cartographie métallogénique et gîtes stratiformes in genesis of stratiform lead-zinc-barite-fluorite deposits. *Econ. Geol. Monogr.*, 3: 227–233.

Lagny, Ph., 1969. Minéralisations plombo-zincifères triasiques dans un paleo-karst (gisement de Salafossa, province de Belluno, Italie). *Compt. Rend. Acad. Sci.*, 268: 1178–1181.

Lahusen, L., 1972. Schicht- und zeitgebunden Antimonit-Scheelit Vorkommen und Zinnober Vererzungen in Kärtnen und Osttirol/Osterreich. *Mineralium Deposita*, 7: 31–60.

Launay, P. and Leenhardt, R., 1959. Les brèches sédimentaires zincifères du Sinémurien du Lot. *Bull. Soc. Géol. Fr., 7e Sér.*, 1 (5): 467–484.

Leleu, M.G., 1966a. Les gisements plombo-zincifères du Laurium. *Sci. Terre*, 11: 293–343.

Leleu, M.G., 1966b. Le karst et ses incidences métallogéniques. *Sci. Terre*, 11 (4): 385–413.

Leleu, M.G., 1969. Essai d'interprétation thermodynamique en métallogénie. Les minéralisations karstiques du Laurium (Grèce). *Bull. B.R.G.M.*, 2e Sér., 4: 1–62.

Lepp, H. and Goldich, S.S., 1964. Origin of precambrian iron formations. *Econ. Geol.*, 59: 1025–1060.

Levorsen, A.I., 1955. Time of petroleum accumulation. *Econ. Geol. (Cinquantenaire, 1905–1955)*, 2: 748–756.

Lindgren, W., 1933. *Mineral Deposits*. McGraw-Hill, New York, N.Y., 930 pp.

Lombard, A., 1953. *Géologie sédimentaire*. Masson, Paris, 722 pp.

Mackay, R.A., 1946. The control of impounding structure on ore deposition. *Econ. Geol.*, 41: 13–46.

Mainguet-Suares, Y. and Samama, J.C., 1969. L'étude des ciments des roches détritiques, outil de prospection des gisements stratiformes enfouis (Trias ardèchois). *Chron. Mines Rech. Min.*, 381: 83–91.

Marmo, V., 1960. Origin of ores. *Neues Jhrb. Mineral.*, 94 (1).

Matsukuma, O. and Korikoshi, E., 1970. Kuroko deposits in Japan, a review. In: T. Tatsumi (Editor), *Volcanism and Ore Genesis*. Tokyo Univ. Press, Tokyo, pp. 153–179.

Maucher, A., 1965. Die Antimon-Wolfram-Quecksilber Formation und ihre Beziehung zu Magmatismus und Geotektonik. *Freib. Forschungsh.*, C 186: 173–188.

Maucher, A., 1966. Discussion paper Maucher and Schneider, in genesis of stratiform lead-zinc-barite-fluorite deposits. *Econ. Geol., Monogr.*, 3: 89.

Maucher, A. and Höll, R., 1968. Die Bedeutung geochemisch-stratigraphischer Bezugshorizonte für die Alterstellung der Antimonlagerstätte von Schlaining in Burgenland, Österreich. *Mineralium Deposita*, 3: 272–285.

McDonald, J.A., 1967. Metamorphism and its effects on sulphide assemblage. *Mineralium Deposita*, 2: 200–220.

McKinstry, H.E., 1955. Structure of hydrothermal deposits. *Econ. Geol. (Cinquantenaire 1905-1955)*, 1: 170–225.

Michard, G., 1967. Signification du potentiel redox dans les eaux naturelles. Conditions d'utilisation des diagrammes (Eh, pH). *Mineralium Deposita*, 2: 34–37.

Miller, L.J., 1960. Massive sulfide deposits in eugeosynclinal belts. *Bull. Geol. Soc. Am.*, 71 (12): 1930.

Monseur, G. and Pel, J., 1973. Reef environments and stratiform ore deposits. In: G.C. Amstutz and A.J. Bernard (Editors), *Ores in Sediments*. Springer, Berlin, pp. 195–207.

Moreau, M., Poughon, A., Puybarreau, Y. and Sanselme, H., 1966. L'uranium et les granites. *Chron. Mines Rech. Min.*, 350: 47–51.

Noble, E.A., 1963. Formation of ore deposits by water of compaction. *Econ. Geol.*, 58: 1145–1156.

Ohle, E.L. and Brown, J.S., 1954. Geologic problems in the southeast Missouri lead district. *Bull. Geol. Soc. Am.*, 45 (4): 201–222.

Padalino, G., Pretti, S., Tamburrini, D., Tocco, S., Uras, L., Violo, M. and Zuffardi, P., 1973. Ore deposition in karst formations with examples from Sardinia. In: G.C. Amstutz and A.J. Bernard (Editors), *Ores in Sediments*. Springer, Berlin, pp. 209–220.

Pedro, E., 1968. Distribution des principaux types d'altération chimique à la surface du globe. Présentation d'une esquisse géographique. *Rev. Géogr. Phys. Géol. Dyn.*, 10: 457–470.

Pelissonnier, H., 1961. Paleoreliefs et minéralisations hydrothermales: la structure paléinsulaire. *Ann. Mines*, 1961: 7–30.

Pelissonnier, H., 1966. Analyse paléohydrogéologique des gisements. In: J.S. Brown (Editor), *Genesis of Stratiform Lead-Zinc-Barite-Fluorite Deposits – Econ. Geol., Monogr.*, 3: 234–252.

Perel'man, A.I., 1967. *Geochemistry of Epigenesis*. Plenum Press, New York, N.Y., 266 pp.

Perrodon, A., 1966. *Géologie du Pétrole*. Presses Universitaires, Paris, 440 pp.

Pospelov, L.G., 1969. Elements of geological resemblance between deposits of oil and fluidogenic ore deposits. *Int. Geol. Rev.*, 11 (7): 751–756.

Puchelt, H., 1973. Recent iron sediment formation at the Kameni Islands, Santorini (Greece). In: G.C. Amstutz and A.J. Bernard (Editors), *Ores in Sediments*. Springer, Berlin, pp. 227–245.

Ramberg, H., 1952. *The Origin of Metamorphic and Metasomatic Rocks*. Univ. Press, Chicago, 317 pp.

Ramdohr, P., 1953. Über Metamorphose und sekundäre Mobilisierung. *Geol. Rundsch.*, 42 (1): 11–19.

Rogers, A.F., 1916. Origin of copper ores of the "Red-Beds" type. *Econ. Geol.*, 11: 366.

Rickard, D.T., 1973. Synsedimentary sulfide ore formation. *Econ. Geol.*, 68 (5): 605–617.

Roberts, W.M.B., Walker, A.L. and Buchanan, A.S., 1969. The chemistry of pyrite in aqueous solution and its relation to the depositional environment. *Mineralium Deposita*, 4: 13–29.

Roukhine, L.B., 1953. *Bases de la Lithologie: Etude des Formations sédimentatires*. Leningrad, 671 pp. (Translated in French by B.R.G.M., 1963).

Routhier, P., 1963. *Les Gisements métallifères: Géologie et Principes de Recherche*. Masson, Paris, 1283 pp.

Rouvier, H., 1971. Minéralisations plombo-zincifères et phénomènes karstiques. Exemple tunisien: le gisement du Djebel Hallouf. *Mineralium Deposita*, 6: 196–208.

Rutten, M.G., 1967. Sedimentary ores of the Early and Middle Precambrian and the history of atmospheric oxygen. In: *Sedimentary Ores: Ancient and Modern* (revised) – *Proc. 15th Inter-Univ. Geol. Congr., Leicester*, pp. 187–195.

Samama, J.C., 1968. Contrôle et modèle génétique de minéralisations en galène de type "red-beds". *Mineralium Deposita*, 3: 261–271.

Samama, J.C., 1969. *Contribution à l'Etude des Gîsements de Type "Red-Beds". Etude et Interprétation de la Géochimie et de la Métallogenèse du Plomb en Milieu continental. Cas du Trias ardèchois et du Gîsement de Largentière.* Thesis, Nancy, 450 pp.

Samama, J.C., 1970. Description et interprétation d'une concentration d'antimoine en milieu lagunaire. L'indice de stibine de Charmes-sur-Rhône (Ardèche). *Bull. B.R.G.M.*, 2e Sér., 2: 1–11.

Samama, J.C., 1973. Ore deposits and continental weathering: a contribution to the problem of geochemical inheritance of heavy metal contents of basement areas and of sedimentary basins. In: G.C. Amstutz and A.J. Bernard (Editors), *Ores in Sediments*. Springer, Berlin, pp., 247–265.

Sato, T., 1971. Physico-chemical environments of kuroko mineralization at Uchinotaï deposit of Kosaka Mine, Akita Prefecture. *Soc. Mining Geol. Japan, Spec. Issue,* 2: 137–144.

Saupe, F., 1973. *La Géologie du Gîsement de Mercure d'Almaden.* Thesis, Nancy, 268 pp.

Schneiderhöhn, H., 1941. *Lehrbuch der Erzlagerstättenkunde.* G. Fisher, Jena, 858 pp.

Shepard, F.P. and Moore, D.G., 1955. Central Texas coast sedimentation; characteristics of sedimentary environment, recent history and diagenesis. *Bull. Am. Assoc. Pet. Geologists*, 39: 1463–1593.

Snyder, F.G. and Odell, J.W., 1958. Sedimentary breccias in the southeast Missouri lead district. *Bull. Geol. Soc. Am.,* 69: 899–925.

Sweeney, R.E. and Kaplan, I.R., 1973. Pyrite framboid formation: laboratory synthesis and marine sediments. *Econ. Geol.,* 68: 618–634.

Tardy, Y., 1969. *Géochimie des Altérations. Etude des Arènes et des Eaux de quelques Massifs cristallins d'Europe et d'Afrique.* Thesis Fac. Sci. Strasbourg, ronéo, 274 pp.

Tardy, Y. and Vine, I.D., 1973. Elements partition ratios in some sedimentary environments. (In press).

Tarr, W.A., 1910. Copper in the "Red-Beds" of Oklahoma. *Econ. Geol.,* 5: 221–226.

Tatsumi, T. and Watanabe, T., 1971. Geological environment of formation of the kuroko-deposits. *Soc. Mining Geol. Japan, Spec. Issue,* 3: 216–220.

Turekian, K.K., 1969. The oceans, streams and atmosphere. In: K. Wedepohl (Editor), *Handbook of Geochemistry.* Springer, Berlin, chapter 10.

Valette, J.N., 1973. Distribution of certain trace elements in marine sediments surrounding Vulcano Island (Italy). In: C.G. Amstutz and A.J. Bernard (Editors), *Ores in Sediments.* Springer, Berlin, pp. 321–338.

Vine, I.D. and Tourtelot, E.B., 1970. Geochemistry of black shales. *Econ. Geol.,* 65 (3): 253–272.

Watanabe, T., 1970. *Volcanism and Ore Genesis.* Tokyo Univ. Press, Tokyo, pp. 423–432.

Watanabe, T., Iwao, S., Tatsumi, T. and Kanehira, K., 1970. Folded ore bodies of the Okuki mine. In: T. Tatsumi (Editor), *Volcanism and Ore Genesis.* Tokyo Univ. Press, Tokyo, pp. 105–117.

Wedepohl, K.H., 1971. Kupferschiefer as a prototype of syngenetic sedimentary ore deposits. *Soc. Min. Geol. Japan,* 3: 268–273.

White, D.E., 1967. Mercury and base-metal deposits with associated thermal and mineral waters. In: H.L. Barnes (Editor), *Geochemistry of Hydrothermal Deposits.* Holt, Rinehart and Winston, New York, N.Y., pp. 575–631.

White, D.E., Hem, J.D. and Waring, G.A., 1963. Chemical composition of subsurface waters. *U.S. Geol. Surv., Prof. Pap.,* 440-F: 67 pp.

White, D.E., Muffler, J.P. and Truesdell, A.H., 1971. Vapour-dominated hydrothermal systems compared with hot-water systems. *Econ. Geol.,* 66: 75–97.

Williams, D., 1962. Further reflections on the origin of the porphyries and ores of Rio Tinto, Spain. *Bull. Inst. Mining Metall., Trans., Lond.,* 71 (8): 492.

Wolf, K.H., 1976. Ore genesis influenced by compaction. In: G.V. Chilingar and K.H. Wolf (Editors), *Compaction of Coarse-grained Sediments, 2.* Elsevier, Amsterdam (in press).